# COMPUTATIONAL METHODS FOR PHYSICS

There is an increasing need for undergraduate students in physics to have a core set of computational tools. Most problems in physics benefit from numerical methods, and many of them resist analytical solution altogether. This textbook presents numerical techniques for solving familiar physical problems, where a complete solution is inaccessible using traditional mathematical methods.

The numerical techniques for solving the problems are clearly laid out, with a focus on the logic and applicability of the method. The same problems are revisited multiple times using different numerical techniques, so readers can easily compare the methods. The book features over 250 end-of-chapter exercises. A website hosted by the author features a complete set of programs used to generate the examples and figures, which can be used as a starting point for further investigation. A link to this can be found at www.cambridge.org/9781107034303.

JOEL FRANKLIN is an Associate Professor in the Physics Department of Reed College. He focuses on mathematical and computational methods with applications to classical mechanics, quantum mechanics, electrodynamics, general relativity, and modifications of general relativity.

# COMPUTATIONAL METHODS FOR PHYSICS

JOEL FRANKLIN

*Reed College*

CAMBRIDGE
UNIVERSITY PRESS

## CAMBRIDGE
### UNIVERSITY PRESS

University Printing House, Cambridge CB2 8BS, United Kingdom

Cambridge University Press is part of the University of Cambridge.

It furthers the University's mission by disseminating knowledge in the pursuit of education, learning and research at the highest international levels of excellence.

www.cambridge.org
Information on this title: www.cambridge.org/9781107034303

© J. Franklin 2013

First published 2013

*A catalogue record for this publication is available from the British Library*

*Library of Congress Cataloguing in Publication data*
Franklin, Joel, 1975–
Computational methods for physics / Joel Franklin, Reed College.
pages   cm
Includes bibliographical references and index.
ISBN 978-1-107-03430-3 (hardback)
1. Mathematical physics.   2. Physics – Data processing.   3. Numerical analysis.   I. Title.
QC20.F735   2013
530.15 – dc23      2013003044

ISBN  978-1-107-03430-3  Hardback

Additional resources for this publication at www.cambridge.org/9781107034303

For Lancaster, Lewis, and Oliver

# Contents

# Preface

This book is meant to teach an advanced undergraduate physics major how to solve problems, that they can easily pose and understand, using numerical methods. It is self-contained (within reason) in its presentation of physical problems,[1] so that a reader need not be previously familiar with all of the physical topics covered. The numerical techniques that are presented are complete, without being exhaustive. I have chosen a set of tools that can be motivated easily – some of these, like Runge–Kutta methods for solving ordinary differential equations, are themselves the "end of the story" for certain problems. In other cases, I choose to present a simplified method that will work for at least some problems, even if it is not what is actually implemented in practice. A homogenous and appropriate level of presentation is the goal. Sometimes that allows for a complete and formal discussion, in other cases, I motivate and inform while ensuring that a successful (if not competitive) method is carefully described.[2]

The chapters are defined by computational technique, with chapters covering "Partial differential equations," "Integration," "Fourier transform," "Matrix inversion," etc. Each chapter introduces a disparate set of physical problems, most of which are "unsolvable" (meaning that there is no closed-form solution expressible in terms of simple functions).[3] I have attempted to draw problems from as wide an array of physical areas as possible, so that in a single chapter, there may be examples from quantum mechanics, E&M, special relativity, etc. All of these physical setups end in a similar analytically intractable problem. That problem then becomes

---

[1] This can lead to some repetition, of, for example, the meaning attached to the solutions of Schrödinger's equation, or the definition of relativistic momentum, etc. But my hope is that repeated local definition allows for easier reading.

[2] To be clear with an example: presenting a fast, research-grade eigenvalue solver requires more mathematics and programming ability than is available at the undergraduate level – that is not, therefore, a compelling target. But, the power method, and simultaneous iteration, which are easy (and delightful) to describe, will do the job in many cases and inform more sophisticated methods.

[3] Of course, many "simple functions" still require numerical evaluation, even sine and cosine have interpolated or otherwise approximated value at most points.

the vehicle for the introduction of a numerical method (or in some cases, a few different methods) in which I focus on the mathematical ideas behind the method, including a discussion of its motivation, its limitations and possible extensions. At the end of each chapter, I include short further reading suggestions,[4] and two types of exercise. The first, called "Problems," are pencil-and-paper exercises. These can involve guiding a student through the analysis of a numerical method presented in the chapter, or setting up analytically solvable limiting cases of problems that will be solved numerically. Then there are "Lab problems," these are physical problems (generally set up at the start of the chapter) that will be solved using the method(s) from the chapter.

The specific numerical methods I discuss in each chapter were chosen for their coverage (of a wide range of physical problems), and presentability. By the latter, I mean that any method should be transparent in its functioning (and implementation), and easy to analyze. Certain types of Monte Carlo "simulations" are omitted, since it is harder to prove (or in some cases even indicate) that they will succeed in solving a particular problem. One can motivate and a posteriori verify that such methods work, and they have a certain physically inspired allure, but my primary goal in choosing methods is to introduce students to ideas whose advantages and limitations can be judged easily. In addition to numerical methods, I discuss "traditional" mathematical methods relevant to the physical sciences. I have chosen specific mathematical methods to present (like separation of variables, shock solutions to time-dependent PDEs, or time-independent perturbation theory) based on the availability of a comparative discussion with numerical methods, and to set up limiting cases so that we can check the functioning of numerical methods on concrete test problems where the answer is known. I hope, then, that the book is more than "just" a numerical methods book, but one that can be used to learn some physics, learn methods for solving physical problems, and build confidence that those methods are working (in the appropriate regime).

## Structure and teaching

Structurally, there are three sections to the book:

1. ODEs, PDEs, and integration. These are the topics covered in the first seven chapters. The idea is to get a bulk of the problem-solving tools out quickly, and most physical problems benefit from an ability to solve ODEs in initial

---

[4] For arguments or examples that come directly from a particular source, I cite the source directly within the text, and those citations refer to the references at the end of the book – suggestions for further reading represents that subset of the full bibliography that provides useful additional information and/or inspiration local to the chapter.

value (Chapter 2) and boundary value form (Chapter 3). There are two chapters on finite difference for PDEs: Chapter 4 covers static, linear operators (mainly the Poisson problem), and in Chapter 5, we see time-dependent PDEs (both linear, like Schrödinger's equation, and nonlinear forms). Chapter 6 develops integration methods from interpolating polynomials, and connects the statistical Monte Carlo integration method to simple box sums. Finally, I count the chapter on the FFT as an integration topic.

2. Numerical linear algebra. Chapters 9–12 discuss numerical techniques for matrix inversion, least squares solution, and the eigenvalue problem. Some of the material in these chapters is required by topics in the first section (for example, solving Poisson's problem in discretized form involves matrix inversion) where we used canned routines. Now we come back and pick up these topics, their motivation coming in part from previous discussions. Chapters 9 and 10 focus on "direct" solution of the matrix inverse and eigenvalue problem, while Chapter 11 introduces iterative techniques (Jacobi, SOR, and Krylov subspace methods).[5] Chapter 12 is not, strictly speaking, a part of linear algebra – but minimization is usefully applied in the nonlinear least squares problem, so must come after the linear form of that problem.

3. Additional topics. The first two sections provide methods that cover many physical problems. The remaining chapters give interesting additional topics, but are slightly different in their coverage from the rest of the book. Chapter 13: "Chaos," for example, has no numerical methods attached to it. The neural network chapter (14) presents these models in terms of their predictive ability for physical phenomena, with emphasis on the promise provided by the Cybenko theorem rather than the physiological inspiration normally encountered. Chapter 15 returns to PDEs and describes the Galerkin approach with some additional nonlinear PDE targets.

Two of the chapters do not fall neatly into these categories – the first chapter provides a review of the minimal programming tools that we need to use to implement the methods presented in the other chapters – it also provides some orienting Mathematica examples of these tools in action. The eighth chapter acts as a pivot between the first and second sections, and is meant to set up our interest in the eigenvalue problem.

The grouping above is informed by the course that inspired this book, and for which it was written – in it, the students present a final project of their own design,

---

[5] Divorcing the finite difference setup from the iterative solutions (Jacobi and SOR) to the matrix inverse problem is not the traditional approach. Typically, authors will combine the discretization of the Laplacian with an iterative update. But I believe that the two pieces are logically separate – discretizing the Laplace operator gives us a matrix inverse problem to solve, how we go about solving it is a separate matter.

and they begin working on this at about the middle of the semester. That is when we have covered the bulk of item one above, so they have seen a number of different kinds of problems and understand numerical methods for solving them. The section on numerical linear algebra is interesting and informative, but students who need to, for example, invert matrices for their final projects can use a built-in routine (few students have made the process of matrix inversion the central idea of their projects) until the relevant material is covered (and even after, in most cases). The additional topics section has then come at a time when students can take advantage of the information in them without relying on it for their project (coupled, driven oscillators can be solved and studied numerically before the nonlinear analysis tools that are typically used to describe them are introduced).

While I use a subset of the programming language of Mathematica both in the book, and in the course, I have found that students can re-render the content in other languages relatively easily, and have had students work out the lab problems using python, Java, C, Sage, and matlab. For those students with less of a programming background, I provide, for each chapter, a Mathematica "notebook" that contains all the commands used to generate the figures and example solutions for the chapter. Then much of the programming can be done by re-working examples from the chapter notebook. My intention is for students who are not as interested in programming to have a way of learning the methods, without worrying about their implementation as much. The computational methods are based on implementation-independent mathematical ideas, and those are the core targets. For students who enjoy the implementation side (I always did, although I was never particularly good at it): optimizing and ordering commands efficiently, don't look at the chapter notebooks, work out your own routines from scratch.

This course is one of my favorites to teach – there are limitless physical problems to set up, so it's a great place for me to learn about new physics. Then the extent to which it is difficult to solve most problems (even simple ones) analytically is always surprising. And, finally, the numerical solutions that allow progress are often simple and fast. I usually cover each chapter in a week with three lectures: on Monday, we just set up physical problems, ones that come from different physics, but all end in a common, fundamental "problem." On Wednesday, we develop a method that solves that problem, and then, during "Friday potpourri," we return to some of the problems from Monday and solve them, and discuss limitations or extensions of the method. The chapter structure mimics these lectures, with physical problems presented first, then methods, then additional items, sometimes additional physics, other times, more in-depth discussion of the method or its variants.

What is notably missing from the weekly lectures is any sort of implementation discussion – in addition to the three lectures, I also run three-hour "labs" every week. Students come to these to work on the lab problems (I assign three or four of these each week), and this is where implementation can be carefully discussed, with the relevant programming ideas built up on an individual basis, or through group discussion. Having a concrete problem to work on (the lab problems) gives natural focus to the programming. This separation between physics, the idea behind the method, and its implementation, is intentional. The goal is to help identify points of confusion as arising from the statement of a physical problem, the proper functioning of a numerical method, or its implementation. Keeping those three separate makes clearing up the confusion easier.

## Website and materials

The book has a website, and on it, students can find all of the chapter notebooks (the `Mathematica` code that generated all the examples and figures in the book), a few sample "lab notebooks," and some project ideas that students have come up with over the years. Some of the lab problems require mocked up data or other information, and those are also available on the website.

## Acknowledgements

It is a pleasure to thank my colleagues in the Reed College physics department, especially Darrell Schroeter, who developed and taught an early version of the computational methods class at Reed with me, Nelia Mann who contributed useful comments and problems, David Latimer, a great sounding board, and David Griffiths for his usual outstanding commentary, criticism and encouragement. Outside the department, Jim Fix gave me some great image rotation advice, Olivia Schelly took the pictures that are used in Chapter 7, and Matt Sayre wrote the Gravitea-Time song for filtering. Simon Capelin and Lindsay Barnes at Cambridge University Press have once again made the publishing process enjoyable, and I thank them for their help at all stages of the project.

I would like to thank Sebastian Doniach and the Stanford University physics department for their hospitality in the winter of 2012: aside from a nice change of scene, and stimulating new environment, I got to teach this material to an entirely new audience, for which I am very grateful. The Writer's Bloc has provided a wonderful setting for revising the manuscript. Finally, a special thanks to the Reed College students who helped me during the process of editing and refining the book: Tom Chartrand, Todd Garon, and Reuven Lazarus.

# 1

# Programming overview

A programming language useful to this book must provide a minimal set of components that can be used to combine numbers, compare quantities and act on the result of that comparison, repeat operations until a condition is met, and contain functions that we can use to input data, and output results. Almost any language will suffice, but I have chosen to use Mathematica's programming environment as the vehicle. The reasoning is that 1. The input/output functions of Mathematica are easy to use, and require little additional preparation[1] 2. We will be focused on the ideas, numerical and otherwise, associated with the methods we study, and I want to draw a clear distinction between those ideas and issues of implementation. This book is not meant to teach you everything you need to know about programming[2] – we will discuss only the bare essentials needed to implement the methods. Instead, we will focus on the physical motivation and tools of analysis for a variety of techniques. My hope is that the use of Mathematica allows us to discuss implementation in a homogeneous way, and our restriction to the basic programming structure of Mathematica (as opposed to the higher-level functionality) allows for easy porting to the language of your choice.

Here, we will review the basic operations, rendered in Mathematica, falling into the broad categories: arithmetic operations, comparisons, loops, and input–output routines. In addition, we must be able to use variable names that can be assigned values, and there is a scoping for these constructions in Mathematica similar to C (and many other languages). Functions, in the sense of C, exist in Mathematica, and we will use a particular (safe) form, although depending on the context, there are faster (and slower) ways to generate functions. We will bundle almost every set of computations into a function, and this is to mimic

---

[1] There are libraries to import audio and video, for example, in C, but the resulting internal representation can be difficult to work with. The details of compiling with those libraries correctly linked is also specific to the language and compiler, issues that I want to avoid.

[2] Although, I will employ good programming practice in the examples and accompanying chapter notebooks.

Arithmetic operations                    Logical operations

In[1]:= **3 + 5**                              In[1]:= **3 < 5**

Out[1]= **8**                                  Out[1]= **True**

In[2]:= **4 * 20**                             In[2]:= **7 ≥ 5**

Out[2]= **80**                                 Out[2]= **True**

In[3]:= **N[Pi / E]**                          In[3]:= **1 < 2 && 2 > 3**

Out[3]= **1.15573**                            Out[3]= **False**

In[4]:= **Sin[.1]**                            In[4]:= **1 < 2 || 2 > 3**

Out[4]= **0.0998334**                          Out[4]= **True**

In[5]:= **1 / 4**                              In[5]:= **3 = 3**

           **1**                               Out[5]= **True**
Out[5]= **—**
           **4**                               In[6]:= **2 ≠ 3**

In[6]:= **1.0 / 4.0**                          Out[6]= **True**

Out[6]= **0.25**

Figure 1.1 Examples of basic arithmetic input and output and logical operations.

good coding practice that is enforced in more traditional languages (for a reason – the logic and readability one gains by breaking calculations up into named constituents cannot be overvalued). Finally, we will look at two important ideas for algorithm development: recursion and function pointers. Both are supported in Mathematica, and these are also available in almost any useful programming language.

Beyond the brief programming overview, there are issues specific to numerical work, like consideration of timing, and numerical magnitude, that provide further introduction into the view of physics that we must take if we are to usefully employ computers to solve problems.

## 1.1 Arithmetic operations

All of the basic arithmetic operations are in Mathematica, and some are shown in Figure 1.1 – we can add and subtract, multiply, divide, even evaluate trigonometric functions (arguments in radians, always). The only occasional hiccup we will encounter is the distinction, made in Mathematica, between an exact quantity and a number – in Figure 1.1, we can see that when presented with $1/4$, Mathematica responds by leaving the ratio alone, since it is already reduced. What we are most interested in is the actual numerical value. In order to force Mathematica to provide real numbers, we can wrap expressions in the N function, as in In[3] on the left of Figure 1.1, or we can insert decimal points, as in In[6]. This is more than an aesthetic point – many simple calculations proceed very slowly if you suggest (accidentally or not) that all numbers are exact quantities (all fraction additions, for example, must be brought to a common, symbolic, denominator, etc.).

Aside from integers and real numbers, Mathematica is aware of most mathematical constants, like $\pi$ and $e$, and when necessary, I'll tell you the name of any specific constant of interest.

## 1.2 Comparison operations

We will use most comparison operations, and Mathematica outputs True or False to any of the common ones – we can determine whether a number is greater than, less than, or equal to another number using >, <, == (notice that equality requires a double equals sign to distinguish it from assignment). We denote "less than or equal to" with <=, and similarly for "greater than or equal to" (>=). The logical "not" operation is denoted !, so that inequality is tested using ! = (not equal). Finally, we can string together the True/False output with "AND" (denoted &&) and "OR" (| |). Some examples are shown on the right in Figure 1.1.

## 1.3 Variables

Unlike C or C++, variables in Mathematica can be instantiated by definition, and do not require explicit typedef-ing. So setting a variable is as easy as typing x = 5.0. The variable has this value (subject to scoping) until it is changed, or cleared (closest to delete that exists in Mathematica) by typing Unset [x].

Variables can take a number of forms: single elements, lists, matrices, etc. For us, variables will be purely numerical (no symbolic variable assignments are allowed – those are generally not available in other languages), and the numbers themselves will be "doubles," i.e. real numbers with maximum precision. We can then define tables and arrays of numbers, again by giving values to a variable name. In Figure 1.2, we see a few different ways of defining variables – first we define and set the variable p to have value 5 – Mathematica will print an output in general, and in the case of defining variables, it prints an output that reminds us of the variable's value. To suppress printing output, we use a semicolon at the end of a line – in the second example on the left in Figure 1.2, we define q to have value 7, and the semicolon tells Mathematica to just set the value without extra verbiage.

We can define variables that are tables of fixed length by specifying the numerical value for each entry, using { . . . }, as in the definition of x on the left in Figure 1.2. The Mathematica command

```
Table[f[k],{k,start,end,step}]
```

can also be used to generate tables that have values related to index number by the function f [k] – in the definition of the array variable y, we use f [k] = k for "iterator" (a dummy name given to the index used to generate the table) k.

Defining variables

In[1]:= **p = 5**

Out[1]= 5

In[2]:= **q = 7;**

In[3]:= **x = {1.0, 2.0, 3.0, 4.0}**

Out[3]= {1., 2., 3., 4.}

In[4]:= **x**

Out[4]= {1., 2., 3., 4.}

In[5]:= **y = Table[k, {k, 1.0, 4.0, .5}]**

Out[5]= {1., 1.5, 2., 2.5, 3., 3.5, 4.}

Setting variables

In[1]:= **q = 9;**

In[2]:= **q**

Out[2]= 9

In[3]:= **q = 10;**

In[4]:= **q**

Out[4]= 10

In[5]:= **x = Table[j^2, {j, 1.0, 10.0, 2.0}]**

Out[6]= {1., 9., 25., 49., 81.}

In[6]:= **x[[2]]**

Out[6]= 9.

In[7]:= **x[[2]] = 4.0**

Out[7]= 4.

In[8]:= **x**

Out[8]= {1., 4., 25., 49., 81.}

Figure 1.2 Examples of defining and setting variable values.

In[1]:= **x = 2;**
        **y = 3;**

In[3]:= **x + y**

Out[3]= 5

In[4]:= **X = Table[Sin[j], {j, 0.0, Pi, Pi / 4}]**

Out[4]= {0., 0.707107, 1., 0.707107, 1.22465 × 10⁻¹⁶}

In[5]:= **Y = Table[2.0 j, {j, 1, 5}]**

Out[5]= {2., 4., 6., 8., 10.}

In[6]:= **X[[2]] * Y[[3]]**

Out[6]= 4.24264

In[7]:= **X - Y**

Out[7]= {-2., -3.29289, -5., -7.29289, -10.}

Figure 1.3 Using arithmetic operations with variables, table elements, and tables.

Once a variable has been defined by giving it a value, the value can be accessed (by typing the name of the variable as input) or changed (using the operator =) as shown on the right in Figure 1.2, where a table x is created, and its second entry set to the value 4.0. The output of such an assignment is the assigned value, if we want to check the full content of x, we can type it as input, as in In[8].

Variables can be used with the normal arithmetic operations, their value replaces the variable name internally, just as in most programming languages. In Figure 1.3, we define x and y, and add them. We can perform operations on elements of lists, or on the lists themselves (so the final example in Figure 1.3 adds each element

| If statement | While statement | For statement |
|---|---|---|
| In[1]:= **x = 4;** | In[1]:= **x = -1;** | In[1]:= **For[x = -1, x ≤ 4, x = x + 1,**<br>        **Print[x];**<br>        **];** |
| In[2]:= **If[x ≤ 4,**<br>        **x = 5;**<br>        **,**<br>        **x = -1;**<br>        **];** | In[2]:= **While[x ≤ 4,**<br>        **Print[x];**<br>        **x = x + 1;**<br>        **];**<br>      −1 | −1<br><br>0<br><br>1<br><br>2<br><br>3<br><br>4 |
| In[3]:= **x**<br>Out[3]= 5 | 0<br><br>1<br><br>2<br><br>3<br><br>4 | |

Figure 1.4  Using Mathematica's If, While, and For.

of the lists X and Y – note that you cannot add together lists of different size). All variable and function names are case-sensitive, so that using x and X as variable names is unambiguous.

## 1.4 Control structures

The most important tools for us will be the if-then-else, while and for constructs. These can be used with logical operations to perform instructions based on certain variable values.

The if-then-else construction operates as you would expect – we perform instructions *if* a certain logical test returns True, and other instructions (*else*) if the test returns False. The Mathematica structure is:

            If[test, op-if-test-true, op-if-test-false]

In Figure 1.4, we define and set the value of x to 4. Then we use the If statement to check the value of x – *if* x is less than or equal to 4, *then* we set x to 5, *else* we set x to −1.

Using While is similar in form – we perform instructions *while* a specified test yields True, and stop when the logical test returns False. The Mathematica command that carries out the While loop is

               While[test, op-if-test-true]

An example in which we set x to −1 and then add one to x if its value is less than or equal to four is shown in Figure 1.4. In this example, we also encounter the i/o function Print[x], which prints the value of the variable x.

Finally, "for loops" perform instructions repeatedly while an *iterator* counts from a specified start value to a specified end value – more generally, the iterator is given some initial value, and a logical test is performed on a function of the iterator – while the logical test is true, operations are executed. We can construct a for loop from a while loop, so the two are, in a sense complementary. In Mathematica, the syntax is:

```
For[j = initialval, f[j], j-update, operations]
```

where j is the iterator, f[j] represents a logical test on some provided function of j, j-update is a rule for incrementing j, and operations is the set of instructions to perform while f[j] returns True – each execution of operations increments j according to j-update. This is easier done than said – an example of using the for loop is shown in Figure 1.4. That example produces the same results as the code in the While example.

## 1.5 Functions

Writing programs requires the ability to break computational instructions into logically isolated blocks – this aids in reading, and debugging. These isolated blocks are called "functions," generically, a name for anything that takes in input and returns output. Mathematica provides a few different ways to define programming functions. We will use the Module form of function definition – the basic structure is:

```
functionname[input1_, input2_]:= Module[{local variables},
                             operations;
                             Return[value];
                           ]
```

An example of Module in action is shown in Figure 1.5 – but the important thing to remember is that we now have a function that can be called with some inputs, returns some output, and has hidden local variables that are not accessible to the "outside world."

In Figure 1.5, we define the function HelloWorld, that takes a single argument called name – the underscore identifies name as an input. The Module is set up with two local variables, one takes the value of name (generally, a string), and the other is set to one. The function itself prints a friendly greeting, and returns the value stored in localvarx (i.e. one). As a check that the variable localvarx really is undefined as far as the rest of the Mathematica "session" is concerned, the last line in Figure 1.5 calls localvarx – the fact that Mathematica returns the variable name, unevaluated, indicates that it is not currently defined.

We can use all of our arithmetic, logical, and control operations inside the function to make it do more interesting things. As an example, the two functions defined in Figure 1.6 are used to sort an array of numbers in increasing order. The first function is Swap – this takes a list, and two numbers, a and b, as inputs, swaps

```
In[1]:= HelloWorld[name_] := Module[{retval, localvarx},
            retval = 1.0;
            localvarx = name;
            Print["Hello ", localvarx];
            Return[retval];
          ]

In[2]:= X = HelloWorld["Dave"]

        Hello Dave

Out[2]= 1.

In[3]:= Y = HelloWorld[33];

        Hello 33

In[4]:= Y

Out[4]= 1.

In[5]:= localvarx

Out[5]= localvarx
```

Figure 1.5 Example of defining, and then calling, a function in `Mathematica` using `Module`.

```
In[1]:= Swap[inlist_, a_, b_] := Module[{holder, outlist},
            outlist = inlist;
            holder = outlist[[b]];
            outlist[[b]] = outlist[[a]];
            outlist[[a]] = holder;
            Return[outlist];
          ]

In[2]:= InsertionSort[inlist_] := Module[{outlist, indexx, indexy, curelm, Nlist},
            outlist = inlist;
            Nlist = Length[outlist];
            For[indexx = 2, indexx ≤ Nlist, indexx = indexx + 1,
              curelm = outlist[[indexx]];
              indexy = indexx - 1;
              While[indexy > 0 && outlist[[indexy]] > curelm,
                outlist = Swap[outlist, indexy, indexy + 1];
                indexy = indexy - 1;
              ];
              outlist[[indexy + 1]] = curelm;
            ];
            Return[outlist]
          ]

In[3]:= InsertionSort[{5, 2, 4, 6, 1, 3}]

Out[3]= {1, 2, 3, 4, 5, 6}

In[4]:= InsertionSort[{-4, 1, 7, 2, 3, -10}]

Out[4]= {-10, -4, 1, 2, 3, 7}
```

Figure 1.6 Definition of the function `InsertionSort` – this function takes a list and sorts the elements of the list in increasing order (from [12]).

the value of the ath and bth elements of the list, and returns the resulting array. For the `InsertionSort` function, we go through the input array, and sequentially generate a sorted list of size `indexx-1`, increasing `indexx` until it is the size of the entire array. This is an inefficient but straightforward way to sort lists of numbers.

## 1.6 Input and output

There are a wide variety of `Mathematica` functions that handle various input and output. We will introduce specific ones as we go, I just want to mention two at the start that are of interest to us. The first we have already seen: `Print[ stuff ]` prints whatever you want, and can be used within a function to tell us what is going on inside the function.

The second output command we will make heavy use of is `ListPlot`. This function takes a table and generates a plot with the table values as heights at locations given by the index. Alternatively, if the table consists of pairs of values, then the plot uses the first of each pair as the $x$ location, and the second provides the $y$ (height). `ListPlot` can be used to visualize arrays of data, or function values. A few examples are shown in Figure 1.7.

## 1.7 Recursion

Most programming languages support a notion of "recursion" – this is the idea that a function can call itself. Recursion can be useful when designing "divide-and-conquer" algorithms. As a simple example of a recursive function, consider `DivideByTwo` defined in Figure 1.8. This function takes a number, and, if it is possible to divide the number by two, calls itself with the input divided by two. If the number cannot be divided by two, the function returns the non-dividable-by-two input. Notice the helper function `IsDivisableByTwo` – this checks divisibility using `Mathematica`'s built-in `Round` command.

An example of the function in action is shown in Figure 1.8 – we are using the `Print` command to see what value the function `DivideByTwo` gets at each call – you can see that it is called four times for the input 88, and returns a concrete result when its input is not divisable by two.

As a more interesting example, we can accomplish the same sorting of numbers idea from `InsertionSort` using recursion. See if you can "sort out" (no pun intended) the recursion in the definition of `MergeSort` shown in Figure 1.9. The helper function `Merge` takes two lists that are already sorted, and combines them to form a sorted list.

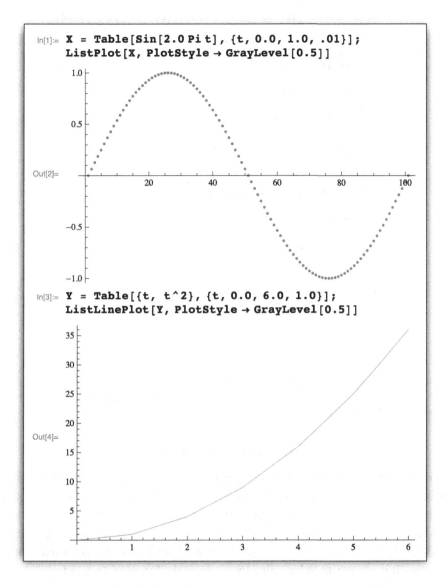

In[1]:= `X = Table[Sin[2.0 Pi t], {t, 0.0, 1.0, .01}];`
`ListPlot[X, PlotStyle → GrayLevel[0.5]]`

In[3]:= `Y = Table[{t, t^2}, {t, 0.0, 6.0, 1.0}];`
`ListLinePlot[Y, PlotStyle → GrayLevel[0.5]]`

Figure 1.7 Plotting – if the table input to `ListPlot` contains single entries, like the first example above, then the $x$-axis is the entry number. If, as in the second case, the table contains pairs of numbers, then the first number is taken to be the $x$ value, the second number the $y$ value. The command `ListLinePlot` is identical to `ListPlot` except that the points are connected by lines.

```
In[1]:= IsDivisableByTwo[inx_] := Module[{retval, div, NZERO},
          NZERO = 10^(-10);
          div = inx / 2;
          If[Abs[div - Round[div]] > NZERO,
            retval = False;
            ,
            retval = True;
          ];
          Return[retval];
        ]

In[2]:= DivideByTwo[num_] := Module[{retval},
          Print[num];
          retval = num;
          If[IsDivisableByTwo[retval] == True,
            retval = DivideByTwo[num / 2];
          ];
          Return[retval];
        ]

In[3]:= DivideByTwo[88]

        88

        44

        22

        11

Out[3]= 11
```

Figure 1.8 Example of recursive function definition – you must provide both the recursive outcome (call the function again with modified input) and the final outcome (the definition of the endpoint).

## 1.8 Function pointers

It is important that functions be able to call other functions – we can accomplish this in a few different ways. One way to make a function you write accessible to other functions is to define it globally, and then call it. That is the preferred method if you have a "helper" function that is not meant to be called by "users." The function Swap in the insertion sort example from Figure 1.6 is such a support function – it is not meant to be called by a user of the function InsertionSort, it is purely a matter of convenience for us, the programmer.

But sometimes, the user must specify a set of functions for use by a program. In this case, we don't know or care what the names of the functions supplied by the user are – they are user-specified, and hence should be part of the argument of any function we write. This "variable" function is known, in C, as a "function pointer" – a user-specifiable routine. Because of the low-key type-checking in Mathematica, we can pass functions as arguments to another function in the same way we pass anything. It is up to us to tell the user what constraints their function must satisfy. As an example, suppose we write a function that computes the time average of some user-specified function $f(t)$ – that is, we want to write a

```
In[1]:= MergeSort[inlist_] := Module[{outlist, oleft, oright, mid},
          outlist = inlist;
          If[Length[outlist] == 1, Return[outlist];];
          mid = Floor[Length[outlist] / 2];
          oleft = MergeSort[Take[outlist, {1, mid}]];
          oright = MergeSort[Take[outlist, {mid + 1, Length[outlist]}]];
          outlist = Merge[oleft, oright];
          Return[outlist];
        ]

In[2]:= Merge[inla_, inlb_] := Module[{outlist, lista, listb,
          index = 1, adex = 1, bdex = 1},
          If[Length[inla] == 0, lista = {inla};, lista = inla;];
          If[Length[inlb] == 0, listb = {inlb};, listb = inlb;];

          outlist = Table[0, {j, 1, Length[lista] + Length[listb]}];
          While[adex ≤ Length[lista] && bdex ≤ Length[listb],
            If[lista[[adex]] ≤ listb[[bdex]],
             outlist[[index]] = lista[[adex]];
             adex = adex + 1;
             ,
             outlist[[index]] = listb[[bdex]];
             bdex = bdex + 1;
            ];
            index = index + 1;
          ];
          If[adex > Length[lista],
           For[bdex = bdex, bdex ≤ Length[listb], bdex = bdex + 1,
            outlist[[index]] = listb[[bdex]];
            index = index + 1;
           ];
           ,
           For[adex = adex, adex ≤ Length[lista], adex = adex + 1,
            outlist[[index]] = lista[[adex]];
            index = index + 1;
           ];
          ];
          Return[outlist];
        ]

In[3]:= MergeSort[{10, -1, 8, 3, 5, 6, 0, -20}]

Out[3]= {-20, -1, 0, 3, 5, 6, 8, 10}
```

Figure 1.9  A recursive sorting algorithm – the implementation is as in [12].

function that takes $f(t)$ as input and computes:

$$\langle f \rangle = \frac{1}{T} \int_0^T f(t)dt. \tag{1.1}$$

We'll use Mathematica's Integrate function for now. Our function, called TimeAverage, takes as input the function $f(t)$ and the period of interest, $T$, and returns the right-hand side of (1.1). The code and two user test cases are shown in Figure 1.10.

## 1.9 Mathematica-specific array syntax

Most programming languages have a notion of memory allocation (whether the responsibility of the programmer, compiler, or operating system) – we need a way

```
In[1]:= TimeAverage[f_, T_] := Module[{retval},
          retval = (1 / T) Integrate[f[t], {t, 0, T}];
          Return[retval];
        ]

In[2]:= userfunction1[t_] := Sin[t]^2

In[3]:= TimeAverage[userfunction1, 2.0 Pi]

Out[3]= 0.5

In[4]:= userfunction2[t_] := Sin[t] Cos[t]

In[5]:= TimeAverage[userfunction2, 2.0 Pi]

Out[5]= 0.
```

Figure 1.10 The function TimeAverage takes a function name as its argument, in addition to the period over which to average.

to store values of various sorts. A table of numbers can be viewed as a vector, but we could also have a table of alphanumeric characters, called a "string," or a table of some more exotic collection of data.

We can make tables of tables (of tables), and use these to store multiple pieces of information. As an example, in Figure 1.11, we see the definition of a table called tXV – each entry contains two elements, one number, and one list of two numbers. This type of table would be useful, for example, in encoding the time (the first element), and position-and-velocity (the second pair) of a particle moving in one dimension. In the example, if we ask for the second element of our list, we get back a table consisting of one number, and a table (as expected) – we can work our way down, so that tXV[[2,1]] gives us back the first element of the table at location 2 of tXV, a "time" (if we like). We could also reference this element using tXV[[2]][[1]]. Finally, we can recover the position via tXV[[2,2,1]], literally "The first element of the second element of the table tXV[[2]]." The built-in Table command can also construct such tables, and in that context, we see at the bottom of Figure 1.11 the same tXV table built automatically. Before moving on to more general considerations, note that Mathematica has a built-in help system – if you see a command you don't recognize, try typing ? and then the command name.

## 1.10 Implementations and pseudo-code

With the exception of the current chapter, which contains pictures of actual Mathematica sessions (meant as a review, and to show what the environment looks like), the algorithms in this book will be displayed in one of two ways. "Implementations" are actual, runnable, Mathematica functions. They are meant to be

In[1]:= **txV = { {0.0, {0.1, 0.0}}, {0.1, {0.11, 0.2}}, {0.2, {0.14, 0.4}}, {0.3, {0.19, 0.6}} }**

Out[1]= {{0., {0.1, 0.}}, {0.1, {0.11, 0.2}}, {0.2, {0.14, 0.4}}, {0.3, {0.19, 0.6}}}

In[2]:= **txV[[2]]**

Out[2]= {0.1, {0.11, 0.2}}

In[3]:= **txV[[2, 1]]**

Out[3]= 0.1

In[4]:= **txV[[2, 2]]**

Out[4]= {0.11, 0.2}

In[5]:= **txV[[2, 2, 1]]**

Out[5]= 0.11

In[6]:= **txV = Table[{t, {.1+ t^2 , 2 t}}, {t, 0, .3, .1}]**

Out[6]= {{0., {0.1, 0.}}, {0.1, {0.11, 0.2}}, {0.2, {0.14, 0.4}}, {0.3, {0.19, 0.6}}}

Figure 1.11 We can define tables with elements consisting of multiple data – txV
is a table whose entries are themselves tables.

simple, with just the minimal elements needed to make a method work (there is no
error checking in these examples, I have stripped out anything that is extraneous to
keep the code short and readable). These code segments are written using only the
subset of Mathematica programming described in this chapter, and so should
be exportable to any other language with minimal changes. In that sense, I am
using the implementations as a sort of pseudo-code. As an example, in Implementation 1.1, we see a function that takes a square matrix $\mathbb{A}$, a vector **b** and returns
the product $\mathbb{A}$**b**.

---

**Implementation 1.1** Matrix-vector multiplication

---

```
MVmul[A_, b_] := Module[{v, i, j, N},
  N = Length[A];
  v = Table[0, {k, 1, N}];
  For[i = 1, i ≤ N, i = i + 1,
   v[[i]] = 0.0;
   For[j = 1, j ≤ N, j = j + 1,
    v[[i]] = v[[i]] + A[[i, j]] b[[j]];
   ];
  ];
  Return[v];
 ]
```

---

The other algorithm display type will be more traditional pseudo-code. I use
this second approach to highlight the logical structure of algorithms that may
have more involved or less clear implementations (especially ones that you might
implement in the lab problems). Sometimes, for example, the order of operations in an implementation can be confusing – in Implementation 1.1, the line:

`v[[i]] = v[[i]] + A[[i,j]] b[[j]]` makes it look as if `v[[i]]` is being defined implicitly in terms of itself. Look at the same matrix-vector multiplication, in pseudo-code form, in Algorithm 1.1. There, the use of $\leftarrow$ makes the order of operations clear. We are updating the elements of a vector **v**, starting at zero, and adding successive terms in the multiplication.

---

**Algorithm 1.1** Matrix-vector multiplication

---

$N \leftarrow$ Length($A$)
**for** $i = 1 \rightarrow N$ **do**
  $v_i \leftarrow 0$
  **for** $j = 1 \rightarrow N$ **do**
    $v_i \leftarrow v_i + A_{ij}b_j$
  **end for**
**end for**
**return** $v$

---

In general, the implementation will involve a full function definition, with arguments and local variables, and you can find the implementation and some examples of its use in the associated chapter notebook. The pseudo-code for the algorithm omits these details, and their inputs and outputs will be described in the text itself.

## 1.11 Timing and operation counts

We will be interested in basic timing for our methods, so we need a notion of counting the amount of time a particular algorithm will take to run. The rules are simple: 1. Every addition, subtraction, multiplication, or division takes the "same" amount of time to execute. 2. We are interested in the timing of an algorithm, up to constants that are machine specific. While a particular computer may add in three microseconds, but take nine microseconds to multiply, these are, up to order of magnitude, the same. Our interest is generally in the scale of the computation, not the details. So, given a problem with $n$ parameters that we know (experimentally, say) can be solved on an iPad in $N$ seconds, our question will always be, how many seconds does it take the iPad to solve the same type of problem with $m$ parameters?

To set the notation, we will learn that a general matrix inversion problem for an $n \times n$ matrix can be solved in $T(n) = \alpha n^3$ seconds, where $\alpha$ is a machine-specific (and therefore uninteresting) constant. So we know what happens to the timing if we double the size of the matrix – we octuple the run-time: $T(2n) = \alpha(2n)^3 = 8T(n)$. It is this scaling that is of interest, constants like $\alpha$ don't really matter. We use the Landau "O" notation to indicate the scaling – if we write $T(n) = O(f(n))$, we mean that $T(n) \leq Af(n)$ for constant $A$, and all $n$. In the matrix inverse timing, we have $T(n) = O(n^3)$, highlighting the scaling of the problem with $n$, and downplaying (by eliminating) the constants of proportionality set by the machine itself (and sometimes, our cleverness of implementation – after all, $0.9n^3$ is not as good as $0.01n^3$).

When counting operations, then, we can simply give the scaling with fundamentally interesting parameters in the problem. As an example, take the dot product of **a** and **b**, both in $\mathbb{R}^n$. We know that:

$$\mathbf{a} \cdot \mathbf{b} = \sum_{i=1}^{n} a_i b_i \tag{1.2}$$

and if we think of how the timing for this calculation goes with $n$, we have:

$$T(n) = \alpha n + \alpha(n-1) \tag{1.3}$$

where $\alpha$ is the constant associated with multiplication/addition (there are $n$ multiplications, and $n-1$ additions needed for the dot product). But, using the $O$ notation, we can simplify:

$$T(n) = 2\alpha n - \alpha = O(n) \tag{1.4}$$

so that fundamentally the time it takes to compute the dot product scales linearly with $n$. This suggests that if you knew, say, $T(2)$ for your computer (in order to get the coefficient out front of $O(n)$), you could compute $T(100)$.

## 1.12 Units and dimensions

We end on a general note about numerical work – on a computer, there is finite precision, so that the smallest number we can represent reliably, called "machine $\epsilon$," is not zero. On most machines circa the turn of the century, the smallest real number that could be used was on the order of $10^{-13}$ (i.e. you got around thirteen digits). That means that you should never compare the result of a calculation, one that is meant to produce small numbers, to zero. It's better to imagine that zero doesn't even exist.[3]

Now that brings us to a point of physics that we should consider prior even to numerical solution – think of a mass falling radially in towards a central body via the force of gravity. We would write the one-dimensional equation of motion as:

$$m\ddot{x} = -\frac{GMm}{x^2} \tag{1.5}$$

where $x$ is the location of the test mass, and $M$ is the mass of the central body ($m$ is the test body mass, but that, famously, does not matter). In the usual units (SI), we

---

[3] Look back at Figure 1.8 – the difference Abs [div - Round[div]] is compared to NZERO, a "numerical" zero, set at $10^{-10}$ in that example.

have $G = 6.67428 \times 10^{-11} \, \text{N(m/kg)}^2$. That is dangerously close to our lower limit, and certainly includes digits below machine $\epsilon$. We could measure all quantities in different units, so that, say, $GM \sim 1$, putting us in the numerically "nice" range (midway between $10^{-13}$ and $10^{13}$)[4]. But why not set up the problem so as to dispense with units entirely? We could render the equation dimensionless (or partially so), and then we get rid of the issue of representation for constants.

For the current case, let $x = \alpha q$, $t = \beta s$ where $\alpha$ will be a length of some sort, then $q$ is dimensionless, and similarly $\beta$ is a time, $s$ is dimensionless. The equation of motion can be written:

$$q'' = -\frac{GM\beta^2}{\alpha^3} \frac{1}{q^2} \tag{1.6}$$

where $q'' \equiv \frac{d^2 q(s)}{ds^2}$. Now we have options: Let $c = \alpha/\beta$, so that we use the fundamental speed of light to relate $\alpha$ to $\beta$. Then set $\alpha = MG/c^2$ to finish the story. We now have to solve:

$$q'' = -\frac{1}{q^2}, \tag{1.7}$$

a dimensionless equation of motion. What we get must, in the end, be translated back to the original units, but this is easy enough to do – given a value of the solution $q(s^*)$, we know that at time $t^* = \frac{MG}{c^3} s^*$, the test particle was located at $x(t^*) = \frac{MG}{c^2} q(s^*)$. Even easier, once you have the numerical result, you just label your axes to reflect the "units" you are using. It is clear that we only need to solve (1.7) once (given initial conditions), while (1.5) looks like it requires numerical solution for all different central masses, all values of $M$ – clarifying the parametric content of an equation is another advantage of the "nondimensionalization" process.

We will encounter a similar idea for each physical system we study, each with its own special constants (think of the constant $\hbar$ appearing in quantum mechanics, for example). When appropriate, we will return to this point, the advantages of removing dimensional consideration from a problem, but file it away now for future reference.

### Further reading

1. Cormen, Thomas H., Charles E. Leiserson, & Ronald L. Rivest. *Introduction to Algorithms*. The MIT Press, 1990.
2. Knuth, Donald. *Art of Computer Programming Vol 1. Fundamental Algorithms*. Addison-Wesley, 1997.

---

[4] Oh yes, there is also no such thing as infinity on a computer – that end seems more obvious than the lack of a true zero.

3. Knuth, Donald. *Art of Computer Programming Vol 3. Sorting and Searching.* Addison-Wesley, 1998.

## Problems

### Problem 1.1
Turn the following `While` statement into a `For` statement that executes identically:

---
**Implementation 1.2** Example While statement
---
```
x = 4.0;
While[Sin[x] ≤ .25,
  Print[x];
  x = x + 1/x;
]
```
---

### Problem 1.2
Referring to Figure 1.8, predict the output of: `DivideByTwo[76]`.

### Problem 1.3
What is the timing for the matrix-vector multiplication, $A\mathbf{v}$ with $A \in \mathbb{R}^{n \times m}$, $\mathbf{v} \in \mathbb{R}^m$? How about matrix-matrix multiplication, $AB$, with $A \in \mathbb{R}^{n \times m}$, $B \in \mathbb{R}^{m \times k}$?

### Problem 1.4
We have $N$ particles interacting under a pair potential $V(r)$ (like the Coulomb or gravitational potential) – we write a function that calculates the force acting on all the particles. How does that function's timing scale with $N$ (i.e. what is $T(N)$ for such a force-calculation)?

### Problem 1.5
In general relativity, it is common to refer to mass with dimension of length – find the appropriate combination of factors of $G$ and $c$ that allows you to take an expression for mass in kilograms, and turn it into an expression for mass in meters, i.e. find $p$ and $q$ such that:

$$M_{\mathrm{m}} = G^p c^q M_{\mathrm{kg}}. \tag{1.8}$$

What is the mass of the sun in kilometers?

### Problem 1.6
For a central charge $Q$ and a test charge $q$ with mass $m$, Newton's second law, together with the Coulomb force reads:

$$m\ddot{x} = \frac{Qq}{4\pi \epsilon_0 x^2} \tag{1.9}$$

(assume that $Q$ and $q$ here have the same sign) for separation $x$. The value of $\epsilon_0$ is $8.85 \times 10^{-12}$ $C^2/(N\,m^2)$. Set $x = \alpha z$ and $t = \beta s$ for $\alpha$ some length, $\beta$ some time, and $z, s$ dimensionless. Rewrite the equation of motion in terms of $z$ and $s$ – what must you set $\beta$ to if your target is $\frac{d^2 z(s)}{ds^2} = \frac{1}{4\pi z(s)^2}$? Notice that we can accomplish the same nondimensionalization by agreeing to "set $\epsilon_0 = 1$, $Q = q = 1$."

### Problem 1.7
Initial and/or boundary conditions must share the dimension of the quantity post-nondimensionalization. In the previous problem, you used $x = \alpha z$ and $t = \beta s$ to write an equation of motion that looks like: $z''(s) = 1/(4\pi z(s)^2)$. If we agree to set $x(0) = R$, and $\dot{x}(0) = 0$, then there is a natural length in the problem, take $\alpha = R$. Write the dimensionless boundary condition for $z(0)$ and $z'(0)$, solve for $z(s)$ and find the amount of time it takes for the particle of charge $q$ to go from $R$ to $2R$ (you must translate back from your dimensionless solution to get an actual time).

### Problem 1.8
For a function $f(x)$, the Taylor series expansion reads:

$$f(x + \Delta x) = \sum_{j=0}^{\infty} \frac{\Delta x^j}{j!} \left( \frac{d^j f(x)}{dx^j} \right) = f(x) + f'(x)\Delta x + \frac{1}{2}f''(x)\Delta x^2 + \dots \quad (1.10)$$

and provides a way to approximate the value of the function $f$ at $x + \Delta x$ (for $\Delta x \ll x$) given its value at $x$ (and the value of all its derivatives there). Let $f(x) = \sin(x)$, and write out the first three terms of the expansion for an arbitrary $x$ and $\Delta x$.

## Lab problems

### Problem 1.9
We can view arrays in a variety of ways. In this problem, we'll make a list of particle locations. Suppose we have one hundred particles to keep track of, each has three spatial coordinates, so we could imagine generating a table with 100 entries, each entry is itself a set of three values. Using the `Table` command, construct a table that spaces the particles evenly along the line from $\mathbf{r}_0 = 0$ to $\mathbf{r}_f = 1\hat{\mathbf{x}} + 1\hat{\mathbf{y}} + 1\hat{\mathbf{z}}$ (so the first particle is at the origin, and the hundredth particle sits at $\{1, 1, 1\}$). If you call your table `Xlist`, what is the output of: `Xlist[[72,2]]`? Provide both the numerical answer to five digits, and describe your interpretation of this entry in words.

### Problem 1.10
Write a function that takes, as input, a list of particle positions:

$$\{\{x_1, y_1, z_1\}, \{x_2, y_2, z_2\} \dots, \{x_n, y_n, z_n\}\}$$

and generates a three-dimensional plot of the individual particle locations. Use the built-in function `Point` to make a table of the points, then use the `Graphics3D` and `Show` commands to render the set of points in the table. Try it out on your list from Problem 1.9 (remember to use `?command` to get help on a built-in `command` within `Mathematica`).

## Problem 1.11

Write a function that approximates the value of sin(0.9) using the Taylor expansion from Problem 1.8 (here, $x + \Delta x = .9$, and you need to choose $x$ to be an angle for which you have a known value for sine). The sine function has an infinite power series expansion that looks like:

$$\sin(x) = \sum_{j=0}^{\infty} (-1)^j \frac{x^{2j+1}}{(2j+1)!}. \tag{1.11}$$

Write a function that computes the partial sum:

$$\sin_N(x) \equiv \sum_{j=0}^{N} (-1)^j \frac{x^{2j+1}}{(2j+1)!}, \tag{1.12}$$

and find the value of $N$ that gives back roughly the same value of sin(0.9) as calculated using the first three terms of the Taylor expansion.

## Problem 1.12

Write a function that takes input parameters `T` and `m`, and output (in a table of length two) that contains: 1. The frequency ($f$, not $\omega$) of a sinusoidal wave that is zero at $t = 0$ and $t = T$ and contains $m$ full cycles, and 2. A plot of this sinusoidal function that has time as its $x$-axis. Give the output of your function for $T = 4.0$ `Exp[1]`, $m = 3$.

## Problem 1.13

Suppose we make a table using the command:

```
ret = Table[{j, Table[Table[m, {m, 1, k}], {k, 1, j}]}, {j, 1, 10}]
```

What will the output of `ret[[8,2,5,3]]` be? (Predict first, then check.)

## Problem 1.14

(a) Write a function that takes two vectors **a** and **b** (both in $\mathbb{R}^n$) and outputs the dot product $\mathbf{a} \cdot \mathbf{b}$ using the `Sum` command. Note that your function should fail if the two vectors are not of the same length (you can check the length of a list using `Length`) – failure should be indicated with a printed statement, and a return value that could *not* be a valid dot-product output (for real-valued vectors).

(b) There is a `Timing` command that can be used to find how many seconds it takes to perform a particular operation – the usage is: `Timing[ ops ]`, and the output is $\{$`timing, return value of ops`$\}$ (i.e. a table with two entries). Using this command, write a function that takes, as its argument, an integer (call it `nsize`), and returns the amount of time it took to compute the dot product of $\mathbf{a} \cdot \mathbf{b}$ for $\mathbf{a}$, $\mathbf{b} \in \mathbb{R}^{nsize}$. Using your function, create a table of the amount of time it takes to compute the dot product for vectors of length 10 000, 20 000, ..., 100 000 (in steps of 10 000) (send in vectors that have all components set to one). From this table, estimate the amount of time it takes your computer (running `Mathematica`) to perform an arithmetic operation – describe your procedure for determining this value.

## Problem 1.15

**(a)** Construct a function that takes as input the height $z$ at which a solid body approaches a sphere of radius $R = 1$, as shown in Figure 1.12, and outputs the angle $\theta$ at which the body "bounces" off the sphere (use angle of incidence equals angle of reflection). What is the angle $\theta$ if $z = 0.25$?

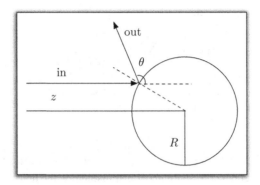

Figure 1.12 A solid body approaches a sphere at height $z$, and bounces off of it at angle $\theta$ (with respect to horizontal).

**(b)** Make a table of angles for a range of heights $z$ (you can use the built-in command `RandomReal[{-1,1}]` to specify a random number from $-1$ to $1$, for example) – using the `Histogram` function, generate a histogram of output angles for 100 000 input heights. Show your histogram on the range $-\pi \longrightarrow \pi$ (make sure you handle the $z < 0$ cases correctly).

# 2

# Ordinary differential equations

We will start off our investigation of numerical techniques by solving ordinary differential equations (ODEs). Our first method is a relatively direct discretization of Newton's second law. This is called the Verlet method, and is an efficient way to find solutions (i.e. a vector function $\mathbf{x}(t)$) given a force that depends on time and position (and appropriate initial data). The method has generalizations, but we will think about it only in the context of equations of motion. As a first method, it has the advantage of being straightforward, while relying in its derivation on ideas that can be extended to develop more involved numerical differential equation solvers.

In the more general context that follows the Verlet method, we note that any ODE can be re-written in the form: $\mathbf{f}'(x) = \mathbf{G}(x, \mathbf{f}(x))$ for vector-valued function $\mathbf{G}$ and vector-valued target function $\mathbf{f}(x)$. The punch-line will be that if we are given initial conditions $\mathbf{f}(0)$, we can use Runge–Kutta methods to find values approximating $\mathbf{f}(x)$ on a grid. First, we'll set up a few ODEs of interest, then introduce the process of discretization and the (special) Runge–Kutta family of methods that can be used to solve the discretized problem.

## 2.1 Physical motivation

There are a variety of problems in physics where ordinary differential equations govern the underlying natural phenomena. We'll start with the most familiar of these, Newton's second law, relating acceleration to forces. After reviewing a few different (relevant) force configurations, we'll look at the modification to Newton's second law that comes from special relativity. This modification presents even more complicated ODEs where, ultimately, numerical techniques are necessary.

### 2.1.1 Newton's second law

Newton's laws are among the first physical laws we encounter. The program of mechanics is, essentially, to find $x(t)$ (in one dimension), the position of a particle as a function of time, given a set of forces that act on the particle. The main computational component of Newton's three laws is the second: $ma = F$, where $m$ is the mass of the particle, and $F$ is the net force on the particle. In this equation, $a$ is the acceleration of the particle, defined to be the second derivative of position:

$$a(t) = \frac{d^2x(t)}{dt^2}. \tag{2.1}$$

So the problem of finding $x(t)$ amounts to solving the second-order differential equation:

$$\frac{d^2x(t)}{dt^2} = \frac{1}{m}F \tag{2.2}$$

given some $F$ (depending, in theory, on both the position itself, $x(t)$, and explicitly on time). In addition to this equation, we must provide data at some known point(s) – as a second-order differential equation, we will need two pieces of information – we could be given the initial position and velocity, $x(0)$ and $v(0)$, or we could be given an initial position and final position, $x(0)$ and $x(T)$, or some other combination. As a physical problem, appropriate for numerical consideration, a differential equation by itself is not enough – we must have the initial data (or some other accessible combination of information).

For most introductory examples of forces, the differential equation can be easily solved. Gravity near the surface of the Earth, for example, has $F = -mg$, a constant, and:

$$\frac{d^2x(t)}{dt^2} = -g \longrightarrow x(t) = -\frac{1}{2}gt^2 + At + B \tag{2.3}$$

where $A$ and $B$ are two arbitrary constants, and are set by the initial data.

The force associated with a simple spring is a little harder, and demands that we know something about solving ordinary differential equations. The force, as a function of position, is $F = -kx$, and our differential equation now reads:

$$\frac{d^2x(t)}{dt^2} = -\frac{k}{m}x(t) \longrightarrow x(t) = A\cos\left(\sqrt{\frac{k}{m}}t\right) + B\sin\left(\sqrt{\frac{k}{m}}t\right). \tag{2.4}$$

So forces can be functions of position. They can also be functions of time – suppose I push on a box of mass $m$ with a time-varying applied force: $F(t) = \alpha t$ – I push harder as time goes on. Then Newton's second law gives us:

$$\frac{d^2x(t)}{dt^2} = \frac{\alpha t}{m} \longrightarrow x(t) = \frac{\alpha t^3}{6m} + At + B, \tag{2.5}$$

again, easily solved in general, with undetermined constants $A$ and $B$ that are set once the rest of the physical configuration has been specified.

We could also imagine a force that depends on speed – various models of friction involve such forces, and it is easy to see why: Let $F = -\gamma v$ for constant $\gamma > 0$ – the larger the speed, the more the force acts. Then the differential equation we need to solve, and its solution, look like

$$\frac{d^2 x(t)}{dt^2} = -\frac{\gamma}{m} \frac{dx(t)}{dt} \longrightarrow x(t) = A - \frac{mB}{\gamma} e^{-\frac{\gamma}{m} t}. \tag{2.6}$$

In this case, we find that the speed of the particle, $v(t) \equiv \frac{dx(t)}{dt}$, is:

$$v(t) = B e^{-\frac{\gamma}{m} t} \tag{2.7}$$

which starts at $B$ and decreases as time goes on.

Friction, or in the absence of an explicit surface, air resistance, is a force that depends on speed, and the geometry of the massive object that is moving. We can approximate a general friction force by assuming we have some function $\mathbf{F}_\mu(\mathbf{v})$ (a vector in three dimensions, hence the bold face), opposing velocity and with magnitude that is dependent on $v$ (the magnitude of the velocity vector $\mathbf{v}$) – by series expansion, we can make some qualitative predictions for the form of $\mathbf{F}_\mu$:

$$\mathbf{F}_\mu(\mathbf{v}) = -\left( a_0 + a_1 v + a_2 v^2 + \cdots \right) \hat{\mathbf{v}}, \tag{2.8}$$

with $\hat{\mathbf{v}} \equiv \mathbf{v}/v$ and $v \equiv \sqrt{\mathbf{v} \cdot \mathbf{v}} \equiv \|\mathbf{v}\|$. The minus sign out front, and the unit vector at the end ensure that this force is always opposing the velocity vector direction of the object. The coefficients $a_0$, $a_1$ and $a_2$ can be computed in certain special cases (spheres in homogenous viscous media, for example), but we can tune them using physical observations as well. For the case of friction in one dimension, the first term, constant $a_0$, represents constant kinetic friction – there, we know the magnitude (again observationally) is $a_0 = \mu_k F_n$ where $F_n$ is the magnitude of the surface normal force. The second term, proportional to $v$, corresponds to the damping encountered in damped harmonic oscillators.

For air resistance, if we work in the regime where $v$ is "large," the squared term will dominate, and our force takes the form: $\mathbf{F}_\mu(\mathbf{v}) = -a_2 v^2 \hat{\mathbf{v}}$. Consider the one-dimensional case, where we'll let $F = -\gamma v^2$ and assume the motion is in the positive $x$ direction. Then the solution in this case, shown below, is different than the damping in (2.6) (associated with $a_1$ in (2.8)):

$$\frac{d^2 x(t)}{dt^2} = -\frac{\gamma}{m} \left( \frac{dx(t)}{dt} \right)^2 \longrightarrow x(t) = A + \frac{m}{\gamma} \log(\gamma t - mB). \tag{2.9}$$

Don't let the long right arrow indicate that we know, immediately, how to solve this ODE – you may or may not.[1] The speed, for this $x(t)$, is:

$$v(t) = \frac{dx(t)}{dt} = \frac{m}{\gamma t - mB} \tag{2.11}$$

so we see that $B$ could be used to set the initial speed.

Finally, we come to a problem that is easy to state, but whose solution is probably not too familiar – how about a box of mass $m$ undergoing friction of the form $F = -\gamma v^2 \hat{v}$ attached to a spring? We need to solve:

$$\frac{d^2 x(t)}{dt^2} = -\frac{k}{m} x(t) - \frac{\gamma v^2}{m} \text{sign}(v) \tag{2.12}$$

subject to the appropriate, physically determined, initial (or boundary) conditions.

Even in this one-dimensional, Newtonian, case, we have encountered a problem that we cannot solve (easily) by brute force. The physical model we have in mind, a mass attached to a spring, sitting in molasses, is neither crazy nor irrelevant.[2] So it would appear to be a reasonable target for study. There is an entire branch of physics (classical mechanics) devoted to the characterization of solutions for this type of scenario. Ideally, we want $x(t)$, but lacking that, there is very sophisticated mathematical machinery that can be employed to gain as much information as possible (maybe we can find $v(t)$, or the energy, or some other physical description that is incomplete but useful). There is also very sophisticated mathematical machinery that can be used to *find* $x(t)$, and this is the realm of computation. A computer can be used, with some care, to provide specific trajectories for almost any force (or more general ODE) you can write down. Classical mechanics gives us the tools to characterize various regimes of solution, and describe useful properties of the motion, even if $x(t)$ is unattainable. Computation will go in the opposite direction, giving us specific trajectories, for specific initial conditions, without allowing us to generalize the behavior of solutions beyond the one currently at hand. The two approaches are complementary, and using both together is a classic "whole is greater than the sum of the parts" parable.

We'll set up some specific targets to focus our attention – there are a variety of applications of Newton's second law that we can use to generate interest in solving various differential equations. Then, we'll start our numerical discussion by studying how to sensibly describe and solve ordinary differential equations

---

[1] If you're feeling good about your ODE skills, try this air resistance model with constant acceleration near the surface of the Earth:

$$\frac{d^2 x(t)}{dt^2} = -g + (\gamma/m)v^2. \tag{2.10}$$

[2] This is apparently a special kind of molasses, since we can easily solve for $x(t)$ provided the molasses exerts only a first order $F_\mu = -\gamma v$.

numerically, and the general scheme here is referred to as "numerical differential equation solving."

## Drag in two dimensions

In two dimensions, we have the vector form of Newton's second law: $m\mathbf{a} = \mathbf{F}$, just two copies of the one-dimensional case. Going back to our model for drag, the dependence of the drag force on $v^2$ refers to the magnitude of the vector $\mathbf{v}(t) = \frac{dx(t)}{dt}$ (squared). The direction of the drag is always opposing the velocity vector (which is in the direction of motion), so we can write:

$$\mathbf{F} = -\gamma v^2 \hat{\mathbf{v}} = -\gamma v \mathbf{v}. \tag{2.13}$$

The force in (2.13) has inherited time dependence through its dependence on $v$. To see the complication once an actual solution to the problem is required, combine drag with gravity, and write the two copies of Newton's second law, for the $x$ and $y$ components separately:

$$
\begin{aligned}
\frac{d^2x(t)}{dt^2} &= -\frac{\gamma}{m}\sqrt{\left(\frac{dx(t)}{dt}\right)^2 + \left(\frac{dy(t)}{dt}\right)^2}\frac{dx(t)}{dt} \\
\frac{d^2y(t)}{dt^2} &= -g - \frac{\gamma}{m}\sqrt{\left(\frac{dx(t)}{dt}\right)^2 + \left(\frac{dy(t)}{dt}\right)^2}\frac{dy(t)}{dt},
\end{aligned}
\tag{2.14}
$$

a pair of coupled, nonlinear ODEs.

## Spinning projectiles

Following [16], we can introduce additional realism by adding spin – suppose we have a ball of radius $r$ that is spinning with constant angular velocity $\omega$ – then the drag force on the top and bottom of the ball will be different – this effect is known as the "Magnus force," and tends to push the ball in a direction perpendicular to its center of mass motion. Referring to the top figure in Figure 2.1 with the ball spinning (about an axis pointing out of the page), think of the rest frame of the center of mass of the ball. Then the air is moving with speed $v_{cm} + \omega r$ at the "top" of the ball, and $v_{cm} - \omega r$ on the "bottom." The air circulates around the ball (assuming the ball is rough, it drags air around with it) – and so there is a force directed towards the center of the ball on an air packet at the top, and another, smaller force (because $v - \omega r < v + \omega r$) directed towards the center of the ball on an air packet at the bottom – the air exerts corresponding forces on the ball via Newton's third law, so that there is a larger force acting up, a smaller force acting down on the ball – the overall effect is a net force upwards. In the lower figure in Figure 2.1, we see the more general setting – the velocity vector $\mathbf{v}_{cm}$ is

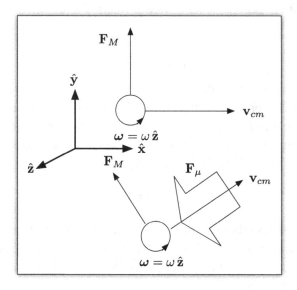

Figure 2.1 Magnus force is perpendicular to the velocity of the center of mass, and the spin axis – shown in the top figure for a specialized axis orientation. The lower figure shows an arbitrary center of mass velocity vector, with the oppositely directed drag force $\mathbf{F}_\mu$ and the Magnus force $\mathbf{F}_M$ given the angular $\boldsymbol{\omega}$ vector as shown.

always opposite the drag force $\mathbf{F}_\mu$, and the Magnus force $\mathbf{F}_M$ is instantaneously perpendicular to both the velocity and spin ($\boldsymbol{\omega}$) vectors.

We can find the direction of the Magnus force given $\mathbf{v}$ (the center of mass velocity) and the vector $\boldsymbol{\omega}$, which we assume does not change. Then the direction of the force is parallel to $\boldsymbol{\omega} \times \mathbf{v}$ – as for the magnitude, again, we need a fudge factor, barring simplified geometry. The simplest assumption to make is that the force is proportional to $\boldsymbol{\omega} \times \mathbf{v}$ – that's sensible, the more spin we have, the bigger the force, and similarly for a faster traveling projectile (i.e. the combinations $v \pm \omega r$ are the relevant ones). Call the proportionality constant $S$, so that $\mathbf{F}_M \equiv S\boldsymbol{\omega} \times \mathbf{v}$, then our equations of motion, augmented from (2.14), read:

$$\frac{d^2x(t)}{dt^2} = -\frac{\gamma}{m}\sqrt{\left(\frac{dx(t)}{dt}\right)^2 + \left(\frac{dy(t)}{dt}\right)^2}\frac{dx(t)}{dt} - \frac{S\omega}{m}\frac{dy(t)}{dt}$$

$$\frac{d^2y(t)}{dt^2} = -g - \frac{\gamma}{m}\sqrt{\left(\frac{dx(t)}{dt}\right)^2 + \left(\frac{dy(t)}{dt}\right)^2}\frac{dy(t)}{dt} + \frac{S\omega}{m}\frac{dx(t)}{dt}.$$

(2.15)

### Electricity and magnetism

In E&M, the primary goal is to find the electric and magnetic fields given their sources (charge and current). What we do beyond that is governed by the Lorentz

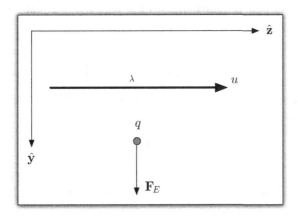

Figure 2.2 An infinite line of charge carrying uniform charge-per-unit-length, $\lambda$, pulled to the right with constant speed $u$ leads to an electric force pointing away from the wire, and magnetic force that depends on the velocity of the charged particle.

force law: $\mathbf{F} = q\mathbf{E} + q\mathbf{v} \times \mathbf{B}$. But given even the simplest electric and magnetic fields, the calculation of a particle trajectory can be difficult. Think of the electric and magnetic field generated by a moving line of charge carrying uniform $\lambda$ (charge per unit length) and moving at constant speed $u$ (to the right, say). In cylindrical coordinates, we would say that the fields are:

$$\mathbf{E} = \frac{\lambda}{2\pi \epsilon_0 s}\hat{\mathbf{s}}$$

$$\mathbf{B} = \frac{\mu_0 \lambda u}{2\pi s}\hat{\boldsymbol{\phi}}. \tag{2.16}$$

Restrict the motion to the $y$-$z$ plane as shown in Figure 2.2, since any particle motion that begins in this plane will remain there. Think of a charge $q$ initially at rest – the electric force will push the particle away from the line of charge. As the particle moves directly away from the line, the magnetic force will tend to push it to the right. While we can qualitatively describe this motion, it is difficult to analytically solve for $\mathbf{x}(t)$. Numerically, however, the problem is relatively simple.

Start with the Lorentz force, $\mathbf{F} = q\mathbf{E} + q\mathbf{v} \times \mathbf{B}$ – in terms of the Cartesian components shown in Figure 2.2,

$$m\ddot{y} = \frac{\lambda q}{2\pi \epsilon_0 y} - \frac{q\dot{z}\mu_0 \lambda u}{2\pi y}$$

$$m\ddot{z} = \frac{q\dot{y}\mu_0 \lambda u}{2\pi y}. \tag{2.17}$$

We can render this pair dimensionless. First take the $y$ equation, which can be written (noting that $\mu_0\epsilon_0 = 1/c^2$)

$$\ddot{y} = \frac{\lambda q}{2\pi m\epsilon_0}\frac{1}{y}\left[1 - \frac{\dot{z}}{c}\frac{u}{c}\right]. \tag{2.18}$$

Define $y = aY$ and $z = aZ$ for $a$ with dimension of length, and take $t = t_0 p$ with $p$ dimensionless. Then

$$Y''(p) = \frac{\lambda q t_0^2}{2\pi m\epsilon_0 a^2}\frac{1}{Y}\left[1 - \frac{a}{t_0}\frac{Z'(p)}{c}\tilde{u}\right] \tag{2.19}$$

defining $\tilde{u} \equiv u/c$. Finally, if we set $c = a/t_0$, then we have:

$$Y''(p) = \underbrace{\frac{\lambda q}{2\pi m\epsilon_0 c^2}}_{\equiv\alpha}\frac{1}{Y(p)}\left[1 - Z'(p)\tilde{u}\right] \tag{2.20}$$

with $\alpha$ dimensionless. Performing these same manipulations on the $z$ equation of motion, we get:

$$Z''(p) = \alpha\frac{1}{Y(p)}Y'(p)\tilde{u}. \tag{2.21}$$

One can simplify the pair (2.20) and (2.21) by noting, for example, that $\frac{Y'(p)}{Y(p)} = \frac{d}{dp}\log(Y(p))$, but as they stand, it is clear that we again have a coupled, nonlinear pair of ODEs, ready for numerical solution.

### 2.1.2 Relativistic mechanics

Newton's second law in the form:

$$\frac{dp(t)}{dt} = F \tag{2.22}$$

holds for relativistic mechanics, provided we understand $p(t)$ as the relativistic momentum:

$$p(t) = \frac{mv(t)}{\sqrt{1 - \frac{v(t)^2}{c^2}}}. \tag{2.23}$$

The list of forces for which this relativistic form can be solved exactly is small[3] – a constant force like $F = -mg$ is possible, but anything beyond that is more difficult.

---

[3] And the list of forces that have a self-consistent relativistic description even smaller.

As a specific problem to think about, consider a charged particle of mass $m$ with charge $q$ moving towards a sphere (centered on the origin) carrying total charge $-Q$ (for $Q > 0$). How long does it take the particle of mass $m$ (our test particle) to reach the origin if it started at $R$ from rest? Assuming the test particle moves along a straight line, we have a one-dimensional ODE to solve:

$$\frac{d}{dt}\left[\frac{mv(t)}{\sqrt{1 - \frac{v(t)^2}{c^2}}}\right] = -\frac{Qq}{4\pi\epsilon_0 x(t)^2},$$ (2.24)

with the initial conditions: $x(0) = R$ and $v(0) = 0$.

One can show that the above ODE can be "integrated once" (a manifestation of conservation of energy) to get:

$$E = \frac{mc^2}{\sqrt{1 - \frac{v(t)^2}{c^2}}} - \frac{Qq}{4\pi\epsilon_0 x(t)}$$ (2.25)

for constant $E$. Now, a change of variables is appropriate – take $x = az$ for some fundamental length $a$ (so that $z$ is dimensionless), and set $t = a/cs$ for dimensionless $s$ (when $s = 1$, $t$ is the time it takes light to go a distance $a$). The natural choice for $a$, given the problem, is $a \equiv Qq/(4\pi\epsilon_0 mc^2)$, so that in this parametrization, we have:

$$\frac{E}{mc^2} = \frac{1}{\sqrt{1 - z'(s)^2}} - \frac{1}{z(s)},$$ (2.26)

with initial conditions: $z(0) = R/a$ and $\frac{dz}{ds}|_{s=0} = 0$. Using these in (2.26), we learn that the initial (and constant) energy is:

$$\frac{E}{mc^2} = 1 - \frac{a}{R}$$ (2.27)

and then the first-order ODE for $z'(s)$ becomes

$$z'(s) = -\sqrt{1 - \left[1 - \frac{a}{R} + \frac{1}{z}\right]^{-2}},$$ (2.28)

with the minus signing indicating that this is *infall*. Note that the speed limit associated with special relativity is automatically enforced here – that limit is $|z'(s)| < 1$ for all $s$, and it is easy to see that this is the case for the expression (2.28).

We can solve this ODE numerically, although there is a solution available by integrating directly, since (2.28) can be written in integral form as

$$\int_{a/R}^{0} \frac{dz}{\sqrt{1 - \left[1 - \frac{a}{R} + \frac{1}{z}\right]^{-2}}} = -\int_{0}^{S} ds = -S$$ (2.29)

where $S$ is the final "time": $T = a/cS$ that was the original target. The left-hand side can be integrated in closed form, or you can wait until Chapter 6 and do it numerically.

### *Relativistic spring*

We can set up the interesting case of a relativistic mass on a spring, just using $F = -kx$ in (2.22), and you will find the solution for this specific case in Problem 2.19. Let's agree to start the mass from rest at an extension of $x(0) = a$. We know, from the non-relativistic solution, that there will be some value of $a$ for which the speed of the mass is greater than $c$, and presumably, the relativistic form of Newton's second law avoids this.

To show that in relativistic dynamics, $v < c$ is enforced for the spring potential, start from the definition of energy (relativistic):

$$E = \frac{mc^2}{\sqrt{1 - \frac{v^2}{c^2}}} + \frac{1}{2}kx^2, \tag{2.30}$$

where the first term on the right is the relativistic generalization of kinetic energy (including the famous rest energy), and we have added the potential, giving us a constant $E$. Then the speed (squared) becomes:

$$v^2 = c^2 \left[ 1 - \left( \frac{mc^2}{\left(E - \frac{1}{2}kx^2\right)} \right)^2 \right] \tag{2.31}$$

and it is clear that the second term in the brackets must be positive, and less than one, so that $v^2 < c^2$.

So we know that the relativistic setup enforces a speed limit, and we can even find the velocity as a function of position using energy conservation. To get $x(t)$, we can integrate $v$ from (2.31) once, 'or we can go back to (2.22) with the spring force in place,

$$\frac{d}{dt}\left[ \frac{m\dot{x}}{\sqrt{1 - \frac{\dot{x}^2}{c^2}}} \right] = -kx \longrightarrow m\ddot{x} = -kx\left(1 - \frac{\dot{x}^2}{c^2}\right)^{3/2}. \tag{2.32}$$

## 2.2 The Verlet method

We'll begin with a simple method that gives us a quick way to solve problems in Newtonian mechanics – our discussion will involve many of the ideas that will be used to develop more involved methods. Rather than a full abstract treatment of the most general problem possible, we'll use our mechanical intuition to generate the method. The starting point is Newton's second law in three dimensions governing the motion of a particle:

$$m\ddot{\mathbf{x}}(t) = \mathbf{F}(t), \tag{2.33}$$

where $\mathbf{x}(t)$ is the (target) position of the particle as a function of time, determined by the (possibly position-dependent force) $\mathbf{F}(t)$.

Suppose we had access to the solution $\mathbf{x}(t)$, then we could Taylor-expand to relate the position at time $t$ to the position(s) at time(s) $t \pm \Delta t$ for some small $\Delta t$. Let $\mathbf{v}(t)$ and $\mathbf{a}(t)$ denote the velocity and acceleration of the particle at time $t$, then:

$$\mathbf{x}(t + \Delta t) = \mathbf{x}(t) + \Delta t \mathbf{v}(t) + \frac{1}{2}\Delta t^2 \mathbf{a}(t) + O(\Delta t^3)$$

$$\mathbf{x}(t - \Delta t) = \mathbf{x}(t) - \Delta t \mathbf{v}(t) + \frac{1}{2}\Delta t^2 \mathbf{a}(t) - O(\Delta t^3). \tag{2.34}$$

---

**Error notation**

We introduced the Landau "O" notation $O(\Delta t^n)$ in Section 1.11 and have used it in (2.34) again – the notation is shorthand for the idea that the leading order corrections to an approximate expression occur at order $\Delta t^n$. Roughly, $O(\Delta t^n)$ refers to a function that has maximum value proportional to $\Delta t^n$ for all $\Delta t$. We know that $\mathbf{x}(t + \Delta t)$ has an infinite Taylor expansion in $\Delta t$, i.e. there is in general an infinite number of coefficients $\{\alpha_j\}_{j=0}^{\infty}$ in

$$\mathbf{x}(t + \Delta t) = \mathbf{x}(t) + \Delta t \dot{\mathbf{x}}(t) + \frac{1}{2}\Delta t^2 \ddot{\mathbf{x}}(t) + \frac{1}{6}\Delta t^3 \dddot{\mathbf{x}}(t) + \cdots$$

$$= \sum_{j=0}^{\infty} \alpha_j \Delta t^j, \tag{2.35}$$

where the values of $\alpha_j$ are related to numerical coefficients like $1/j!$ and the derivatives of the function $\mathbf{x}(t)$ evaluated at $t$. These coefficients don't change if we modify the value $\Delta t$, and what we are interested in is the behavior of different terms under a change in $\Delta t$.

When we write $O(\Delta t^3)$ in (2.34), we mean "some constant times $\Delta t^3$ plus an infinite sum of terms of the form constant $\times \Delta t^j$ with $j > 3$." Using this error order notation allows us to write equalities in Taylor expansions, and focuses our attention on terms that are relevant to our immediate purpose. We could, equivalently, write the first equation in (2.34) as

$$\mathbf{x}(t + \Delta t) \approx \mathbf{x}(t) + \Delta t \mathbf{v}(t) + \frac{1}{2}\Delta t^2 \mathbf{a}(t) \tag{2.36}$$

where we use the symbol $\approx$ to indicate that this is an approximate expression, but it is nice to use the reminder $O(\Delta t^3)$, together with equality, to reinforce the order of the remaining terms.

The point, summed up in an example, is: If you don't care about the form of the coefficient sitting in front of $\Delta t^{15}$, or any successive coefficients in front of higher powers of $\Delta t$, you can just write $+O(\Delta t^{15})$, and we will.

We can add the equations in (2.34) to give an expression for $\mathbf{x}(t + \Delta t)$ given $\mathbf{x}(t - \Delta t)$, $\mathbf{x}(t)$ and $\mathbf{a}(t)$ (available through the force, $\mathbf{a}(t) = \mathbf{F}(t)/m$):

$$\mathbf{x}(t + \Delta t) \approx 2\mathbf{x}(t) - \mathbf{x}(t - \Delta t) + \Delta t^2 \mathbf{a}(t). \tag{2.37}$$

So, given two previous values for $\mathbf{x}$, we can construct an approximation to $\mathbf{x}(t + \Delta t)$, propagating forward in time. The velocities can be constructed, as needed, via (this time subtracting the two expansions in (2.34)):

$$\mathbf{v}(t) \approx \frac{\mathbf{x}(t + \Delta t) - \mathbf{x}(t - \Delta t)}{2\Delta t}. \tag{2.38}$$

The method itself consists of stringing together positions updated using (2.37) – given a force, we know that $\mathbf{a}(t) = \frac{1}{m}\mathbf{F}(t)$, so the physical input is the acceleration. We also need initial data, $\mathbf{x}(0)$ and $\mathbf{v}(0)$. That poses a slight problem, since it is most natural to start the method off by specifying $\mathbf{x}(-\Delta t)$ (set $t = 0$ in (2.37)). But $\mathbf{x}(-\Delta t)$ and $\mathbf{v}(0)$ are related – take the Taylor expansion of $\mathbf{x}(-\Delta t)$:

$$\mathbf{x}(-\Delta t) \approx \mathbf{x}(0) - \mathbf{v}(0)\Delta t + \frac{1}{2}\mathbf{a}(0)\Delta t^2. \tag{2.39}$$

Now, given $\mathbf{x}(0)$ and $\mathbf{v}(0)$ we calculate $\mathbf{a}(0)$, and have $\mathbf{x}(-\Delta t)$. Then use that to proceed to find $\mathbf{x}(\Delta t)$, then $\mathbf{x}(2\Delta t)$ (requiring both the value $\mathbf{x}(0)$ and $\mathbf{x}(\Delta t)$), and so on, up to any desired final time $T = N\Delta t$ for integer $N$. An example of the implementation of the Verlet method is shown in Implementation 2.1 – physically, we need to provide a function to calculate the acceleration at time $t$, location $\mathbf{x}$ (called a in the argument list of the function in Implementation 2.1), and give the initial data (x0 and v0). In addition, we must specify the time step and number of steps to take (dt and Ns). Beyond that, the implementation functions as you'd expect, starting with our initial position and moving forward in time until the required number of steps have been taken.

---

**Implementation 2.1** The Verlet method

---

```
Verlet[a_, x0_, v0_, dt_, Ns_] := Module[{xout, xnext, xcur, xprev, t, index},
  xout = Table[0.0, {j, 1, Ns}];
  xcur = x0;
  xprev = x0 - dt v0 + (1/2) a[x0, 0.0] dt^2;
  t = 0.0;
  xout[[1]] = {t, x0};
  For[index = 2, index ≤ Ns, index = index + 1,
    xnext = 2.0 xcur - xprev + dt^2 a[xcur, t];
    t = t + dt;
    xout[[index]] = {t, xnext};
    xprev = xcur;
    xcur = xnext;
  ];
  Return[xout];
]
```

---

The advantage of Verlet is its speed, only one force evaluation is necessary per update.[4] It is also "time-reversible," up to the initial data, owing to its symmetric derivative approximation. That implies that energy should be relatively well conserved using the method. With our first (usable) method in place, we are ready to think about the more abstract issues associated with solving general ordinary differential equations. We will develop the Runge–Kutta class of methods in that context, and highlight some of the philosophical and moral issues involved in numerical solution. Aside from presentation, Runge–Kutta methods are different from the Verlet method in that they do not depend on values beyond the current one to perform their update.

## 2.3 Discretization

There are two sides to "solving" differential equations numerically: 1. the operation of discretization (which is our main topic) and 2. the mathematics of convergence. The point is, we can discretize (put onto a grid) almost anything, and that is the idea we will discuss, but a related question is how good solving an equation on a grid *is* in terms of approximating a continuous real solution to a continuous equation. That is a fundamental question in numerical analysis – the existence, and uniqueness of solutions to differential equations themselves (function theory) as well as existence and uniqueness of discrete solutions (algebra), and the "finishing move": if everything exists and is unique, how close can one make the discrete to the real? We'll return to that briefly at the end of the chapter, and set the question aside for now to focus on the practical act of discretization.

---

[4] The Runge–Kutta methods we will see later on require multiple evaluations of the force function per step – if that function is expensive, like many-body Coulomb forces, for example, then Verlet is a better method to use.

As our model problem, we will start with the first-order ODE in one dimension, together with a boundary condition:

$$\frac{df(x)}{dx} = G(x) \quad f(a) = f_a, \tag{2.40}$$

where $x$ is some generic independent variable (could be position, time, etc.), $f$ represents a function of that variable, and $G(x)$ is a given function of $x$. Because the differential equation is first order, we must provide the value of $f$ at some point to produce a unique solution, and we have taken $x = a$ with $f(a)$ given. The goal of a numerical solution to the above is to generate a set of values approximating $f(x)$ at some definite locations guided by the ODE itself (i.e. not just a random set of values at random locations).

Operationally, discretization procedures, whether on large spatial grids (as for partial differential equations) or simple forward propagation of variables (ODEs), come from Taylor series expansion. Any (well-behaved) function can be expanded close to a known point $x$ by

$$f(x + \epsilon) = f(x) + \epsilon f'(x) + \frac{1}{2}\epsilon^2 f''(x) + O(\epsilon^3) \tag{2.41}$$

for small $\epsilon$. This expansion says, among other things, that if you know the value of a function and all its derivatives at a point, then you can find its value at nearby points. As a corollary, if you know the value of a function and *some* of its derivatives at a point, you can find an *approximation* to the value of the function at nearby points.

As we go, keep in mind the following conventions – unbarred functions refer to the continuous form $f(x)$. In order to compare with our (at this point necessarily discrete) approximation, we project these continuous functions onto a grid $\{f_n \equiv f(x_n)\}_{n=1}^{N}$ for some set of points $\{x_n\}_{n=1}^{N}$. For example, we might space the grid points equally, $x_n \equiv n\Delta x$ for a provided $\Delta x$. Finally, our numerical approximation (which will be compared with $f_n$) is denoted $\bar{f}_n$ or its variants (tildes, hats, etc.). The numerical methods we develop will generate sequences of values, $\{\bar{f}_n\}_{n=1}^{N}$ (as the Verlet method did) – these have no continuum analogue; they are, in a sense, "just" sequences of numbers. The identification of $\bar{f}_n$ with $f_n$ comes from an analysis of the numerical method. This strict distinction between the "Platonic, continuous function" $f(x)$, its projection onto a grid, $f_n$, and the numerical approximation to $f_n$, called $\bar{f}_n$, will be relaxed as we move forward – I want to stress that there are three *different* objects here, but in practice we will generally rely on context to distinguish between the numerical approximation and the projected true solution.

### 2.3.1 Euler's method

Take a fixed grid, with $x_j = a + j\Delta x$ for $j = 0 \longrightarrow N$. The first point ($j = 0$), then, is the boundary point in (2.40), where we know the value of $f$ prior to any discretization or approximation. We'll use the Taylor expansion (2.41) to make an approximate statement about the sequence $\{f_n\}_{n=0}^N$, the projection of the continuous solution $f(x)$ onto our grid. Then we'll use that statement to define a numerical method that generates a sequence $\{\bar{f}_n\}_{n=0}^N$.

The relation between $f(x_{n+1})$ to $f(x_n)$ (and its derivatives), from (2.41), is:

$$f(x_{n+1}) = f(x_n + \Delta x) = f(x_n) + \Delta x f'(x_n) + \frac{1}{2}\Delta x^2 f''(x_n) + O(\Delta x^3). \quad (2.42)$$

We'll truncate this expression at $\Delta x$, using $f_{n+1} \equiv f(x_{n+1})$, we can write

$$f_{n+1} = f_n + \Delta x f'(x_n) + O(\Delta x^2). \quad (2.43)$$

Since we are given the function $G(x)$, the derivative of $f(x)$, we can express the right-hand side in terms of $G(x)$ evaluated at $x_n$ (written unambiguously as $G_n \equiv G(x_n)$):

$$f_{n+1} = f_n + \Delta x G_n + O(\Delta x^2). \quad (2.44)$$

So far, everything is exact – we have simply chosen specific values for $x$ and $\epsilon$ in (2.41), and used the ODE.

Now the method implied by this observation, called Euler's method, is used to generate a sequence $\{\bar{f}_n\}_{n=0}^N$ recursively given the starting point $\bar{f}_0$, which we will take to be $f_a$. Starting from $n = 0$, update via:

$$\boxed{\begin{aligned} \bar{f}_{n+1} &= \bar{f}_n + \Delta x G_n \\ \bar{f}_0 &= f_a. \end{aligned}} \quad (2.45)$$

Notice, there is no approximation sign, no error notation; this pair of equations defines a sequence, period. The properties of that sequence are what we will analyze.

First, we can define the "local truncation error" of the method. Suppose we start with the value $f_n \equiv f(x_n)$, i.e. the value of the true solution $f(x)$ evaluated at $x_n$, and take one step using the method (2.45), call the result $\tilde{f}_{n+1}$:

$$\tilde{f}_{n+1} = f_n + \Delta x G(x_n) = f_n + \Delta x f'(x_n), \quad (2.46)$$

Now subtract the Taylor expansion of $f(x_n + \Delta x)$ in (2.43) (i.e. $f_{n+1}$) from $\tilde{f}_{n+1}$: $\tilde{f}_{n+1} - f_{n+1} = O(\Delta x^2)$, the difference is $O(\Delta x^2)$. So taking a single step of (2.45), starting from $f_n$ yields $f_{n+1}$ up to an error of order $\Delta x^2$.

In general, if we take a single step of a method taking us from $x_n$ to $x_n + \Delta x$ starting with $f_n$, and subtract the Taylor expansion of $f(x_n + \Delta x)$, we will get an error $O(\Delta x^p)$, and the method is then said to have local truncation error order $p$. This is an error per-step. If our problem is defined on a grid $x \in [0, 1]$, say (which would be typical), then $\Delta x = 1/N$, and we make $O(\Delta x^p)$ error per step for $N$ steps, so the global truncation error will be $O(\Delta x^p)N = O(\Delta x^{p-1})$. The global truncation error for the Euler method is $O(\Delta x)$.

**Example**

Here's a simple example, where the recursion (2.45) is solvable, and the exact solution (and hence its projection onto a grid) is known: Take $G(x) = x$, and $f(0) = \alpha$ in (2.40), and let $x_j = j\Delta x$ for fixed $\Delta x$. The continuous solution and its projection are:

$$f(x) = \alpha + \frac{1}{2}x^2$$

$$\bar{f}_n = \alpha + \frac{1}{2}x_n^2 = \alpha + \frac{1}{2}n^2 \Delta x^2. \tag{2.47}$$

Using (2.45), we have:

$$\tilde{f}_{n+1} = \tilde{f}_n + \Delta x x_n = \tilde{f}_n + n \Delta x^2 \tag{2.48}$$

and writing out a few terms allows us to simplify the recursion,

$$\tilde{f}_n = \tilde{f}_{n-1} + (n-1)\Delta x^2$$

$$= \tilde{f}_{n-2} + ((n-2) + (n-1))\,\Delta x^2$$

$$\vdots \tag{2.49}$$

$$= \tilde{f}_{n-j} + \left(\sum_{k=1}^{j}(n-k)\right)\Delta x^2.$$

From the final line, we can set $j = n$, and sum, to get:

$$\tilde{f}_n = \tilde{f}_0 + \sum_{k=1}^{n}(n-k)\Delta x^2 = \alpha + \frac{1}{2}n^2\Delta x^2 - \frac{1}{2}n\Delta x^2. \tag{2.50}$$

We can form the residual, for any $n$, $r_n \equiv \tilde{f}_n - \bar{f}_n$, since we have explicit expressions for both, and

$$r_n = \frac{1}{2}n\Delta x^2 = \frac{1}{2}x_n\Delta x. \tag{2.51}$$

It is clear that the error at any step will go to zero as $\Delta x \longrightarrow 0$, so the method is "consistent." But, for a fixed step size $\Delta x$, the error depends on how many steps we take – the maximum value of $r_n$ is determined by the number of points we include in our sequence, $N$.

### 2.3.2 Improved accuracy

There are two approaches to improved accuracy – one is to use more derivative information from the Taylor series (2.41) to generate a method that has, at each step, a smaller error – the other (related) idea is to modify the numerical method directly by, for example, evaluating the function $G(x)$ in $f'(x) = G(x)$ at some new point. Let's explore both of these options.

We can retain terms up to and including $\epsilon^2$ in the expansion (2.41) – for the projection $f_n$ with our grid defined as above, this gives:

$$f_{n+1} = f_n + \Delta x G(x_n) + \frac{1}{2}\Delta x^2 G'(x_n) + O(\Delta x^3), \qquad (2.52)$$

where we used $f'(x) = G(x) \longrightarrow f''(x) = G'(x)$ to replace $f''(x_n)$. Define the *method*, then, by:

$$\bar{f}_{n+1} = \bar{f}_n + \Delta x G(x_n) + \frac{1}{2}\Delta x^2 G'(x_n)$$

$$\bar{f}_0 = f_a. \qquad (2.53)$$

The local truncation error again comes from comparing the action of this update on $f_n$ to $f_{n+1}$. We have:

$$\tilde{f}_{n+1} = f_n + \Delta x G(x_n) + \frac{1}{2}\Delta x^2 G'(x_n) \qquad (2.54)$$

and $\tilde{f}_{n+1} - f_{n+1} = O(\Delta x^3)$, so the method has local accuracy $O(\Delta x^3)$ (or "is third-order accurate").

Now for an interesting observation. Consider the Taylor expansion of $G(x_n + \frac{1}{2}\Delta x)$ – since $G(x)$ is a continuous (assume, nice) function, the Taylor expansion exists and

$$G\left(x_n + \frac{1}{2}\Delta x\right) = G(x_n) + \frac{1}{2}\Delta x G'(x_n) + O(\Delta x^2). \qquad (2.55)$$

Compare the right-hand side with (2.52) – which can then be written as

$$f_{n+1} = f_n + \Delta x G\left(x_n + \frac{1}{2}\Delta x\right) + O(\Delta x^3), \qquad (2.56)$$

the two expressions (2.56) and (2.52) are *identical* up to errors of order $\Delta x^3$. That means that the method

$$\bar{f}_{n+1} = \bar{f}_n + \Delta x G\left(x_n + \frac{1}{2}\Delta x\right)$$

$$\bar{f}_0 = f_a \qquad (2.57)$$

is equivalent to (2.53) (at each update, the two methods give the same value up to $O(\Delta x^3)$).

---

**Example**

Let's return to our test case, $G(x) = x$ with $f(0) = \alpha$, and compare the solutions to both (2.53) and (2.57) to the exact solution. The derivative of $G(x)$ is $G'(x) = 1$ here. If we iterate using (2.53), then our sequence is:

$$\bar{f}_n = \bar{f}_{n-1} + x_{n-1}\Delta x + \frac{1}{2}\Delta x^2 = \bar{f}_{n-1} + \Delta x^2\left(n - \frac{1}{2}\right)$$

$$= \bar{f}_{n-2} + \Delta x^2\left[\left(n - \frac{1}{2}\right) + \left(n - \frac{3}{2}\right)\right]$$

$$\vdots$$

$$= \bar{f}_{n-j} + \Delta x^2\sum_{k=1}^{j}\left(n - k + \frac{1}{2}\right),$$

(2.58)

and performing the sum, with $j = n$,

$$\bar{f}_n = \bar{f}_0 + \frac{1}{2}n^2\Delta x^2 = \alpha + \frac{1}{2}x_n^2.$$

(2.59)

The sequence $\{\bar{f}_n\}_{n=0}^N$ is identical to the projection from (2.47). That's no surprise – the higher-order terms in (2.52) all vanish since the second derivative of $G(x) = x$ is zero.

Now, using (2.57) to generate the sequence $\{\bar{f}_n\}_{n=0}^N$, we have the update

$$\bar{f}_n = \bar{f}_{n-1} + \Delta x G\left(x_{n-1} + \frac{1}{2}\Delta x\right) = \bar{f}_{n-1} + \Delta x\left((n-1)\Delta x + \frac{1}{2}\Delta x\right)$$

(2.60)

$$= \bar{f}_{n-1} + \Delta x^2\left(n - \frac{1}{2}\right),$$

identical to (2.58), the resulting sequence will also be identical. Again, since there are no errors of order $\Delta x^3$ in this example, the methods (2.53) and (2.57) generate the same sequence, and each is identical to the grid-projection of the solution to the problem.

---

We'll move on to more general ODEs in the next section, but the pattern for generating methods and improving accuracy will be the same – we'll take the "shifted evaluation" point of view represented by (2.57) to define the class of Runge–Kutta methods.

## 2.4 Runge–Kutta methods

The most interesting applications of ODE solving start with the more general form:

$$\frac{df(x)}{dx} = G(x, f(x)) \quad f(a) = f_a, \tag{2.61}$$

where the derivative of $f(x)$ depends on both $x$ and $f(x)$. This is still a first-order ODE, but we'll see in Section 2.4.2 how, if we view (2.61) as a *vector* equation, any ODE can be written in this form, so the setup is generic. We'll generate methods appropriate to a grid $x_j = a + j \Delta x$ for $j = 0 \longrightarrow N$ as before.

Start with the extension of Euler's method to (2.61):

$$\boxed{\begin{aligned} \bar{f}_{n+1} &= \bar{f}_n + \Delta x G(x_n, \bar{f}_n) \\ \bar{f}_0 &= f_a. \end{aligned}} \tag{2.62}$$

The motivation is the truncation of the Taylor expansion (2.43), with $f'(x)$ replaced by $G(x, f(x))$, so in discrete form, $G(x_n, f_n)$. You can easily verify that the method has local error order $O(\Delta x^2)$.

To improve the accuracy, we will again include more terms from the Taylor expansion, as in (2.52). Now, however, the derivative of $G(x, f(x))$ with respect to $x$ is more involved:

$$\frac{dG(x, f(x))}{dx} = \frac{\partial G}{\partial x} + \frac{\partial G}{\partial f}\frac{df}{dx} = \frac{\partial G}{\partial x} + \frac{\partial G}{\partial f}G \tag{2.63}$$

since $G$ depends on $x$ both explicitly, and implicitly through $f$. Notice that we have used the ODE itself in the second equality, replacing $\frac{df}{dx}$ with $G$.

The truncated expansion (2.52) now reads

$$\begin{aligned} f_{n+1} = f_n &+ \Delta x G(x_n, f_n) \\ &+ \frac{1}{2}\Delta x^2 \left[ \frac{\partial G(x_n, f_n)}{\partial x} + \frac{\partial G(x_n, f_n)}{\partial f} G(x_n, f_n) \right] + O(\Delta x^3). \end{aligned} \tag{2.64}$$

We can once again associate terms on the right-hand side with the expansion of $G$ evaluated at a modified point, consider

$$\begin{aligned} G &\left( x_n + \frac{1}{2}\Delta x, f_n + \frac{1}{2}\Delta x G(x_n, f_n) \right) \\ &= G(x_n, f_n) + \frac{1}{2}\Delta x \left[ \frac{\partial G(x_n, f_n)}{\partial x} + \frac{\partial G(x_n, f_n)}{\partial f} G(x_n, f_n) \right] + O(\Delta x^2) \end{aligned} \tag{2.65}$$

so that we can write (2.64) in terms of the shifted evaluation

$$f_{n+1} = f_n + \Delta x G\left(x_n + \frac{1}{2}\Delta x, \, f_n + \frac{1}{2}\Delta x G(x_n, f_n)\right) + O(\Delta x^3). \quad (2.66)$$

The method that we get from this approach is called the "midpoint Runge–Kutta" method, and we can write it in the transparent form:

$$
\begin{aligned}
k_1 &\equiv \Delta x G(x_n, \bar{f}_n) \\
k_2 &\equiv \Delta x G\left(x_n + \frac{1}{2}\Delta x, \, \bar{f}_n + \frac{1}{2}k_1\right) \\
\bar{f}_{n+1} &= \bar{f}_n + k_2 \\
\bar{f}_0 &= f_a.
\end{aligned}
\qquad (2.67)
$$

From its motivation, it should be clear that the local error is $O(\Delta x^3)$. We have written the method to highlight the "shifted evaluation" of $G$ as the accuracy-improving mechanism, although we know that it is just shorthand for keeping additional terms from the Taylor series for $f(x)$.

The presence of the modified locations for the evaluation of $G(x, f(x))$ is the hallmark of a Runge–Kutta method, and any higher-order "RK" method can be derived by introducing additional terms like $k_1$ and $k_2$ – these are chosen by tuning coefficients to kill off higher and higher order terms in the Taylor series expansion of $f(x + \Delta x)$. The code implementing the midpoint method is shown in Implementation 2.2. To use the function RKODE, we provide the function G that computes the right-hand side of the ODE, the initial value of the independent variable, x0, and the initial value of the function, f0 (i.e. $f(x_0)$). This information is enough to specify the mathematical problem, but there are numerical details,

---

**Implementation 2.2** Midpoint Runge–Kutta

```
RKODE[G_, x0_, f0_, dx_, Ns_] := Module[{k1, k2, retvals, index, xn, fn},
  retvals = Table[{0.0, 0.0}, {i, 1, Ns}];
  xn = x0;
  fn = f0;
  retvals[[1]] = {x0, f0};
  For[index = 2, index ≤ Ns, index = index + 1,
    k1 = dx G[xn, fn];
    k2 = dx G[xn + dx / 2, fn + k1 / 2];
    fn = fn + k2;
    xn = xn + dx;
    retvals[[index]] = {xn, fn};
  ];
  Return[retvals];
  ]
```

we need to give the spacing in $x$, called dx and the number of steps to take, Ns. Because of the way in which Mathematica treats vectors, the code shown in Implementation 2.2 will work for both scalar and vector values of $f$ and $G$, provided they match correctly (if $f$ is a vector of length $p$, then $G$ must be a vector of length $p$).

A popular combination with the same flavor as the midpoint Runge–Kutta method is the fourth-order Runge–Kutta method, defined by the update:

$$
\begin{aligned}
k_1 &= \Delta x\, G(x_n, \bar{f}_n) \\[4pt]
k_2 &= \Delta x\, G\left(x + \frac{1}{2}\Delta x, \bar{f}_n + \frac{1}{2}k_1\right) \\[4pt]
k_3 &= \Delta x\, G\left(x + \frac{1}{2}\Delta x, \bar{f}_n + \frac{1}{2}k_2\right) \\[4pt]
k_4 &= \Delta x\, G(x + \Delta x, \bar{f}_n + k_3) \\[4pt]
\bar{f}_{n+1} &= \bar{f}_n + \frac{1}{3}\left(\frac{k_1}{2} + k_2 + k_3 + \frac{k_4}{2}\right) \\[4pt]
\bar{f}_0 &= f_a
\end{aligned}
\tag{2.68}
$$

with local truncation error $O(\Delta x^5)$ (the method is called "fourth order" in reference to its global error). Again, there is no mystery here – just judicious choices.

### 2.4.1 Adaptive step size

We have been considering a grid of fixed $\Delta x$ – but what if we change $\Delta x$? The nice thing about Runge–Kutta methods is that typically, one set of evaluations can be used to generate two different approximations – for example, referring to (2.67), we can construct both the $\bar{f}_{n+1}$ there, as well as the Euler approximation (denoted $\tilde{f}_{n+1}$ here to distinguish it), out of the components $k_1$ and $k_2$ with no additional evaluations of $G$:

$$
\begin{aligned}
\bar{f}_{n+1} &= \bar{f}_n + \Delta x_n G\left(x_n + \frac{1}{2}\Delta x_n, \bar{f}_n + \frac{1}{2}\Delta x_n G(x_n, \bar{f}_n)\right) + O\left(\Delta x_n^3\right) \\[4pt]
\tilde{f}_{n+1} &= \bar{f}_n + \Delta x_n G(x_n, \bar{f}_n) + O\left(\Delta x_n^2\right)
\end{aligned}
\tag{2.69}
$$

where $\Delta x_n$ is the step size for use at the $n$th step. Suppose we start with $\bar{f}_n = \tilde{f}_n$, so we consider the same initial point – then the difference between these two approximations after one step can be written:

$$
r_{n+1} \equiv \bar{f}_{n+1} - \tilde{f}_{n+1} = O\left(\Delta x_n^2\right) = \alpha \Delta x_n^2
\tag{2.70}
$$

for some constant $\alpha$. We have an automatic bound on the error – this information can be used as follows: Fix a tolerance $\epsilon$ for each step and *find* $\Delta x$ that satisfies this tolerance. So rather than fixed $\Delta x$, we have fixed error-per-step $\epsilon$. Suppose we take a step with $\Delta x_n$ (specified) – if $|r_{n+1}| < \epsilon$, then we're done, but we can "suggest" a $\Delta x$ for the next step (i.e. $\Delta x_{n+1}$) by noting that we *could* have taken a step of size $\Delta x_n$ with error: $\alpha \Delta x_n^2 = \epsilon$, but instead we took $\alpha \Delta x_n^2 = |r_{n+1}|$ so the next step could be made larger:

$$\frac{\Delta x_{n+1}^2}{\Delta x_n^2} = \frac{\epsilon}{|r_{n+1}|} \rightarrow \Delta x_{n+1} = \Delta x_n \sqrt{\frac{\epsilon}{|r_{n+1}|}}. \tag{2.71}$$

If instead, we find $|r_{n+1}| > \epsilon$, then we took too large a step to satisfy our fixed tolerance – we must *reduce* $\Delta x_n$ and take the step again – in that case, we use the above formula to re-set the step size and try again:

$$\Delta x_n^{\text{new}} = \Delta x_n \sqrt{\frac{\epsilon}{|r_{n+1}|}} \tag{2.72}$$

this time, with $\epsilon < |r_{n+1}|$, our new $\Delta x_n^{\text{new}}$ is smaller than the original.

The same basic procedure holds for higher-order RK methods – for example, RK54 coefficients (this is the Cash–Karp embedding you will see in Problem 2.9) can be combined to form a method of order five and four – two different approximations of different order, and then we use that difference to set $\Delta x$ (see [40], there the power will be $\sim \epsilon^{1/5}$ instead of the $1/2$ we obtained here).

### 2.4.2 Vectors

The equation we are using RK methods to solve,

$$\frac{df(x)}{dx} = G(x, f(x)) \quad f(a) = f_a \tag{2.73}$$

does not have to be a single equation – we can just as easily view the problem defined with $\mathbf{f}(x), \mathbf{G}(x, \mathbf{f}(x)) \in \mathbb{R}^p$:

$$\frac{d\mathbf{f}(x)}{dx} = \mathbf{G}(x, \mathbf{f}(x)) \quad \mathbf{f}(a) = \mathbf{f}_a, \tag{2.74}$$

where now a vector of initial values must be provided for $x = a$.

The vector form is useful when we have multiple, coupled, ODEs. For example, given:

$$\frac{df_1(x)}{dx} = -f_2(x)$$
$$\frac{df_2(x)}{dx} = f_1(x), \tag{2.75}$$

we could write the vector equation

$$\frac{d}{dx} \underbrace{\begin{pmatrix} f_1(x) \\ f_2(x) \end{pmatrix}}_{\equiv \mathbf{f}(x)} = \underbrace{\begin{pmatrix} -f_2(x) \\ f_1(x) \end{pmatrix}}_{\equiv \mathbf{G}(x,\mathbf{f})}. \tag{2.76}$$

But we can also turn higher-order ODEs into multiple coupled ODEs. Suppose we were given the second-order differential equation

$$\frac{d^2 h(x)}{dx^2} = J(x, h(x), h'(x)), \tag{2.77}$$

for some function $J$. Let $f_1(x) = h(x)$, $f_2(x) = h'(x) = f_1'(x)$ and then we could write the first-order, two-dimensional vector equation

$$\frac{d}{dx} \underbrace{\begin{pmatrix} f_1(x) \\ f_2(x) \end{pmatrix}}_{\equiv \mathbf{f}(x)} = \underbrace{\begin{pmatrix} f_2(x) \\ J(x, f_1(x), f_2(x)) \end{pmatrix}}_{\equiv \mathbf{G}(x,\mathbf{f})} \tag{2.78}$$

where the top equation enforces the definition $f_1'(x) = f_2(x)$ and the bottom equation sets $h''(x) = J(x, h(x), h'(x))$, expressing the ODE (2.77).

In classical mechanics, the move from the second-order equation of motion (Newton's second law) to a pair of first-order equations is accomplished by Hamilton's formulation of the equations of motion. Given

$$m\ddot{x} = F(x) \tag{2.79}$$

in one dimension, define $p = m\dot{x}$, so that $\frac{dp}{dt} = F(x)$ and $\frac{dx}{dt} = p/m$ and we can write the pair as a vector

$$\frac{d}{dt} \begin{pmatrix} x \\ p \end{pmatrix} = \begin{pmatrix} \frac{p}{m} \\ F(x) \end{pmatrix}. \tag{2.80}$$

As an example with all the pieces in place, take a damped harmonic oscillator,

$$m\ddot{x}(t) = -m\omega^2 x(t) - 2m\gamma\omega\dot{x}(t), \tag{2.81}$$

with frequency $\omega$, and damping coefficient $\gamma$. This second-order ODE can be turned into the first-order vector system (we'll use $v$ instead of $p$, but the sentiment is precisely that of (2.80)):

$$\frac{d}{dt} \underbrace{\begin{pmatrix} x \\ v \end{pmatrix}}_{\equiv \mathbf{f}} = \underbrace{\begin{pmatrix} v \\ -\omega^2 x - 2\gamma\omega v \end{pmatrix}}_{\equiv \mathbf{G}}. \tag{2.82}$$

If we were going to use our code from Implementation 2.2 to solve this system, we would have to define $\mathbf{G}$, explicitly, in terms of the components of $\mathbf{f}$. Since we have associated $f_1$ with $x$ and $f_2$ with $v$, we would define **G** as:

$$\mathbf{G} = \begin{pmatrix} f_2 \\ -\omega^2 f_1 - 2\gamma\omega f_2 \end{pmatrix}. \tag{2.83}$$

Note that the ordering of the vector **f** is arbitrary, we could associate $f_1$ with $v$ and $f_2$ with $x$ – as long as we are consistent, in this case we would have:

$$\mathbf{G} = \begin{pmatrix} -\omega^2 f_2 - 2\gamma\omega f_1 \\ f_1 \end{pmatrix}. \tag{2.84}$$

Once you've chosen an embedding, stick with it. Now we have the vector **G**, and all we need to do in order to use any of the RK variants, is provide the initial data in vector form (ensuring that it is correctly encoded) and let it go.

## 2.5 Stability of numerical methods

We will discuss the issue of stability briefly, just enough to introduce the basic idea. We'll use a linear model problem, the simple harmonic oscillator with vectorized (first-order) equations as in (2.82) ($\gamma = 0$),

$$\frac{d}{dt}\begin{pmatrix} x \\ v \end{pmatrix} = \begin{pmatrix} v \\ -\omega^2 x \end{pmatrix}. \tag{2.85}$$

Call the vector on the left **f** (dropping the bars, this is a purely numerical issue, we do not need to compare with the continuum form), then the forward Euler method updates according to:

$$\mathbf{f}_{n+1} = \mathbf{f}_n + \Delta t \begin{pmatrix} 0 & 1 \\ -\omega^2 & 0 \end{pmatrix} \mathbf{f}_n = \underbrace{\begin{pmatrix} 1 & \Delta t \\ -\omega^2 \Delta t & 1 \end{pmatrix}}_{\equiv \mathbb{D}} \mathbf{f}_n, \tag{2.86}$$

so that given the initial conditions, $\mathbf{f}_0$, the $n$th step of the method can be written in terms of $\mathbb{D}$ as $\mathbf{f}_n = \mathbb{D}^n \mathbf{f}_0$.

We shall see, during our discussion of the eigenvalue problem in linear algebra (in Chapter 10), that this equation (for almost all $\mathbf{f}_0$) implies that $\mathbf{f}_n$ goes like $\rho(\mathbb{D})^n$ for large $n$, where $\rho(\mathbb{D})$ is the "spectral radius" of the matrix $\mathbb{D}$: $\rho(\mathbb{D}) \equiv \max_i(|\lambda_i|)$ the maximum (in magnitude) eigenvalue of $\mathbb{D}$. For the matrix $\mathbb{D}$ defined in (2.86), $\rho(\mathbb{D}) = \sqrt{1 + \omega^2 \Delta t^2}$, meaning that the norm of the vector $\mathbf{f}_n$ will grow (since $\rho(\mathbb{D}) > 1$) with each successive application of $\mathbb{D}$, *for all values of* $\Delta t$. This growth will render any solution useless after some number of iterations. An example is shown in Figure 2.3.

A method with this sort of intrinsic growth is called an unstable method. The lack of stability has nothing to do with the local error order; forward Euler is correct

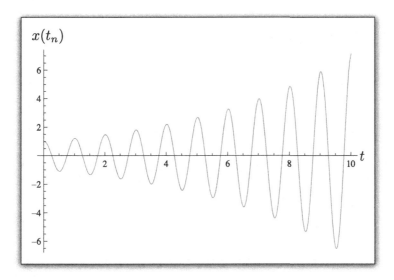

Figure 2.3 Forward Euler applied to the model problem (2.85) with $\omega = 2\pi$, $\Delta t = 0.01$. The initial condition implies a solution with amplitude 1, but each successive peak of the numerical solution is larger than the previous.

to $O(\Delta t^2)$ at each step, so that is not the problem. The growth is coming from the definition of the method itself – errors are accumulating and adding together in an unfortunate, but quantifiable way.

We can fix this behavior by modifying the method. The "backward" Euler method is defined as:

$$\mathbf{f}_{n+1} = \mathbf{f}_n + \Delta t \begin{pmatrix} 0 & 1 \\ -\omega^2 & 0 \end{pmatrix} \mathbf{f}_{n+1}, \tag{2.87}$$

and has the same accuracy as the "forward" case (2.86) (i.e. it is also $O(\Delta t^2)$ for each step). All we have done is evaluate $\mathbf{G}(\mathbf{f})$ (still the right-hand side of $\mathbf{f}'(x)$) in the above at $\mathbf{f}_{n+1}$ rather than at the current value as in $\mathbf{G}(\mathbf{f}_n)$. This introduces additional computation since we can write the method again as a matrix iteration, but with a matrix inversion:

$$\underbrace{\begin{pmatrix} 1 & -\Delta t \\ \omega^2 \Delta t & 1 \end{pmatrix}}_{\tilde{\mathbb{D}}} \mathbf{f}_{n+1} = \mathbf{f}_n. \tag{2.88}$$

The method has $n$th iterate given by: $\mathbf{f}_n = \left(\tilde{\mathbb{D}}^{-1}\right)^n \mathbf{f}_0$. The evaluation of the right-hand side at the updated time and subsequent inversion give this type of approach its name, it is known as an "implicit" method. Implicit methods are generally stable (if the corresponding explicit method is unstable), since the matrix inverse

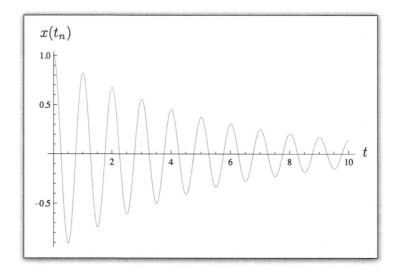

Figure 2.4 Backward Euler applied to the model problem (2.85) with $k = 2\pi$, $\Delta t = 0.01$. This time, the solution decays with iteration.

has "inverted" eigenvalues.[5] Again, the growth of the solution is determined by the spectral radius, this time for $\tilde{\mathbb{D}}^{-1}$, which is: $\rho(\tilde{\mathbb{D}}^{-1}) = \left(1 + \omega^2 \Delta t^2\right)^{-1/2} < 1$ for $\omega > 0$. Now the solution will decrease in amplitude from step to step, but that is better than the alternative. What we have in effect done is introduce an artificial "numerical" viscosity into the problem. The initial condition sets the initial magnitude, and the magnitude of the numerical solution decays as times goes on; this is shown in Figure 2.4.

The reason the decaying solution provided by an implicit method is viewed more favorably than the growth behavior comes from the physical viability of solutions that decay in time. The "numerical viscosity" that is a feature of the method can be used to mimic, enhance, or replace physical viscosity – for our oscillating example, neither growth nor decay appears to be acceptable, but in real situations we might have damping or some other energy-loss mechanism that could be co-opted (at least in part) by numerical viscosity.

Aside from that physically inspired benefit, methods that are both consistent (their global error goes to zero as $\Delta x \to 0$) and stable converge to the true solution as $\Delta x \to 0$. These, then, are the only methods that will correctly solve problems posed on the continuum side. In practice, it is easy to establish consistency using

---

[5] Not literally, of course – the hope is that if the original matrix has eigenvalues with magnitude greater than one, the inverse has eigenvalues with magnitude less than one.

Taylor expansion. Stability is more difficult – we have just explored stability in the context of a linear problem, and that is certainly not the only type of problem we might try to solve. But, if you can get a method that is both consistent and stable (even if only for some parameter values, choices of $\Delta x$, say), then the discrete numerical solution will be a good approximation to the projection of the continuous solution.

## 2.6 Multi-step methods

The Runge–Kutta style method is an example of a "one step" method: it takes information from the current step, and uses it to generate a new value for the solution vector. We can also imagine methods that depend not just on $\bar{f}_n$ to compute $\bar{f}_{n+1}$, but a whole set of previous values as well, $\{\bar{f}_j\}_{j=0}^n$. Updates that rely on more than just one previous value are called "multi-step" methods. Verlet, defined in (2.37), is an example: it requires the current and previous values of the solution variable to define the next value.

Many multi-step methods can be generated by turning the target ODE into an integral equation, and approximating the integral. Take a first-order ODE, $\frac{df(x)}{dx} = G(x, f(x))$ (with $f(a) = f_a$ given) as usual – formally, we know the solution to this ODE can be written in integral form as:

$$f(x) = f_a + \int_a^x G(\bar{x}, f(\bar{x}))d\bar{x} \tag{2.89}$$

given the starting point $a$. The integral itself depends on all values of $G$ for $\bar{x} = a \longrightarrow x$, and so it is not surprising that a discrete method based on the above integral might involve multiple grid points. We could even imagine an approximation of the form:

$$\bar{f}_{j+1} = f_a + \int_a^{x_j} G(\bar{x}, f(\bar{x}))d\bar{x}. \tag{2.90}$$

We will think about numerical approximations to integrals later on in Chapter 6, but one way to approximate an integral, for functions that have values only on a grid of points, is to use sums – that would make the above integral form of the solution look like:

$$\bar{f}_{j+1} = f_a + \sum_{k=1}^{j} \alpha_k G(x_k, \bar{f}_k) \tag{2.91}$$

for some set of coefficients $\{\alpha_k\}_{k=1}^{j}$. Choosing the coefficients with an eye towards accuracy and stability defines a numerical method relating $\bar{f}_j$ to (potentially) all previous values.

## Further reading

1. Allen, M. P. & D. J. Tildesley. *Computer Simulations of Liquids*. Oxford University Press, 2001.
2. Arfken, George B. & Hans J. Weber. *Mathematical Methods for Physicists*. Academic Press, 2001.
3. Giordano, Nicholas J. *Computational Physics*. Prentice Hall, 1997.
4. Goldstein, Herbert, Charles Poole, & John Safko. *Classical Mechanics*. Addison-Wesley, 2002.
5. Isaacson, Eugene & Herbert Bishop Keller. *Analysis of Numerical Methods*. Dover, 1994.
6. Koonin, Steven E. & Dawn C. Meredith. *Computational Physics: Fortran Version*. Westview Press, 1990.
7. Press, William H., Saul A. Teukolsky, William T. Vetterling, & Brian P. Flannery. *Numerical Recipes in C*. Cambridge University Press, 1996.

## Problems

### Problem 2.1

Find the time it takes for a projectile fired at an angle $\theta$ with speed $v$ to return to its firing height (use only gravity, no drag or Magnus force).

### Problem 2.2

Using the self-force associated with a uniformly charged spherical shell (with total charge $Q$, total mass $M$, also uniformly distributed), find the amount of time it takes for a shell of radius $R$, initially at rest, to expand to twice its starting radius. Write this time in terms of the dimensionless variables you set up in Problem 1.6, with $\alpha = R$.

### Problem 2.3

For a mass on a spring (with constant $k$) in Newtonian mechanics, if we start the mass ($m$) from rest at an extension $a$, what is its maximum speed? For what value of $a$ will this maximum speed equal $c$?

### Problem 2.4

For relativistic mechanics:
(a) It is possible to identify an "effective" non-relativistic force from (2.22). Using the relativistic momentum there, rewrite as $m\dot{v} = F_{\text{eff}}$ – what is $F_{\text{eff}}$ in terms of $F$ and $v$?
(b) Show that (2.24) implies (2.25).

**Problem 2.5**

There are other options available to set constants in a general solution to an ODE – solve:

$$m\ddot{x} = -kx \tag{2.92}$$

with $x(0) = x_0$ and $\dot{x}(T) = v_T$, i.e. given the initial position, and velocity at time $T$.

**Problem 2.6**

The approximation (2.37) could be written as an equality with the addition of $O(\Delta t^p)$ – find the value of $p$ relevant to that approximation (don't just quote the easy, but correct, $O(\Delta t^3)$).

**Problem 2.7**

Why would it be difficult to include magnetic forces in the Verlet method as written in (2.37)?

**Problem 2.8**

(a) For the equation of motion $\ddot{x} = -g$, associated with gravity near the Earth with initial conditions $x(0) = h$, $\dot{x}(0) = 0$, define vectors $\bar{\mathbf{f}}$ and $\mathbf{G}$ so that $\frac{d\bar{\mathbf{f}}}{dt} = \mathbf{G}$ as in Section 2.4.2. Now apply the Euler method to your first-order vector equation, i.e.

$$\bar{\mathbf{f}}_{n+1} = \bar{\mathbf{f}}_n + \Delta t \mathbf{G}_n. \tag{2.93}$$

Solve this recursion equation for $\bar{\mathbf{f}}_n$ in terms of $n$, $\Delta t$, $g$, and $h$.

(b) In this case, we can compare the projection of the true solution with the numerical solution explicitly. Write the continuous function $x(t)$ solving the ODE with the provided initial conditions, generate the sequence of discrete values $\{x_n \equiv x(t_n)\}_{n=1}^N$ associated with a grid of spacing $\Delta t$. Finally, construct the residual $r_n \equiv x_n - \bar{x}_n$.

**Problem 2.9**

The so-called "Cash–Karp" embedding for Runge–Kutta is defined by (from [40]):

$$\bar{f}_{n+1} = \bar{f}_n + \frac{37}{378} k_1 + 0 k_2 + \frac{250}{621} k_3 + \frac{125}{594} k_4 + 0 k_5 + \frac{512}{1771} k_6 \tag{2.94}$$

with

$$k_\ell = \Delta x G\left(x + a_\ell \Delta x, \bar{f}_n + \sum_{j=1}^{\ell-1} b_{\ell j} k_j\right), \tag{2.95}$$

and coefficients:

$$a_i \doteq \begin{pmatrix} 0 \\ \frac{1}{5} \\ \frac{3}{10} \\ \frac{3}{5} \\ 1 \\ \frac{7}{8} \end{pmatrix} \qquad b_{\ell j} \doteq \begin{pmatrix} 0 & 0 & 0 & 0 & 0 \\ \frac{1}{5} & 0 & 0 & 0 & 0 \\ \frac{3}{40} & \frac{9}{40} & 0 & 0 & 0 \\ \frac{3}{10} & -\frac{9}{10} & \frac{6}{5} & 0 & 0 \\ -\frac{11}{54} & \frac{5}{2} & -\frac{70}{27} & \frac{35}{27} & 0 \\ \frac{1631}{55296} & \frac{175}{512} & \frac{575}{13824} & \frac{44275}{110592} & \frac{253}{4096} \end{pmatrix}. \tag{2.96}$$

Verify, using `Mathematica`'s `Series` command that these coefficients result in a method that has local truncation error $O(\Delta x^6)$.

## Problem 2.10
We can write the forward and backward Euler methods applied to the model problem: $\dot{x}(t) = G(x)$ (so that $G$ is a function only of $x$, and not of $t$):

$$x_{n+1} = x_n + \Delta t G(x_n) \qquad x_{n+1} = x_n + \Delta t G(x_{n+1}) \tag{2.97}$$

where the left(right) is the forward(backward) method appropriate for a grid with spacing $\Delta t$. Show that these two methods have the same local error order.

## Problem 2.11
Turn the following third-order ODE into a vector first-order ODE, be careful to define the entries of $\mathbf{f}$ and refer to them correctly in $\mathbf{G}(x, \mathbf{f})$:

$$\frac{j'''(x)}{x^2} - j'(x)j''(x) + j(x) = F(x). \tag{2.98}$$

## Problem 2.12
Suppose that Newton's second law looked like: $\alpha \ddot{x}(t) = G(x)$ (for some force-like analogue, $G(x)$).
**(a)** Develop the one-dimensional analogue of the Verlet method to numerically solve this equation given $G(x)$, indicate the local truncation error in your derivation.
**(b)** Write the vector form of the equation you would use in an RK-type solver.

## Problem 2.13
Show that:

$$\frac{d}{dx} \int_{x_0}^{x} h(\bar{x}) d\bar{x} = h(x). \tag{2.99}$$

## Problem 2.14

Solve the following ODEs preparatory to comparison with numerical solutions:

**(a)**

$$f'(x) = x^{10} - 5x^2 \quad f(0) = 1. \tag{2.100}$$

**(b)**

$$f'(x) = -xf \quad f(0) = 1. \tag{2.101}$$

**(c)**

$$f''(x) = -29 f(x) - 4f'(x) \quad f(0) = 1 \quad f'(0) = 0. \tag{2.102}$$

## Lab problems

## Problem 2.15

The Verlet method does a good job at conserving energy (when the force is derivable from a potential). We can see this property, in comparison with the midpoint Runge–Kutta method for the linear harmonic oscillator problem.

**(a)** Solve

$$\ddot{x} = -(2\pi)^2 x \quad x(0) = 1 \quad \dot{x}(0) = 0 \tag{2.103}$$

using Verlet – take $N = 1000$ steps with $\Delta t = 20/(N-1)$. Calculate the total energy for each step (use (2.38) for the velocities). Plot the energy, calculate the mean, and find the difference between the largest and smallest energy values.

**(b)** Calculate and plot the energies for the solution found via midpoint Runge–Kutta – use the same $N$ and $\Delta t$, and observe the basic qualitative difference between the RK energies and those coming from Verlet.

## Problem 2.16

Set up a discrete shell of charge ($N$ points, each with charge $q = Q/N$, and mass $m = M/N$) – use the conversion from spherical coordinates to Cartesian coordinates,

$$x = r \sin\theta \cos\phi \quad y = r \sin\theta \sin\phi \quad z = r \cos\theta \tag{2.104}$$

to note an initial configuration of point charges, with $\Delta\theta = \pi/20$ and $\Delta\phi = 2\pi/10$ (make sure not to double-count the charges at $\phi = 0$ and $\phi = 2\pi$, and omit the particles at $\theta = 0, \pi$). Using the Verlet solver, solve Newton's second law for this system, using the pair force $\mathbf{F} = \frac{1}{4\pi r_{ij}^2}\hat{\mathbf{r}}_{ij}$. Starting with a shell of radius $R = 1.0$, integrate from (dimensionless time) $s = 0 \to 1$ in steps of $\Delta s = 0.01$. Find the approximate time at which the shell's radius has doubled and compare with the result from Problem 2.2.

**Problem 2.17**

Modify Implementation 2.2 so that it executes the fourth-order method from (2.68). Solve the ODEs in Problem 2.14 numerically for $x = 0 \longrightarrow 1$ with $\Delta x = 0.001$ and the initial conditions shown using the midpoint integrator from the chapter notebook, and your new RK4 integrator (with fixed step-size). In each case, calculate (and plot) the residual with respect to the exact result (i.e. the exact solution evaluated on your grid minus your numerical approximation), compare with the residual from the midpoint RK solution. Finally, find the value of $\bar{f}_n$ at $x = 0.95$ for the final ODE (Problem 2.14c) from your two numerical solutions.

**Problem 2.18**

Calculate the test charge's trajectory $\mathbf{x}(t)$ for the moving line of charge from (2.20) and (2.21), assuming the charge $q$ starts from rest at $Y = 1$, $Z = 0$. Take $\alpha = 10.0$ and use $\tilde{u} = 0.25$. Plot the trajectory of the test particle in the $Y - Z$ plane, make sure to run your solver out far enough to capture the interesting (and expected) behavior.

**Problem 2.19**

Start from the second-order ODE in (2.32) for the relativistic spring equation of motion. Using your fourth-order RK routine:

**(a)** Solve the relativistic spring equation for a spring with constant $k = 12$ N/m, with attached mass $m = 1$ kg starting with $x(0) = 1$ m, $v(0) = 0$ for $t = 0 \longrightarrow 10$ s with $\Delta t = 0.05$ s. Plot this with the non-relativistic solution on top of it. Qualitatively, then, does the relativistic solution approximate the non-relativistic one?

**(b)** For a spring with constant $k = 12$ N/m and a mass $m = 1$ kg attached to it, calculate the initial displacement that would lead classically to a maximum velocity of $c$, use this (and $v(0) = 0$) as your initial displacement, and solve for $t = 0 \longrightarrow 10$ – does your relativistic spring achieve $v = c$? Provide a plot of $x(t)$ and $v(t)$ for $t = 0 \longrightarrow 10$.

**(c)** Use the same parameters as above, but this time, start the mass off at five times the critical displacement (again, include a plot of $x(t)$ and $v(t)$ for $t = 0 \longrightarrow 10$). Describe the motion of the mass qualitatively.

**Problem 2.20**

Solve the drag problem (2.15) for the parameters specified below. In each case, provide a plot of the motion in $x - y$ space – run each solution long enough for the projectile to touchdown (after that, the $y$-component of position will be negative, since we aren't counting the ground in the problem):

**(a)** Start with $m = 1.0$ kg, an initial speed of 1000 m/s and angle $\theta = 30°$ – set $\omega = 0$ and $\gamma = 0$, so this motion is purely projectile. Adjust your step size so that you obtain the correct touchdown time to two digits after the decimal (in the true solution, from Problem 2.1). You can use this step size in the remaining parts of this problem.

**(b)** Use the same parameters as above, but include $\gamma/m = 5 \times 10^{-5}$ 1/m.

**(c)** A golf ball travels at around 100 mph (120 if you're Tiger Woods). Take an initial speed of 100 mph $\approx$ 45 m/s, with angle of 10°, use $\gamma/m = 5 \times 10^{-5}$ 1/m and backspin such that $S\omega/m \approx 0.5$ 1/s. The typical mass of a golf ball is 0.045 kg.

# 3

# Root-finding

Finding the "roots" of a function of one variable, $F(x)$, means finding the set of values $\{\bar{x}_i\}_{i=1}^N$ such that

$$F(\bar{x}_i) = 0. \tag{3.1}$$

There are a number of problems in which this type of calculation is important – we will discuss a few cases in which what we actually want are the roots of some polynomial or other function. There are also examples in which the ability to find roots allows us to turn boundary-value problems into initial value problems, an important conversion given our numerical ODE solvers.

## 3.1 Physical motivation

There are three distinct types of physical problem that involve root-finding. We'll start by considering problems that are solved directly by finding roots, of, say, polynomial or transcendental functions. Then there are two flavors of purely numerical root-finding associated with "shooting" methods.

The first of these takes a boundary-value formulation of an ODE and turns it into an initial value form. If we think of Newton's second law, the boundary-value form is: "Given $x(0) = x_0$ and $x(T) = x_T$, find $x(t)$ satisfying $m\ddot{x} = F$." We will turn this into an initial value problem: "Given $x(0)$ find $\dot{x}(0)$ such that $x(T) = x_T$ and $x(t)$ solves $m\ddot{x} = F$." Our ODE solvers (whether RK or Verlet-style) require initial value formulations of ODE problems, and turning a boundary-value form into an initial value form amounts to finding the roots of a (numerically defined) function.

The second type of shooting method applies to "eigenvalue" ODE problems. The typical setup comes from quantum mechanics, where we are trying to solve

Schrödinger's equation,

$$-\frac{\hbar^2}{2m}\psi''(x) + V(x)\psi(x) = E\psi(x) \tag{3.2}$$

for both $E$ and $\psi(x)$. We are usually given two boundary conditions, $\psi(0) = \psi(a) = 0$, for example (where 0 and $a$ are often replaced by $-\infty$ and $\infty$). The difference between this type of problem and Newton's second law is that the derivative of the wave function, $\psi'(x)$, is undetermined numerically: if Schrödinger's equation is solved by $\psi(x)$, then it is also solved by $A\psi(x)$ for arbitrary constant $A$. That means that $\psi'(0)$ can be anything, and the shooting approach does not turn a boundary-value problem into an initial value problem in this case. Instead, we end up solving the following: "Given $\psi(0) = 0$, find $E$ such that $\psi(a) = 0$," and it is $E$ that is variable now, not the derivative of the wave function. Once again, we will identify a numerically defined function whose roots provide precisely these "allowed" energies.

### 3.1.1 Roots of functions

First, we can consider physical problems in which the root of some function is of direct interest. There are examples coming from a variety of setups. As we go, you will see that there is little distinction between solving an "equation," and finding the roots of a function. This is because it is always possible to turn an equation (whose solutions are of interest) into a function whose roots *are* the solutions. As an example, suppose we want to find $x$ satisfying the equation:

$$\sin(x)\cos(x) = x. \tag{3.3}$$

Define the function $F(x) \equiv \sin(x)\cos(x) - x$, then the roots of $F(x)$ are precisely the solutions to (3.3).

#### Orbital motion

For orbital gravitational motion in the $\theta = \frac{1}{2}\pi$ plane, we have the total energy, written in spherical coordinates:

$$E = \underbrace{\frac{1}{2}m(\dot{r}^2 + r^2\dot{\phi}^2)}_{=\frac{1}{2}mv^2} - \frac{GMm}{r}. \tag{3.4}$$

Here, $m$ is the mass of the orbiting body, and $M$ is the mass of the central body – Newtonian gravity is providing the force that causes the orbital motion.

We know, from the equations of motion (or otherwise), that $mr^2\dot{\phi} = J_z$, a constant (the $\hat{\mathbf{z}}$ component of angular momentum). So, we can solve the above

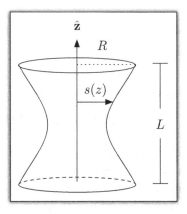

Figure 3.1 We want to find the function $s(z)$, the radius as a function of height, associated with a surface connecting two rings of radius $R$, separated a distance $L$, that has minimal surface area.

energy equation for $\dot{r}^2$, giving us a polynomial in $r$:

$$\dot{r}^2 = \frac{2}{mr^2}\left[Er^2 + GMmr - \frac{J_z^2}{2m}\right]. \tag{3.5}$$

Physically, the zeroes of $\dot{r}^2$ correspond to turning points of the motion, the points of furthest and closest approach. In this case, then, we'd like to know how to set $E$ and $J_z$ to get specific zeroes, allowing us to generate orbits of known elliptical character.

You might think this is a joke – and in this case, I'll admit that the utility of a numerical method for solving such a problem is . . . unclear. But imagine additional pieces to the potential – suppose we considered perturbing bodies (additional gravitational contributions to the potential, say, that do not have such a simple structure). Another way to obtain additional terms is to take a central body that is not spherically symmetric – the dipole, quadrupole, and higher moments of the central body contribute to the potential. Finally, in general relativity, we typically end up with a quartic (because of the four coordinates there) for the analagous equation, and it is simply faster to find the roots numerically than to write them down from the general formula.

### *Area minimization*

There are also cases in which the function of interest is not just polynomial. Many "minimization" problems end in functions that require numerical root-finding. As a simple example of this type of problem, consider a surface connecting two rings of equal radii, $R$, separated a distance $L$ as shown in Figure 3.1. We want to find the

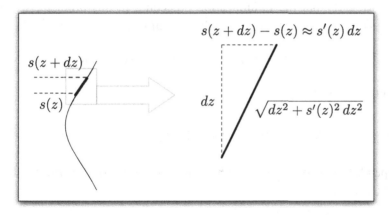

Figure 3.2 Calculating the infinitesimal hypotenuse for the platelet extending from $z \longrightarrow z + dz$.

surface with *minimal* area – soap films find these minimal surfaces automatically. There the soap film is taking advantage of a minimal energy configuration, leading to a stable equilibrium.

The immediate goal is a function $s(z)$ that gives the radius of the surface as a function of height. We'll take $z = 0$ at the bottom ring, then $z = L$ is the height at the top ring. Our area expression follows from the azimuthal symmetry – for a platelet extending from $z$ to $z + dz$ and going around an infinitesimal angle $d\phi$, the area is:

$$dA = sd\phi\sqrt{dz^2 + s'(z)^2 dz^2}, \tag{3.6}$$

as can be seen in Figure 3.2.

If we integrate this expression for the area in both $\phi$ and $z$, we get the total area of the surface:

$$A = 2\pi \int_0^L s(z)\sqrt{1 + s'(z)^2}dz. \tag{3.7}$$

This formula is nice, but it proceeds from a *given* function of $s(z)$. There is a general method for taking such a functional (here $A$ is a number that depends on the function $s(z)$, so $A$ is itself a function of the function $s(z)$ – we call those functionals) and minimizing it – the result is an ODE for $s(z)$ that can be used to *find* $s(z)$.[1]

---

[1] This procedure should be familiar to you from classical mechanics, it is an application of variational calculus.

When we carry out the minimization procedure in this problem, we get the following ODE, with appropriate boundary conditions:

$$1 + s'^2 - ss'' = 0 \quad s(0) = R \quad s(L) = R. \tag{3.8}$$

The general solution is:

$$s(z) = \alpha \cosh\left(\frac{z - \beta}{\alpha}\right), \tag{3.9}$$

for independent real constants $\alpha$ and $\beta$ (with what dimensions?). Now for the $z = 0$ boundary, we have:

$$s(0) = \alpha \cosh\left(\frac{\beta}{\alpha}\right) = R \tag{3.10}$$

and we must simultaneously solve:

$$s(L) = \alpha \cosh\left(\frac{L - \beta}{\alpha}\right) = R. \tag{3.11}$$

From the first equation, we can write $\beta = \alpha \cosh^{-1}\left(\frac{R}{\alpha}\right)$, and then the second equation becomes:

$$\alpha \cosh\left(\frac{L}{\alpha} - \cosh^{-1}\left(\frac{R}{\alpha}\right)\right) = R. \tag{3.12}$$

Define the function:

$$F(\alpha) \equiv \alpha \cosh\left(\frac{L}{\alpha} - \cosh^{-1}\left(\frac{R}{\alpha}\right)\right) - R. \tag{3.13}$$

It should be clear that the roots of this function solve (3.12), and define the final constant of integration for our solution. In this area minimization example, there is a solution not captured by our setup, where we assume $s(z)$ is a continuous function. It could be that the area minimizing solution is really two flat surfaces, one on the top, and one on the bottom ring (this happens all the time when you actually try to make a soap film extend between two rings – if you pull the rings apart far enough, the connecting bubble breaks, and there is a flat soap film covering the top and bottom rings). That solution would not be captured by the roots of $F(\alpha)$, and in that case (i.e. for a certain separation $L$), you would find that $F(\alpha)$ has *no* roots in the physically relevant domain. A coarse plot of the function $F(\alpha)$ will tell you whether or not a solution exists – you should not try to find the roots of a function that doesn't have any.

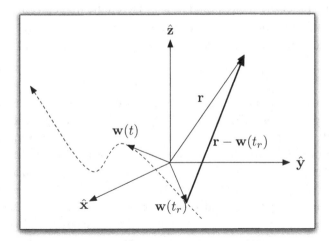

Figure 3.3 A charged particle travels along the dashed trajectory, its location at time $t$ is given by $\mathbf{w}(t)$. Field information travels from $\mathbf{w}(t_r)$ to $\mathbf{r}$ in time $t - t_r$, traveling at the speed of light, so the electric field at $\mathbf{r}$ is informed by the location (and velocity, acceleration) of the charged particle at $t_r$.

### Retarded time in E&M

Another natural place for root-finding is in the retarded time condition of E&M. Suppose we have a particle moving along a given curve $\mathbf{w}(t)$ (so that the vector $\mathbf{w}(t)$ points from the origin to the location of the moving charge at time $t$), and we want to know the electric field at a particular location, $\mathbf{r} = x\hat{\mathbf{x}} + y\hat{\mathbf{y}} + z\hat{\mathbf{z}}$ at time $t$. Because the electromagnetic field information travels with speed $c$, the field at $\mathbf{r}$ will be informed by the moving charge's location at some earlier time – we take into account the time-of-flight for the field information itself. This earlier time, called the "retarded time," and denoted $t_r$, is defined by the equation:

$$c(t - t_r) = \|\mathbf{r} - \mathbf{w}(t_r)\|. \tag{3.14}$$

The vector $\mathbf{r} - \mathbf{w}(t_r)$ points from the retarded location of the charge to the point $\mathbf{r}$ as shown in Figure 3.3. The length of this vector represents the distance field information must travel (at speed $c$) to reach the location $\mathbf{r}$ at time $t$. Then (3.14) just equates the travel time (multiplied by $c$) with the distance travelled. It's an equation that is easy to understand geometrically, but difficult to work with since $t_r$ appears on both the left and right.

We can turn (3.14) into a root-finding project by defining the function:

$$F(x) = c(t - x) - \|\mathbf{r} - \mathbf{w}(x)\| \tag{3.15}$$

whose roots are precisely the retarded times.[2] Given the trajectory $\mathbf{w}(t)$, a location of interest $\mathbf{r}$, and the time $t$, we can use root-finding on $F(x)$ in (3.15) to find $t_r$, the time at which the charge's motion effected the electric field at $\mathbf{r}$.

---

**Example**

We can combine the root-finding required by retarded time with the rest of the (relativistic) dynamics of particle motion. Suppose we have two equal and opposite charges $\pm q$ that start from rest at $\pm x_0$ on the $x$-axis (as you will see in a moment, we must also assume that the particles were at rest for all $t < 0$). How long does it take for the two charges to halve the initial distance between them (we could ask how long it takes for the two particles to collide, but the forces as the charges approach each other become arbitrarily large)? Call $x(t)$ the position of the charge that starts at $x_0$, then from our symmetrical setup, we know that $-x(t)$ is the position of the charge starting at $-x_0$. There is no magnetic field along the line connecting the two charges, so the only force is electric. The electric field at $x(t)$ due to the charge at $-x(t)$ is (see, for example [22]):

$$E = -\frac{q}{4\pi \epsilon_0 (x(t) + x(t_r))^2} \left[ \frac{1 - \dot{x}(t_r)/c}{1 + \dot{x}(t_r)/c} \right] \tag{3.16}$$

where the position and velocity of the particle at $-x(t)$ are evaluated at the retarded time. The full problem (equation of motion, retarded time and initial conditions) is

$$\frac{d}{dt} \left[ \frac{m\dot{x}(t)}{\sqrt{1 - \frac{\dot{x}(t)^2}{c^2}}} \right] = -\frac{q^2}{4\pi \epsilon_0 (x(t) + x(t_r))^2} \left[ \frac{1 - \dot{x}(t_r)/c}{1 + \dot{x}(t_r)/c} \right]$$

$$c(t - t_r) = |x(t) + x(t_r)| \tag{3.17}$$

$$x(0) = x_0$$

$$\dot{x}(0) = 0.$$

Now we want the time $\bar{t}$ at which $x(\bar{t}) = x_0/2$ – this problem must be solved numerically, and then at each "step" of the numerical recursion, the retarded time must be found using a root-finding procedure (note that the discretization will make $x(t)$ available only at grid locations, so that the accuracy of any method is limited by our ability to solve the retarded time condition). At time $t = 0$, the right-hand side of the ODE in (3.17) relies on $t_r < 0$, and so we must know the past history of the charges – the simplest assumption is that prior to $t = 0$, the particles were pinned in place by some external agent (so have $\dot{x} = 0$ for $t < 0$).

---

[2] There is no a priori reason that an arbitrary trajectory $\mathbf{w}(t)$ will be in "causal contact" (meaning that (3.14) is satisfied, and $t_r \leq t$) with $\mathbf{r}$ at only one point in its past, although for actual particle trajectories, the equation (3.15) will have only one root for $t < t_r$.

## Finite square well

In quantum mechanics, the time-independent Schrödinger equation (3.2) provides a connection between a potential $V(x)$ and the stationary wave function $\psi(x)$. For a particle of mass $m$, moving under the influence of a potential $V(x)$, the solution, $\psi(x)$ has a statistical interpretation: The probability of finding the particle within $dx$ of the location $x$ is $\psi(x)^*\psi(x)dx$, and we call $\psi(x)^*\psi(x)$ the "probability density." The variable $E$ is then the energy of the particle – if you measured the energy of a particle in the "state" $\psi(x)$, you would get $E$. Solutions fall into two categories: bound and scattering states. Bound states typically have energies that are quantized, while scattering states admit solutions with a continuum of energies. If the potential goes to zero at spatial infinity, then bound states have negative energy, and we can find the energy associated with bound states, in some cases, by solving a transcendental equation (equivalently, finding the roots of a transcendental function).

The time-independent Schrödinger equation (3.2) can be applied to a finite square well. The potential we'll use is

$$V(x) = \begin{cases} -V_0 & 0 < x < a \\ 0 & \text{else} \end{cases}. \tag{3.18}$$

There are scattering states, corresponding to a particle coming in, interacting with the potential, and going out. And there are bound states, in which $\psi(x)$ is small outside the well, dropping to zero at $\pm\infty$. The scattering states are not localized in this sense. Bound states have negative energy for this well, and scattering states have energy greater than zero – so we can choose to focus on one set or the other by taking $E$ to be either negative or positive.

There are three regions of space defined by the potential (3.18): region I, with $x < 0$, region II, $0 < x < a$, and region III, $x > a$. The approach will be to solve Schrödinger's equation in each of these three regions, and then stitch together the full solution by requiring that $\psi(x)$ and $\psi'(x)$ be continuous everywhere. Since we are looking for bound states, take $E = -f$ with $f > 0$ to be the energy. Then Schrödinger's equation in region I is:

$$-\frac{\hbar^2}{2m}\psi_1''(x) = -f\psi_1(x), \tag{3.19}$$

and we'll take the final step (following [23], for example), define $z_0^2 \equiv 2mV_0/\hbar^2$, and $z^2 \equiv z_0^2 - 2mf/\hbar^2$, so that $2mf/\hbar^2 = z_0^2 - z^2$. Then the solution to Schrödinger's equation, in region I is:

$$\psi_1(x) = Ae^{\sqrt{z_0^2-z^2}x} + Be^{-\sqrt{z_0^2-z^2}x}, \tag{3.20}$$

with constants $A$ and $B$. The same solution holds for $\psi_{\text{III}}(x)$ (with a different set of integration constants). In region II, we have:

$$-\frac{\hbar^2}{2m}\psi_{\text{II}}''(x) - V_0\psi_{\text{II}}(x) = -f\psi_{\text{II}}(x) \longrightarrow \psi_{\text{II}}''(x) = -z^2\psi_{\text{II}}(x) \tag{3.21}$$

so that

$$\psi_{\text{II}}(x) = C\cos(zx) + D\sin(zx). \tag{3.22}$$

We can tabulate our solution, piece by piece:

$$\psi_{\text{I}}(x) = Ae^{\sqrt{z_0^2 - z^2}\,x} + Be^{-\sqrt{z_0^2 - z^2}\,x}$$
$$\psi_{\text{II}}(x) = C\cos(zx) + D\sin(zx) \tag{3.23}$$
$$\psi_{\text{III}}(x) = Ee^{\sqrt{z_0^2 - z^2}\,x} + Fe^{-\sqrt{z_0^2 - z^2}\,x}.$$

We want our wave function to vanish at $x \longrightarrow \pm\infty$ (no chance of finding the particle way out there), and this requirement forces $B = 0$, and $E = 0$. Finally, we must impose continuity for $\psi(x)$ and its derivative at $x = 0$ and $a$. That requirement gives the following equation (as you will show in Problem 3.6), which must be solved for $z$ (remember, that will determine the bound-state energies, since $z^2 = z_0^2 - 2mf/\hbar^2$, and $f$ is our goal):

$$\tan(za) = \frac{2\sqrt{\left(\frac{z_0}{z}\right)^2 - 1}}{2 - \left(\frac{z_0}{z}\right)^2}. \tag{3.24}$$

This equation is not analytically solvable, but we can turn it into a root-finding problem, and then solve that numerically. Here, we have $z_0$ and $a$ given, and the function whose roots are of interest is:

$$F(z) = \tan(za) - \frac{2\sqrt{\left(\frac{z_0}{z}\right)^2 - 1}}{2 - \left(\frac{z_0}{z}\right)^2}. \tag{3.25}$$

Those roots will define the bound-state energies, the allowed values of an energy measurement (for a localized particle).

### *3.1.2 Shooting for Newton's second law*

For simple forces, we can easily solve Newton's second law given either a particle's initial position and initial velocity, or its initial position and final position. Take constant forcing in one dimension: $F = F_0$. For a particle of mass $m$ acted on by this force, we know that:

$$m\frac{d^2x(t)}{dt^2} = F_0 \longrightarrow x(t) = \frac{1}{2}\frac{F_0}{m}t^2 + At + B, \tag{3.26}$$

for arbitrary constants $A$ and $B$. As a second-order differential equation, it is appropriate to recover two "constants of integration." Physically, they give us the freedom to choose our description based on, say, experimental data. For example, if we know the initial position $x_0$ and initial velocity $v_0$ of the particle, then we can set $A$ and $B$ and write the solution as:

$$x(t) = \frac{1}{2}\frac{F_0}{M}t^2 + v_0 t + x_0. \tag{3.27}$$

This case, where we are told the initial values of $x(t)$ and its derivative, is called an "initial value" formulation of the problem.[3]

If instead, we know that the particle started at $x_0$, and at time $T$, was observed at $x_T$, we could write:

$$x(t) = \frac{1}{2}\frac{F_0}{M}t^2 + \frac{1}{T}\left((x_T - x_0) - \frac{1}{2}\frac{F_0}{M}T^2\right)t + x_0. \tag{3.28}$$

We are given the values of the particle's position on the boundaries of a temporal domain of interest, and this type of data is referred to as "boundary-value" data, and the problem (Newton's second law plus boundary data) is in boundary-value form. In either of these cases, we are rendering the mathematical constants of integration $A$ and $B$ from (3.26) into physically relevant form.

Now suppose we ask the question: What value $v_0$ should we choose in order to get a particle that starts at $x_0$ to be at $x_T$ at time $T$? The answer is clear from a comparison of (3.27) and (3.28):

$$v_0 = \frac{1}{T}\left((x_T - x_0) - \frac{1}{2}\frac{F_0}{M}T^2\right), \tag{3.29}$$

and we have turned a boundary-value problem into an initial value problem. It is precisely this inversion that we will need to carry out in a more general setting, where the force may or may not have an analytic solution.

---

**Example**

As another example of turning arbitrary constants into physically meaningful information, consider a spring (with constant $k$) with a mass $m$ attached to it. The force exerted on the mass by the spring is:

$$F = -kx(t) \tag{3.30}$$

---

[3] I use the word "problem" in a technical sense. Newton's second law is an ordinary differential equation, the "problem" of solving for $x(t)$ is only completely specified when either initial or boundary values are given.

and the solution to Newton's second law is:

$$m\frac{d^2x}{dt^2} = -kx \longrightarrow x(t) = A\cos\left(\sqrt{\frac{k}{m}}t\right) + B\sin\left(\sqrt{\frac{k}{m}}t\right). \qquad (3.31)$$

Again, we have two arbitrary constants. Suppose we were given $x(0) = x_0$ and $\dot{x}(0) = v_0$ – our solution, at $t = 0$ is:

$$x(0) = A \qquad (3.32)$$

so we naturally associate $A = x_0$. The velocity of the particle is given by:

$$v(t) = \frac{dx(t)}{dt} = \sqrt{\frac{k}{m}}\left(-A\sin\left(\sqrt{\frac{k}{m}}t\right) + B\cos\left(\sqrt{\frac{k}{m}}t\right)\right) \qquad (3.33)$$

so that:

$$v(0) = \sqrt{\frac{k}{m}}B = v_0 \longrightarrow B = v_0\sqrt{\frac{m}{k}}, \qquad (3.34)$$

and we can write out the solution, tailored to the initial value formulation of the problem, as

$$x(t) = x_0\cos\left(\sqrt{\frac{k}{m}}t\right) + v_0\sqrt{\frac{m}{k}}\sin\left(\sqrt{\frac{k}{m}}t\right). \qquad (3.35)$$

While it is easy to write complete solutions like (3.26) or (3.31) in terms of either initial values or boundary values, the problems of interest to us do not have simple closed-form solutions. Both numerical methods we worked out to solve for $\mathbf{x}(t)$ for nonlinear or otherwise complicated forces, Verlet and Runge–Kutta, apply *only* to an initial value formulation of a problem. How, then, can we solve a problem that is posed in terms of boundary values, numerically? And how does this relate to root-finding? Formally, our goal is to solve Newton's second law given a force:

$$m\ddot{\mathbf{x}} = \mathbf{F} \qquad (3.36)$$

and boundary values: $\mathbf{x}(0) = \mathbf{x}_0$ and $\mathbf{x}(T) = \mathbf{x}_T$. When we use an RK method, we need to specify $\mathbf{x}_0$ and $\mathbf{v}_0$ – the plan is to guess a value for $\mathbf{v}_0$, see where the particle is at time $T$ using our RK solution, then, using that information, modify the initial velocity. We do this until we are within an acceptable range of $\mathbf{x}_T$. This process is called "shooting," in reference to the problem (shown in Figure 3.4): "Given a cannon firing a slug of mass $m$ with muzzle speed $v_0$, what angle $\theta$ should the cannon barrel have so as to hit a target located a distance $R$ away?"

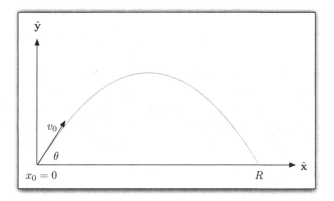

Figure 3.4 An object of mass $m$ travels along the parabolic trajectory shown here. Given $v_0$ and $\theta$, we can find $R(v_0)$, and the idea behind the shooting method is that (by inverting this relationship) given $R$, we can find $v_0$ (or $\theta$): $v_0(R)$.

**Range formula**

In two dimensions, the parabolic motion that occurs for cannon slugs has initial velocity given by both $v_0$ (the muzzle speed) and the angle $\theta$ (elevation). The equations of motion, together with initial conditions, are:

$$m\ddot{x} = 0 \quad m\ddot{y} = -mg$$
$$x(0) = 0 \quad \dot{x}(0) = v_0 \cos\theta \quad y(0) = 0 \quad \dot{y}(0) = v_0 \sin\theta. \tag{3.37}$$

The solutions for $x(t)$ and $y(t)$ are:

$$x(t) = v_0 \cos\theta t \quad y(t) = v_0 \sin\theta t - \frac{1}{2}gt^2. \tag{3.38}$$

For the boundary-value formulation of this problem: "Given $v_0$, find the angle $\theta$ for which $y = 0$ when $x = R$," the temporal parametrization in the equations of motion is inefficient – it is better to parametrize the motion in terms of $x$. The solution for $x(t)$ in (3.38) can be inverted, so we solve for $t$ in terms of $x$: $t = x/(v_0 \cos\theta)$. Inserting this into the $y(t)$ equation, we get vertical motion parametrized by $x$:

$$y(x) = \tan\theta x - \frac{1}{2}g\left(\frac{x}{v_0 \cos\theta}\right)^2. \tag{3.39}$$

Now we can generate a function $F(\theta)$ whose roots will correspond to angles $\theta$ enforcing $y(R) = 0$

$$0 = \tan\theta R - \frac{1}{2}g\left(\frac{R}{v_0 \cos\theta}\right)^2, \tag{3.40}$$

by defining

$$F(\theta) = \tan\theta R - \frac{1}{2}g\left(\frac{R}{v_0\cos\theta}\right)^2. \tag{3.41}$$

In the current case, a numerical approach is overkill. We can find the desired angle easily using the quadratic formula. But what if the force was more complicated than the constant gravitational force near the surface of the Earth? For any force, we can re-parametrize the equations of motion in terms of $x$ as above. If we take $y(t) \equiv y(x(t))$, then $\frac{dy}{dt} = \frac{dy(x)}{dx}\frac{dx}{dt} \equiv y'(x)v_x$, so that we send $\frac{d}{dt} \longrightarrow v_x\frac{d}{dx}$, then

$$F_x = mv_x\frac{dv_x}{dx}$$

$$F_y = mv_x\left[\frac{dv_x}{dx}\frac{dy}{dx} + v_x\frac{d^2y}{dx^2}\right] = F_x\frac{dy}{dx} + mv_x^2\frac{d^2y}{dx^2}, \tag{3.42}$$

and in this form, we can start with almost any force, and develop the ODE version of the range formula with height $y$ parametrized by $x$.

As a check, take $F_x = 0$, and $F_y = -mg$, then (3.42) reads:

$$\frac{dv_x}{dx} = 0 \quad y''(x) = -\frac{g}{v_x^2}, \tag{3.43}$$

and $v_x$ is a constant, equal to $v\cos\theta$ from the initial condition, so

$$y''(x) = -\frac{g}{v^2\cos^2\theta} \longrightarrow y(x) = \tan\theta x - \frac{1}{2}g\frac{x^2}{v^2\cos^2\theta}, \tag{3.44}$$

as before. This time, the $\tan\theta$ term comes up naturally, since the $x$ derivative is related to the time derivative at zero: $\dot{y}(0) = y'(0)v_x(0)$, and $v_x(0) = \dot{x}(0) = v\cos\theta$, so we have $\dot{y}(0) = y'(0)\dot{x}(0)$, and then $y'(0) = \dot{y}(0)/\dot{x}(0) = \tan\theta$.

### 3.1.3 Shooting for eigenvalues

Now for the eigenvalue problem – we want to solve:

$$-\frac{\hbar^2}{2m}\psi''(x) + V(x)\psi(x) = E\psi(x) \tag{3.45}$$

for $\psi(x)$ and $E$. This is called an eigenvalue problem because we have a linear *operator* (the familiar form would be a matrix) acting on a function (vector) producing a number times the function, so that if you squint:

$$\left[-\frac{\hbar^2}{2m}\frac{d^2}{dx^2} + V(x)\right]\psi(x) = E\psi(x) \tag{3.46}$$

looks like the more familiar $\mathbb{A}\mathbf{v} = \lambda\mathbf{v}$ type of eigenvalue problem.

Regardless, here we have an ODE and natural boundary values, but we can't develop an equation that is meant to determine the initial value $\psi'(0)$ (as we did in the previous section), since that is arbitrary. Instead we want to find the eigenvalue, $E$, in this case. We'll do this both for bound and scattering states.

### Bound states in one dimension

Let's see how the bound states, $\psi(x)$, and associated energies, for a given potential, can be found numerically using an RK solver applied to (3.46) and a root-finder. Take the infinite square well, defined by the potential:

$$V(x) = \begin{cases} 0 & 0 < x < a \\ \infty & \text{else} \end{cases}. \tag{3.47}$$

In this case, the wave function must be zero at both $x = 0$ and $x = a$.[4] Schrödinger's equation for values of $x$ between zero and $a$ reads:

$$-\frac{\hbar^2}{2m}\frac{d^2\psi(x)}{dx^2} = E\psi(x). \tag{3.48}$$

In the interests of ridding ourselves of unwieldy numerical constants, suppose we take $x = aq$ where $a$ is the length of the well, then a change of variables gives:

$$\frac{d^2\psi(q)}{dq^2} = -\frac{2mEa^2}{\hbar^2}\psi(q), \tag{3.49}$$

and we can define $\tilde{E} \equiv 2mEa^2/\hbar^2$, a dimensionless form of the energy, to get $\psi''(q) = -\tilde{E}\psi(q)$. Solving this equation subject to the boundary condition $\psi(0) = 0$ gives:

$$\psi(q) = A\sin(\sqrt{\tilde{E}}q). \tag{3.50}$$

The second boundary condition requires $\sin(\sqrt{\tilde{E}}) = 0$, and this has solutions at $\tilde{E} = n^2\pi^2$ for integer $n$. The boundary condition is the mathematical source of the energy quantization.

Suppose we didn't know the answer – how would we approach this problem? In the shooting setup, the boundary conditions are fixed, and the goal is to numerically determine $\tilde{E}$. What we have is an ODE, which we will solve using our Runge–Kutta

---

[4] This requirement comes from the continuity of $\psi(x)$ – the probability of finding a particle outside of the infinite square well is zero, so $\psi(x) = 0$ outside the well. Then $\psi(0) = 0$ and $\psi(a) = 0$ allow the interior solution to smoothly match up with the exterior one.

routine, and a pair of boundary conditions: $\psi(0) = \psi(1) = 0$ (for $\psi(q)$). In terms of the Runge–Kutta setup, what we are missing is $\psi'(0)$ and $\tilde{E}$. The derivative value is arbitrary, since that only serves to set the scale for $\psi(q)$ itself, and that scale is arbitrary (absent normalization requirement), so take $\psi'(0) = 1$. Our vectorized first-order ODE reads

$$\frac{d}{dq}\begin{pmatrix} \psi(q) \\ \frac{d\psi(q)}{dq} \\ \tilde{E} \end{pmatrix} = \begin{pmatrix} \frac{d\psi(q)}{dq} \\ -\tilde{E}\psi(q) \\ 0 \end{pmatrix}. \tag{3.51}$$

We can solve this ODE numerically for any value of $\tilde{E}$ – define PsiE to be the numerical value at the endpoint $q = 1$ given some $\tilde{E}$. Then the function whose roots we want is:

$$F(\tilde{E}) = \text{PsiE}, \tag{3.52}$$

so that when $F(\tilde{E}) = 0$, the boundary condition at $q = 1$ is satisfied. In Implementation 3.1, we see an example of defining the function $F(\tilde{E})$ – we assume that the function RKODE from Implementation 2.2 (or an equivalent, more accurate version) is available.

---

**Implementation 3.1** Example function for root-finding

```
F[Et_] := Module[{Nsteps, G, nvals, PsiE},
  Nsteps = 100;
  G[x_, f_] := {f[[2]], -f[[3]] f[[1]], 0};
  nvals = RKODE[G, 0.0, {0.0, 1.0, Et}, 1.0/(Nsteps - 1), Nsteps];
  PsiE = nvals[[Length[nvals], 2, 1]];
  Return[PsiE];
]
```

---

Of course, in this case, the root-finding has been done analytically above – after all, what we really have (dropping the numerical subterfuge, and just setting $F(\tilde{E}) = \psi(1)$ from (3.50)) is

$$F(\tilde{E}) = A\sin(\sqrt{\tilde{E}}), \tag{3.53}$$

and the roots are located at $\tilde{E} = n^2\pi^2$ for integer $n$. As a check, we can plot the function defined in Implementation 3.1 and verify that the zero-crossings are approximately correct – the plot is shown in Figure 3.5.

### Hydrogen spectrum

Using the shooting technique, we can find the numerical spectrum of hydrogen. We have a proton and an electron interacting electrostatically with potential energy:

$$V(r) = -\frac{e^2}{4\pi\epsilon_0 r} \tag{3.54}$$

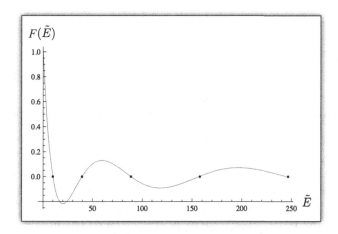

Figure 3.5 A plot of the function $F(\tilde{E})$ defined numerically in Implementation 3.1. The points are the first five values of $n^2\pi^2$.

where $r$ is the radial location of the electron. The problem is intrinsically three dimensional, but because the potential is spherically symmetric, the radial and angular portions separate, leaving us with the usual spherical harmonics (functions of $\theta$ and $\phi$), and a radial equation that is one dimensional. In terms of the separation ansatz: $\psi(r, \theta, \phi) = R(r)Y_\ell^m(\theta, \phi)$, the radial part of Schrödinger's equation is (see [23])

$$\frac{d}{dr}\left[r^2\frac{dR}{dr}\right] - \frac{2mr^2}{\hbar^2}[V(r) - E]R = \ell(\ell + 1)R, \qquad (3.55)$$

subject to the boundary conditions: $R(0) = A_\ell$ (a constant, that could be zero) and $R(\infty) = 0$. We know, from the (bound-state) solutions themselves, that when $\ell$ is zero, $A_\ell$ is a non-zero constant used to set the normalization of the radial portion of the wave function. When $\ell$ is non-zero, $A_\ell = 0$. In this case, there is an obvious parallel between the hydrogen problem and the infinite square well problem: We want $R$ to vanish at 0 and $\infty$ (in the square well case, we wanted $\psi$ to vanish at zero and $a$). The only issue is the numerical representation of infinity. But if you think about what "zero" means, to a computer, a number on the order of $10^{-13}$, and you remember, again from the radial solutions, that $R(r) \sim e^{-nr/a}$ (for $a = 4\pi\epsilon_0\hbar^2/(me^2)$ the Bohr radius associated with hydrogen, $a \approx 0.5$ Å), then we are motivated to choose a finite value for "infinity." Take, for example, $n = 1$, and note that $e^{-20} \sim 10^{-9}$. Then $r = 20a$ gives a solution that is very close to "zero." Put infinity where you like (it changes the resulting root structure); we'll start with $r_\infty = 20a$ but you should move this value around and see how the solutions (the energies) respond.

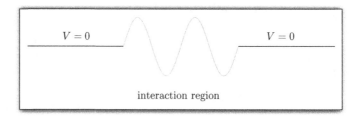

Figure 3.6 For some region of space, the potential is non-zero.

Again, we want the various constants, $m$, $e$ (the electron mass and charge), and $\hbar$, soaked into variables that allow us to work in a reasonable range. Take $r = aq$, this time using the Bohr radius itself as the fundamental length scale. Then (3.55) becomes

$$\frac{d}{dq}\left[q^2\frac{dR}{dq}\right] - [-2q + \ell(\ell+1)]R = -\frac{2E}{\alpha^2 mc^2}q^2 R \qquad (3.56)$$

where $\alpha = \hbar/(mac)$ is the dimensionless fine structure constant, $\alpha \approx 1/137$. If we define $\tilde{E} \equiv 2E/(\alpha^2 mc^2)$, then our final differential equation reads, written in vector first-order form:

$$\frac{d}{dq}\begin{pmatrix} R \\ \frac{dR}{dq} \\ \tilde{E} \end{pmatrix} = \begin{pmatrix} \frac{dR}{dq} \\ -\frac{2}{q}\frac{dR}{dq} + \left(-\frac{2}{q} + \frac{\ell(\ell+1)}{q^2} - \tilde{E}\right)R \\ 0 \end{pmatrix}. \qquad (3.57)$$

With these conventions, $\tilde{E} = -1/n^2$ for integer $n$ is the correct result.

We now define the function $F(\tilde{E})$ that is zero when the boundary condition is satisfied. Let `PsiE` be the numerically determined value approximating $\psi(20)$ (remember that our "infinity" is $r_\infty = 20a$, so that $q_\infty = 20$ is the right-hand endpoint of our solution region), then $F(\tilde{E}) = $ `PsiE`, and we can find zeroes of this function, giving us the bound-state energies.

### Scattering in one dimension

In one dimension, we can define the general scattering problem. Given a potential that is localized to a region $0 < x < a$ (so, zero potential outside) as in Figure 3.6, we imagine a "free particle" (no potential) of mass $m$ coming in from the left, inter-acting with the potential, and emerging on the right (with some portion reflected). Operationally, we will solve (3.45) for $x < 0$, $0 < x < a$ and $x > a$, just as we did for the bound states of the finite well, and then use continuity and derivative continuity to fix all constants in the resulting solutions.

Define the three wave functions of interest: $\psi_{\mathrm{I}}(x)$ for $x < 0$, $\psi_{\mathrm{II}}(x)$ for $0 \leq x \leq a$, and $\psi_{\mathrm{III}}(x)$ for $x > a$. We'll use the dimensionless energy $\tilde{E} \equiv \frac{2ma^2 E}{\hbar^2}$ and similarly $\tilde{V}(x) = \frac{2ma^2}{\hbar^2} V(x)$; then our three equations are:

$$\frac{d^2\psi_{\mathrm{I}}}{dx^2} = -\frac{1}{a^2}\tilde{E}\psi_{\mathrm{I}}(x)$$

$$\frac{d^2\psi_{\mathrm{II}}}{dx^2} = -\frac{1}{a^2}\left(\tilde{E} - \tilde{V}(x)\right)\psi_{\mathrm{II}}(x) \qquad (3.58)$$

$$\frac{d^2\psi_{\mathrm{III}}}{dx^2} = -\frac{1}{a^2}\tilde{E}\psi_{\mathrm{III}}(x).$$

We can eliminate the final constant in the problem, $a$, by taking a dimensionless spatial variable $x = aq$. Then, letting primes refer to $q$-derivatives, we have

$$\psi_{\mathrm{I}}''(q) = -\tilde{E}\psi_{\mathrm{I}}(q)$$

$$\psi_{\mathrm{II}}''(q) = -\left(\tilde{E} - \tilde{V}(q)\right)\psi_{\mathrm{II}}(q) \qquad (3.59)$$

$$\psi_{\mathrm{III}}''(q) = -\tilde{E}\psi_{\mathrm{III}}(q).$$

The left and right solutions do not depend on $V(q)$, those are:

$$\psi_{\mathrm{I}}(q) = Ae^{i\sqrt{\tilde{E}}q} + Be^{-i\sqrt{\tilde{E}}q}$$

$$\psi_{\mathrm{III}}(x) = Ce^{i\sqrt{\tilde{E}}q} + De^{-i\sqrt{\tilde{E}}q}. \qquad (3.60)$$

From the full time-dependent form of Schrödinger's equation, it is correct to associate $e^{i\sqrt{\tilde{E}}q}$ with right-traveling solutions, and $e^{-i\sqrt{\tilde{E}}q}$ with left-traveling ones. So in terms of the physical configuration, if a particle "comes in from the left," we expect $A$ to set the size of the right-traveling portion, $B$ will then tell us about the part of $\psi_{\mathrm{I}}(x)$ that is "reflected" from the potential. On the right, we expect only a right-traveling solution, so set $D = 0$. The scattering solutions here are not normalizable, precisely because of the oscillatory solutions on the left and right. But, we still have the freedom to set, say $A = 1$, since even though the solutions are not normalizable, we still have eigenfunctions that can be scaled by any number.

The continuity conditions are potential-independent

$$\psi_{\mathrm{I}}(0) = \psi_{\mathrm{II}}(0) \quad \psi_{\mathrm{I}}'(0) = \psi_{\mathrm{II}}'(0) \quad \psi_{\mathrm{II}}(1) = \psi_{\mathrm{III}}(1) \quad \psi_{\mathrm{II}}'(1) = \psi_{\mathrm{III}}'(1), \qquad (3.61)$$

(remember that $x = a$ corresponds to $q = 1$) and these give us constraints on the constants in $\psi_{\mathrm{II}}(q)$. If we write out all four of these, we obtain two relations between the value of $\psi_{\mathrm{II}}$ and its derivative, one at each boundary point:

$$\psi_{\mathrm{II}}(0) = 1 + B$$

$$\psi'_{\mathrm{II}}(0) = i\sqrt{\tilde{E}}(1 - B) = i\sqrt{\tilde{E}}(2 - \psi_{\mathrm{II}}(0))$$

$$\psi_{\mathrm{II}}(a) = C e^{i\sqrt{\tilde{E}}} \tag{3.62}$$

$$\psi'_{\mathrm{II}}(a) = i\sqrt{\tilde{E}}\left(C e^{i\sqrt{\tilde{E}}}\right) = i\sqrt{\tilde{E}}\psi_{\mathrm{II}}(1).$$

So the equations we need to solve are (once we are given $V(q)$):

$$\psi''_{\mathrm{II}}(q) = -\left(\tilde{E} - \tilde{V}(q)\right)\psi_{\mathrm{II}}(q)$$

$$\psi'_{\mathrm{II}}(0) = i\sqrt{\tilde{E}}(2 - \psi_{\mathrm{II}}(0)) \tag{3.63}$$

$$\psi'_{\mathrm{II}}(1) = i\sqrt{\tilde{E}}\psi_{\mathrm{II}}(1),$$

one second-order ODE, and two boundary conditions. Remember that for scattering states, the energy is not quantized, so unlike the bound-state case, here we will *input* an energy $\tilde{E}$, and find the associated wave function. From a shooting point of view, we have a single unknown: $\psi_{\mathrm{II}}(0)$ (but it has real and complex parts), and we must choose this value so that the third equation in (3.63) is satisfied. The boundary condition at $q = 0$ will be enforced, given $\psi_{\mathrm{II}}(0)$, by setting the initial value of the first derivative to $\psi'_{\mathrm{II}}(0) = i\sqrt{\tilde{E}}(2 - \psi_{\mathrm{II}}(0))$.

We'll use the output of our Runge–Kutta routine to obtain numerical values for $\psi_{\mathrm{II}}(1)$ and $\psi'_{\mathrm{II}}(1)$ (call these output numerical values: Psi, Psip, respectively) given the input complex value $\psi_{\mathrm{II}}(0)$ (denoted Psi0), then we can define the function,

$$u(\mathrm{Re}(\mathrm{Psi0}), \mathrm{Im}(\mathrm{Psi0})) = |\mathrm{Psip} - i\sqrt{\tilde{E}}\mathrm{Psi}|^2. \tag{3.64}$$

The function $u$ depends on *two* variables, the real and imaginary parts of Psi0, and returns a real number, the magnitude (squared) of the boundary requirement. This is different – our original shooting problem found the values of $x$ for which a function $F(x) = 0$ – now we have a function $u(x, y)$, and we want to find the values of $x$ *and* $y$ such that $u(x, y) = 0$. We'll introduce the steepest descent method in Section 3.2.4 to find the root of (3.64). The method is a minimization algorithm

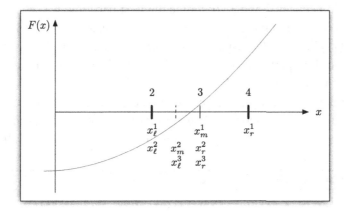

Figure 3.7 The successive bisections for the function $F(x)$. Here, $x_\ell^n$ and $x_r^n$ refer to the left and right endpoints of the interval for the $n$th iteration of the bisection.

(discussed in detail in Chapter 12), but for $u(x, y)$ defined by (3.64), the minimum is also a "root" (of sorts).

## 3.2 Finding roots

Our first job is to find the zeroes of a function $F(x)$ – here, we mean at least a subset of all values $\{\bar{x}_i\}_{i=1}^N$ such that $F(\bar{x}_i) = 0$. Our function could be polynomial, or transcendental; the basic idea is the same either way.

### 3.2.1 Bisection

The bisection method is straightforward, and reliable. We start with an interval defined by its endpoints, $x_\ell$ and $x_r$, such that the product $F(x_\ell)F(x_r) < 0$, i.e. somewhere in between the left and right points, the function $F(x)$ has a zero (possibly many). We then split the interval in half by computing $x_m = \frac{1}{2}(x_\ell + x_r)$. If the product $F(x_\ell)F(x_m) < 0$, then the zero of $F(x)$ lies between $x_\ell$ and $x_m$, otherwise, the zero must lie between $x_m$ and $x_r$. Whichever case is relevant, we define a new left and right endpoint for an interval of half the size, and repeat until $|F(x_m)| \leq \epsilon$, a user-specified tolerance.

A picture of the process for an arbitrary function is shown in Figure 3.7. In that picture, we start with the interval bounded by $x_\ell^1 = 2$, $x_r^1 = 4$ – the product of $F(x_\ell^1)F(x_r^1) < 0$ since the evaluation on the right endpoint is positive, and the value of $F$ at the left endpoint is negative. So the condition for a zero to exist between the two points is met. We form $x_m^1 = \frac{1}{2}(2+4) = 3$ and see that $F(x_m^1)F(x_\ell^1) < 0$, so the zero lies in between 2 and 3. To define the next partition, we set $x_\ell^2 = x_\ell^1 = 2$, and $x_r^2 = x_m^1 = 3$. The midpoint between these two is $x_m^2 = \frac{1}{2}(2+3) = 2.5$, and this time $F(x_\ell^2)F(x_m^2) > 0$, indicating that the zero lies between $x_m^2$ and $x_r^2$ – so we take, as our third partition: $x_\ell^3 = x_m^2 = 2.5$, and $x_r^3 = x_r^2 = 3$. This process will continue until we are (numerically) satisfied.

Bisection can be naturally implemented in a recursive form, and the code to take a function $F(x)$ and apply bisection, given the endpoints of an interval containing the root (xl and xr), until $|F(x)| \leq$ eps, is shown in Implementation 3.2. It should be

---

**Implementation 3.2** Recursive bisection

```
Bisect[F_, xl_, xr_, eps_] := Module[{val, vam, xm, retval},
  xm = .5 (xl + xr);
  val = F[xl];
  vam = F[xm];
  If[Abs[vam] ≤ eps,
   Return[xm];
  ];
  If[val vam < 0,
   retval = Bisect[F, xl, xm, eps];
   ,
   retval = Bisect[F, xm, xr, eps];
  ];
  Return[retval];
 ]
```

---

clear that any of our polynomial or transcendental functions can be provided as the function F in the Bisect routine from Implementation 3.2, the return value is the location of "the" (read "a") numerical zero lying in between the initial endpoints. It is therefore important that there *be* a root in that initial interval, otherwise the function will never return. The downside of bisection is its timing – the method uses no information about the function itself to determine the size of the step it should take. That means that the overall time is determined only by the size of the initial interval, and the target tolerance $\epsilon$.

### 3.2.2 Newton's method

Let's look at how using information about the function $F(x)$ can speed up the search for its zeroes. Suppose a function $F(x)$ has a root at $x = x_0$, and you are within $\delta$ of that root at $x = x_0 - \delta$, the question is: "How far away from the root are you?" – i.e. what is $\delta$? Using Taylor expansion, we can write:

$$F(x_0) = F(x + \delta) \approx F(x) + \delta F'(x) + O(\delta^2), \tag{3.65}$$

and then, setting $F(x_0) = 0$ by assumption, we can make a linear approximation to $\delta$, call it $\Delta$:

$$\Delta \equiv -\frac{F(x)}{F'(x)}. \tag{3.66}$$

As an example, suppose we take the linear function $F(x) = Ax - B$, so that there is a root at $x_0 = B/A$. Our estimate, in this case, will be exact – at any point $x$, we are a distance:

$$\Delta = -\frac{F(x)}{F'(x)} = -\frac{Ax + B}{A} = -\left(x - \frac{B}{A}\right) \tag{3.67}$$

from the root, and as a check, we have $x + \Delta = B/A = x_0$ as desired. The distance is exact since the function is linear, so that the $O(\delta^2)$ term in (3.65) goes away (that term has a coefficient $F''(x_0)$ which will vanish for linear $F(x)$).

Clearly, for nonlinear functions, $\Delta$ determined according to (3.66) will be approximate rather than exact – take $F(x) = (x - 1)(x - 2)$, and let $x = 3/4$. How far from the root $x_0 = 1$ are we? According to (3.66),

$$\Delta = -\frac{\frac{5}{16}}{-\frac{3}{2}} = \frac{5}{24}, \tag{3.68}$$

and $\delta = 1/4 = 6/24$ is the exact answer. It's a small error in this case, and suggestive of the method itself. The idea is to start at some location $x$, and try to *step towards* the root using $\Delta$ in (3.66) to guide the choice of step size. For a function $F(x)$, and a starting point $x_1$, we update $x$ according to:

$$x_2 = x_1 + \left(-\frac{F(x_1)}{F'(x_1)}\right)$$

$$x_3 = x_2 + \left(-\frac{F(x_2)}{F'(x_2)}\right)$$

$$\vdots \tag{3.69}$$

$$x_j = x_{j-1} + \left(-\frac{F(x_{j-1})}{F'(x_{j-1})}\right).$$

The iteration stops when $|F(x_j)| < \epsilon$, a user-defined tolerance.

When you think of the geometry of this method, the derivative defines the slope of the line tangent to the function $F(x)$ at the point $x$ – we construct this line and then step down it to its zero. We know that if the function $F(x)$ were

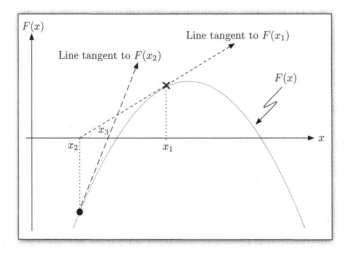

Figure 3.8 Starting at $x_1$, we find the slope of the line tangent to the point $(x_1, F(x_1))$ – that is given by $F'(x_1)$. Then we find the root of this line, i.e. the point at which this line crosses zero, call that $x_2$, and repeat the process.

*actually* linear, $\Delta$ would represent the distance to the zero, but for our arbitrary function $F(x)$, we must iterate to achieve refinement. The first few steps of the process are shown in Figure 3.8. The code to execute Newton's method, given a function F and its derivative DF, a starting point x0 and tolerance eps, is shown in Implementation 3.3.

---

**Implementation 3.3** Newton's method

---

```
Newton[F_, DF_, x0_, eps_] := Module[{x, deltax, val},
  x = x0;
  val = F[x];
  While[Abs[val] ≥ eps,
    deltax = -F[x]/DF[x];
    x = x + deltax;
    val = F[x];
  ];
  Return[x];
]
```

---

The problem with the method is that in order to generate the iteration (3.69), we assumed that we were close to a particular root. If we are not close to a root, then the Taylor expansion that motivated the method is inaccurate, higher-order terms contribute heavily, and our iterates will run off in one direction or another. Without an explicit check for this behavior, the code in Implementation 3.3 will not necessarily run to completion.

### *3.2.3 Newton's method and shooting*

Let's return to the shooting problem – in a general setting, we have the following RK-oriented setup:

$$\frac{d}{dt}\underbrace{\begin{pmatrix} x \\ v \end{pmatrix}}_{\equiv \mathbf{f}} = \underbrace{\begin{pmatrix} v \\ a \end{pmatrix}}_{\equiv \mathbf{G}}$$

$$x(0) = x_0$$

$$x(T) = x_T$$

(3.70)

where we are given the acceleration $a$, and initial and final values $x_0$ and $x_T$, and our goal is to find the value of $v(0)$ that allows $x(T) = x_T$.

What can we get out of the numerical solution on a grid with uniform spacing: $t_j = j\Delta t$ for $j = 0 \longrightarrow N$, given $x(0) = x_0$ and $v(0) = v_0$? Let $\texttt{xofT}(v_0)$ be the value of $x(T)$ as determined by whatever Runge–Kutta routine you like, given $v_0$. Then the function we want to find the zero of is:

$$F(v_0) = \texttt{xofT}(v_0) - x_T = 0$$

(3.71)

and the value of interest here will be $v_0$. You can use bisection to do the root finding, in which case the function $F(v_0)$ is enough. But in order to use Newton's method, we must provide the function to minimize, $F(v_0)$ and its derivative. Now, since we don't know the analytical solution, we'll generate the derivative itself, numerically. In fact, we can approximate the derivative of $F(v_0)$ with the finite difference (dropping the "limit" from the formal definition of derivative to obtain the approximation):

$$F'(v_0) \approx \frac{F(v_0 + \Delta v) - F(v_0 - \Delta v)}{2\Delta v},$$

(3.72)

where we choose $\Delta v$ in some appropriate fashion (it should be "small").

Since for many problems of interest, there will not be a clear analytic expression for the derivative, we may as well define a version of Newton's method that automatically approximates the derivative, so that all we do is provide the function F. In Implementation 3.4 we have a purely numerical Newton's method – we give it F, a starting point, x0, and a tolerance eps as usual, but in addition we provide a step for the numerical derivative approximation: delta (the generalization of $\Delta v$ in (3.72)). This approach, in which a finite difference approximation appears for the derivative in Newton's method, is referred to as the "secant method."

---

**Implementation 3.4** Newton's method with numerical derivative

```
nNewton[F_, delta_, x0_, eps_] := Module[{x, deltax, val, DF},
  DF[x_] := (F[x + delta] - F[x - delta]) / (2. delta);
  x = x0;
  val = F[x];
  While[Abs[val] ≥ eps,
    deltax = -F[x] / DF[x];
    x = x + deltax;
    val = F[x];
  ];
  Return[x];
]
```

---

Comparing Implementation 3.3 with Implementation 3.4, we can see that the only difference is the numerical determination of DF in the latter. This allows us to use Newton's method for all of the numerically defined shooting and eigenvalue problems we have encountered so far.

### 3.2.4 Steepest descent

Related to the Newton method is the notion of "steepest descent." In one dimension, if we know the derivative of a function at a point, then we know whether to step "left" or "right" to achieve a decrease in the function's value (just check the sign of the derivative at the point). If, however, we have a function $u$ that returns a real value given multiple arguments, then we have to choose a direction in a high-dimensional space in order to decrease the value of $u$. In the end, rather than the single derivative, we end up taking a gradient, and this gives us a direction in which to travel.

In two dimensions, we have a function $u(x, y) \in \mathbb{R}$. Our goal is to start at some initial point $(x_0, y_0)$ and make a move to a new point $(x_0 + \Delta x, y_0 + \Delta y)$ such that:

$$|u(x_0 + \Delta x, y_0 + \Delta y)| < |u(x_0, y_0)|. \tag{3.73}$$

Then by iteratively stringing together such moves, we will end up at a minimum of some sort. In the setting of this chapter, our functions will have minima at zero, so this is still a root-finding exercise, and we'll return to functions with non-zero minima in Chapter 12.

How do we pick the update directions $\Delta x$ and $\Delta y$ so as to satisfy the inequality of (3.73)? Think of the Taylor expansion of the argument on the left of (3.73); the value of $u$ at the "new point" is:

$$u(x_0 + \Delta x, y_0 + \Delta y) \approx u(x_0, y_0) + \frac{\partial u}{\partial x} \Delta x + \frac{\partial u}{\partial y} \Delta y \tag{3.74}$$

where the partial derivatives are taken, and then evaluated at $(x_0, y_0)$. From vector calculus, we recognize a gradient on the right-hand side:

$$u(x_0 + \Delta x, y_0 + \Delta y) \approx u(x_0, y_0) + \nabla u \cdot d\mathbf{x} \tag{3.75}$$

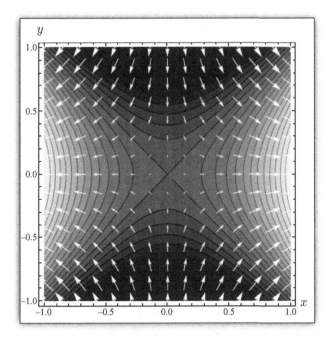

Figure 3.9 A contour plot of a two-dimensional $u(x, y)$ with smaller values shown darker. On top of the plot is the $\nabla u$ vector field.

where we set $d\mathbf{x} = \Delta x \hat{\mathbf{x}} + \Delta y \hat{\mathbf{y}}$. It is clear that if we want to get a smaller value for the new point, we should take $d\mathbf{x} \sim -\nabla u$, i.e. go in the direction *opposite* the gradient of $u$ at the point $(x_0, y_0)$. In Figure 3.9, we see the gradient of a two-dimensional function plotted on top of a contour representation of the function – dark areas correspond to minima, notice that $\nabla u$ is pointing *away* from the minima.

In the context of finding roots, what we will do is take a function $u(x, y)$ that has a zero for some value of $(x, y)$, start "nearby," and then tune $x$ and $y$ so as to find values of $u$ that are within $\epsilon$ of zero. The algorithm is shown in Algorithm 3.1 (given starting values $x_0$ and $y_0$, and some small parameter $\eta$), where we use $\eta$ to control how far in the current steepest descent direction we will go.

---

**Algorithm 3.1** Steepest descent in two dimensions

---

$x = x_0$
$y = y_0$
$r = u(x, y)$
**while** $|r| > \epsilon$ **do**
    $x \leftarrow x - \eta \frac{\partial u}{\partial x}|_{x,y}$
    $y \leftarrow y - \eta \frac{\partial u}{\partial y}|_{x,y}$
    $r = u(x, y)$
**end while**

---

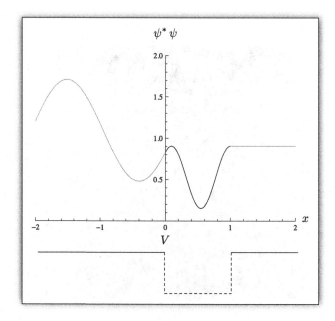

Figure 3.10 The probability density associated with an incoming plane wave – the potential is a finite well, in this case.

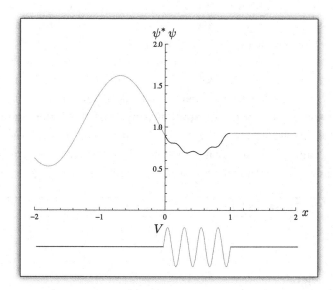

Figure 3.11 Probability density for scattering from a localized, spatially oscillating potential.

*Scattering solution*

Referring to (3.64), we have a system to which Algorithm 3.1 pertains. Notice that we will need to compute the gradient of $u$, and this will have to be done numerically, by varying the real and imaginary parts of `Psi0` separately to construct the descent direction. Remember that the end result of the steepest descent here will be the initial value of the wave function, $\psi_{\text{II}}(0)$, given an energy $\tilde{E}$. We still need to solve the ODE (3.63) once we have our potential. The resulting $\psi(x)^* \psi(x)$ (probability density) for a finite square well is shown in Figure 3.10. To show that the method is robust, we also solve for an oscillatory potential in the region $0 < x < a$ (as in our scattering setup); the probability density for that solution is shown in Figure 3.11.

## Further reading

1. Giordano, Nicholas J. *Computational Physics*. Prentice Hall, 1997.
2. Goldstein, Herbert, Charles Poole, & John Safko. *Classical Mechanics*. Addison-Wesley, 2002.
3. Griffiths, David J. *Introduction to Electrodynamics*. Prentice Hall, 1999.
4. Griffiths, David J. *Introduction to Quantum Mechanics*. Pearson Prentice Hall, 2005.
5. Koonin, Steven E. & Dawn C. Meredith. *Computational Physics: Fortran Version*. Westview Press, 1990.
6. Press, William H., Saul A. Teukolsky, William T. Vetterling, & Brian P. Flannery. *Numerical Recipes in C*. Cambridge University Press, 1996.
7. Stoer, J. & R. Bulirsch. *Introduction to Numerical Analysis*. Springer-Verlag, 1993.

## Problems

### Problem 3.1

Take the general harmonic oscillator solution:

$$x(t) = A \cos\left(\sqrt{\frac{k}{m}}t\right) + B \sin\left(\sqrt{\frac{k}{m}}t\right), \tag{3.76}$$

and put it in a form appropriate to "boundary-value" data, i.e. given $x(0) = x_0$ and $x(T) = x_T$. Find the function $v_0(T)$ that gives the initial speed needed to achieve $x(T) = x_T$.

### Problem 3.2

We can view the harmonic oscillator problem as an eigenvalue problem, too – suppose we were given $x(0) = 0$, $x(T) = 0$ – what would the solution you got in the previous problem look like (if you literally inserted $x_0 = x_T = 0$)? It is possible to satisfy these boundary conditions for specific values of $k/m$ – find those values by solving the eigenvalue problem: $\ddot{x} = -\frac{k}{m}x$ with $x(0) = 0$ and $x(T) = 0$ – what happens to the two constants of integration we have in the general solution?

## Problem 3.3

Generate the range formula that applies to the modified problem (start from the solution (3.39)): We want the projectile to land on a hill whose height is given by $h(x)$ (a monotonically increasing function of $x$), a distance $R$ away. Write the function $F(\theta)$ whose roots provide the starting angle $\theta$, given a muzzle speed $v_0$.

## Problem 3.4

From our work on projectile motion in two dimensions in Section 3.1.2, we had:

$$\frac{dy(x(t))}{dt} = \frac{dy}{dx}\frac{dx}{dt}. \tag{3.77}$$

Verify this expression for the case $x(t) = At^2$ (with $x > 0$ for $t > 0$), $y(t) = y_0 \sin(Bt)$ (these expressions do not, of course, come from a particular projectile problem, they are only meant to provide practice). Verification here means: rewrite $y$ in terms of $x$, then form the right-hand side of (3.77), write that right-hand side entirely in terms of $t$, and then compare with the result of taking the derivative $\frac{dy(t)}{dt}$ directly.

## Problem 3.5

Using (3.42) and the appropriate forces, generate the equations of motion, in $x$-parametrization, appropriate to projectile motion with drag and the Magnus force (the equations of motion in temporal parametrization are in (2.15)).

## Problem 3.6

Using the general solution for the bound states of the finite square well (3.23), show that continuity and derivative-continuity at both boundaries yields (3.24).

## Problem 3.7

In order to set up the bound states of the quantum mechanical finite square well as a shooting problem, we need relations equivalent to the second two conditions in (3.63) starting from the general solution (3.23).

**(a)** Show that, for bound states (where solutions on the left and right of the non-zero potential must go to zero as $x \longrightarrow \pm\infty$),

$$\psi_{\mathrm{II}}'(0) = \sqrt{z_0^2 - z^2}\,\psi_{\mathrm{II}}(0) \quad \psi_{\mathrm{II}}'(a) = -\sqrt{z_0^2 - z^2}\,\psi_{\mathrm{II}}(a), \tag{3.78}$$

using continuity of $\psi$ and its derivative at $x = 0$ and $a$.

**(b)** We are free to choose a non-zero value for $\psi_{\mathrm{II}}(0)$, and then $\psi_{\mathrm{II}}'(0)$ is fixed by the first equation in (3.78). The second equation is the one that we will need to enforce – let psif($z$) and psipf($z$) be numerically determined values for $\psi_{\mathrm{II}}(a)$ and $\psi_{\mathrm{II}}'(a)$, respectively given some $z$. Write a function $F(z)$ in terms of these whose zeroes enforce the second equation in (3.78).

## Problem 3.8

The bisection method described in Section 3.2.1 has a run-time that is basically independent of the function $F(x)$ (whose roots we want to find). Let $\ell \equiv x_r - x_\ell$ be the length of the initial interval, the one provided in the first call to the routine. Find the order of the timing of the method in terms of $\ell$ and the desired tolerance $\epsilon$. You should assume that in the vicinity of the root the function $F(x)$ is linear (that's reasonable, given a small enough interval), so that $F(x) \sim \alpha(x - \bar{x})$ for constant $\alpha$ near the root $\bar{x}$.

## Problem 3.9

Implement the bisection method using a `While` loop rather than recursion. Your function should take the same inputs as the one shown in Implementation 3.2; we will be using this bisector for the lab problems.

## Problem 3.10

We can take the idea behind Newton's method one step further, starting from (3.65), write the $O(\delta^2)$ term explicitly, and use that to make a more accurate step approximation, $\Delta$.

## Lab problems

### Problem 3.11

It is important that we provide your bisection routine from Problem 3.9 with an initial interval that brackets a root. Given a function $F(x)$, we can generate this interval numerically. Write a function `BisectStart` that takes steps of size $\Delta x$ starting at $x$, and returns an interval containing (at least) one root of a function $F(x)$. Your function should take, as its arguments, $x$, $\Delta x$, and $F$, and return $\{x + (N - 1)\Delta x, x + N\Delta x\}$ for some $N$ such that $F(x + (N - 1)\Delta x)F(x + N\Delta x) < 0$. Write a final function `Bisection` that first calls `BisectStart` and then, using the initial bracketing, calls your bisection routine from Problem 3.9. In the end, then, your function `Bisection` takes, as its arguments: the function to bisect, `F`, the starting point `xstart`, a step size `deltax`, and a tolerance `eps`. This is the function you should use for bisection in the following problems.

### Problem 3.12

How many iterations of your bisection loop are required to find the zero of the function:

$$F(x) = \sin(x^2 - 2x - 1) \tag{3.79}$$

located on $x \in [0, 3]$ with $\epsilon = 10^{-10}$ (use $x_\ell = 0.0$, $\Delta x = 1.0$ in your `Bisection` routine from the previous problem)? How many iterations of the Newton method yield the same result (starting from 2.0)?

**Problem 3.13**

A charged particle moves along a spiral trajectory about the $z$ axis – the vector that points from the origin to the particle location at time $t$ is:

$$\mathbf{w} = R\sin(\omega t)\hat{\mathbf{x}} + R\cos(\omega t)\hat{\mathbf{y}} + vt\hat{\mathbf{z}}, \tag{3.80}$$

with $R = 2$ m, $\omega = 10$ Hz and $v = 0.1c$. An electric field is generated by the particle. Find the location of the moving charge that informs its electric field at the location $\mathbf{r} = 1\hat{\mathbf{x}} + 1\hat{\mathbf{y}} + 1\hat{\mathbf{z}}$ at time $t = 1$ s. To do this, you will need to find the retarded time, $t_r$, defined by:

$$c(t - t_r) = \|\mathbf{r} - \mathbf{w}(t_r)\|. \tag{3.81}$$

The difference $t - t_r$ represents the time it takes light to travel from $\mathbf{w}(t_r)$ to $\mathbf{r}$. Note that you may need to scale the function $F(t_r)$ whose roots define the retarded time – that's fine, multiplying by a constant doesn't change the location of the root.

**Problem 3.14**

Find the first three finite square well energies for $a = 1$, $z_0 = 10$ by solving the transcendental equation (3.24) (i.e. finding the roots of (3.25)). Use Newton's method to find the roots with $\epsilon = 10^{-9}$.

**Problem 3.15**

For the arbitrary height function range formula you developed in Problem 3.3, find the angle $\theta$ given $v = 100$ m/s, and a target range of $R = 100$ m, use $h(x) = \frac{1}{1000}x^2$ as your height function (use $\epsilon = 10^{-5}$ in your bisection routine). What happens if you instead set $v = 10$ m/s? What is your physical interpretation of this phenomenon?

**Problem 3.16**

Using the equations you set up in Problem 3.5, write a function RKODE($\theta$) (where $\theta$ is the initial angle for velocity) that returns the value of $y$ at $x = R$. Bisect (at $\epsilon = 10^{-4}$) this function to find the initial angle we should use if a ball is to hit the ground at $R = 200.0$ m assuming a ball of mass $m = 0.045$ kg, drag given by $\gamma/m = 5 \times 10^{-5}$ 1/m, with spin parameter $S\omega/m = 0.3$ 1/s, and initial speed $v_0 = 45$ m/s.

**Problem 3.17**

In this problem, we'll solve for the first three bound-state energies of a finite square well (set up in Section 3.1.1) by shooting. To remain as general as possible, use your RKODE routine to solve Schrödinger's equation in region II: $\psi_{\mathrm{II}}''(x) = -z^2\psi_{\mathrm{II}}(x)$ using the initial conditions $\psi_{\mathrm{II}}(0) = 1$, $\psi_{\mathrm{II}}'(0) = \sqrt{z_0^2 - z^2}\psi_{\mathrm{II}}(0)$ (the first equation in (3.78)) with $z_0 = 10$

setting the depth of the well, and $z$ the quantity of interest. Take $\Delta x = 0.01$. You should output your numerically determined value of $\psi_n(a)$ and $\psi_n'(a)$ for $a = 1$, then using these values, find the first three roots of the function you defined in part (b) of Problem 3.7. Those roots, $z$, are related to the first three observable energies for the system: $E = -\frac{\hbar^2}{2m}(z_0^2 - z^2)$. Compare your energies with the results from Problem 3.14.

Once you have the first three energies, plot the associated wave functions (just the portion in $x = 0 \rightarrow a$). Note that we already know what these look like from the solution (3.23), but with your numerical setup, you can handle any localized potential (not just constant steps).

## Problem 3.18

Write a function that shoots the bound states for hydrogen as outlined in Section 3.1.3 – start by solving (3.57) for $\ell = 0$ given an energy $\tilde{E}$ using your RKODE routine to get PsiE and then use bisection to find the first three roots of this function (corresponding to the first three allowed values of an energy measurement). Use $R(0) = 0$ and $R'(0) = 1$ for your initial conditions – note that you cannot start your solver at exactly $q = 0$, since the right-hand side is infinite there, so move away from zero a "little bit" (start at, for example, the first grid location $\Delta q$).

You have two "knobs" in your numerical solution, the size of the grid spacing, $\Delta q$, and the location of the effective "infinity." If you put your numerical value for infinity in close, you will lose roots associated with higher energy states (higher energy eigenstates of hydrogen have larger spatial extent). If you move infinity out, there will indeed be more roots in your function, but you'll need a smaller value of $\Delta q$ to hone in on their location accurately. Start with $q_\infty = 20$ as the value of infinity (where the boundary condition must be met, as in the setup from Section 3.1.3) and $\Delta q = 0.002$ – find the first three eigenvalues and compare with $-1/n^2$. You should try changing both $q_\infty$ and $\Delta q$ to see how the spectrum changes.

# 4

# Partial differential equations

So far, we have been working with ordinary differential equations, functions of a single variable, satisfying a relation between the function and its derivative(s). But in many physical contexts, what we have is a function of multiple variables. The relevant differential equations then depend on derivatives with respect to each of the variables. The first example one encounters in physics is in E&M, where the electrostatic potential $V(x, y, z)$ is determined by a distribution of source charges $\rho(x, y, z)$ via:

$$\left[\frac{\partial^2 V}{\partial x^2} + \frac{\partial^2 V}{\partial y^2} + \frac{\partial^2 V}{\partial z^2}\right] = -\frac{\rho}{\epsilon_0}. \tag{4.1}$$

We use the "Laplace operator", $\nabla^2$, as shorthand for the derivatives appearing on the left, $\nabla^2 \equiv \frac{\partial^2}{\partial x^2} + \frac{\partial^2}{\partial y^2} + \frac{\partial^2}{\partial z^2}$, so we can write, compactly,

$$\nabla^2 V = -\frac{\rho}{\epsilon_0}. \tag{4.2}$$

When boundary conditions are provided, this differential equation has a unique solution, and is an example of the "Poisson problem."

We'll take the generic setup:

$$\nabla^2 f(x, y, z) = s(x, y, z) \tag{4.3}$$

for a "source" $s(x, y, z)$ and $f(x, y, z)$ given on the boundary of some volume, as the definition of the Poisson problem. Its "source-free" form, $\nabla^2 f = 0$ (with boundary conditions), is referred to as the Laplace problem. In either case, the goal is to find $f(x, y, z)$, given source function $s(x, y, z)$ (possibly zero), in some region $\Omega$, with $f(x, y, z)$ on the boundary of that region, $\partial\Omega$, matching a provided function.

There are many other examples of *linear* differential operators – e.g. the Helmholtz equation:

$$\left(\nabla^2 - \mu^2\right) f = s, \tag{4.4}$$

appropriate for fields with "mass" $\mu$ (see Section 15.1.2). The operator is linear because it acts on $f$ (and not $f^2$ or $\sin(f)$, say). The goal is the same: Find $f$ given source $s$ and boundary values.

Numerically, we exploit the linearity of these differential operators. When we introduce a grid of spatial values, and approximate the derivatives on that grid, we can define an unknown *vector* of values $\mathbf{f}$ (the projection of $f$ onto the grid) and a matrix operator $\mathbb{D}$ such that:

$$\mathbb{D}\mathbf{f} = \mathbf{s} - \mathbf{b} \tag{4.5}$$

where $\mathbf{s}$ is the projection of the source onto the grid and $\mathbf{b}$ provides boundary information. The matrix $\mathbb{D}$ depends on the operator in question, we will focus on this "finite difference" approximation to $\nabla^2$. Now the problem of finding $\mathbf{f}$ is one of matrix inversion. For now, we'll use built-in routines to perform this inversion; that's not cheating because we will return to the problem of solving an equation like (4.5) numerically in Chapter 9 (and in an iterative, approximate, manner in Chapter 11).

## 4.1 Physical motivation

The Poisson problem

$$\nabla^2 f = s \tag{4.6}$$

is familiar as the time-independent field equation for (gauge-fixed) quantities of interest from both E&M and gravity. For a given charge density $\rho$, the electrostatic potential is determined by $\nabla^2 V = -\rho/\epsilon_0$. Given a mass density $\rho$, we have gravitational potential governed by $\nabla^2 V = 4\pi G\rho$.

Combining $V$ with $\mathbf{A}$ in $A^\mu \doteq (V/c, \mathbf{A})^T$, the static form of Maxwell's equations can be written as

$$\nabla^2 A^\mu = -\mu_0 J^\mu, \tag{4.7}$$

with source $J^\mu \doteq (c\rho, \mathbf{J})^T$, four copies of Poisson's equation. We have the similar construct:

$$\nabla^2 h_{\mu\nu} = -16\pi \frac{G}{c^4} T_{\mu\nu} \tag{4.8}$$

for time-independent gravitational problems – this is the field equation of static, linearized general relativity. In this latter case, there are more (of the same) equations to solve. But the point remains, if you can solve (4.6), you can easily solve it four or ten times.

Our primary focus will be on linear partial differential equations in two and three dimensions typified by (4.6), but it is interesting to note the one-dimensional reduction of the problem, where we have a general second-order ODE with "source." In this setting, the finite difference method we will apply amounts to a generalization of the Verlet method (a multi-step method), and can be used to solve boundary-value problems directly (i.e. without the shooting we encountered in the previous chapter).

### 4.1.1 One dimension

In one dimension, what we are doing is solving a second-order ODE – we have seen a variety of examples, and methods, for doing this. Think of the damped, driven harmonic oscillator – that is represented by an ODE of the form:

$$\ddot{x}(t) = -\omega^2 x(t) - 2 \underbrace{\gamma \omega}_{\equiv b} \dot{x}(t) + F(t)/m. \tag{4.9}$$

As a second-order differential equation, we know that the solution $x(t)$ requires either an initial position and velocity (we provide $x(0)$ and $\dot{x}(0)$), or an initial and final position (we provide $x(0)$ and $x(T)$ for some time of interest $T$).

When we are given initial data, we can use Runge–Kutta methods easily to solve the ODE. When boundary data are provided, we can use RK with shooting as in Chapter 3 to solve for the initial velocity $\dot{x}(0)$ that leads to a specified $x(T)$. The discrete setup we will develop in this section is most easily applied to the boundary-value formulation. For completeness, we record the nondimensionalized form of the problem.

---

**Damped driven harmonic oscillator**

A damped, driven harmonic oscillator is described (using Newton's second law) by the equation of motion:

$$\ddot{x} = -\omega^2 x - 2b\dot{x} + F(t)/m \tag{4.10}$$

for a particle of mass $m$ attached to a spring with constant $k$ (then $\omega^2 \equiv k/m$), driven with time-varying force $F(t)$ and experiencing a damping force governed by $b$ (which then has dimension of $1/\text{time}$). We will render this equation dimensionless in the usual way, by setting $x = x_0 q$ and $t = t_0 s$. Then

$$q''(s) = -t_0^2 \omega^2 q(s) - 2bt_0 q'(s) + \frac{F(s)t_0^2}{mx_0}. \tag{4.11}$$

Choose $t_0 = 1/(2b)$, and let $\tilde{\omega}^2 \equiv \omega^2/(4b^2)$, we have

$$q''(s) = -\tilde{\omega}^2 q(s) - q'(s) + \frac{F(s)t_0^2}{mx_0}, \tag{4.12}$$

and we can define the dimensionless forcing: $\tilde{F}(s) \equiv F(s)/(4mx_0 b^2)$. Our final ODE is

$$q''(s) = -\tilde{\omega}^2 q(s) - q'(s) + \tilde{F}(s). \tag{4.13}$$

We know how to solve this problem analytically once $\tilde{F}(s)$ is provided, but in this form, it is also easy to apply the finite difference approach as we shall see in Section 4.2.

### 4.1.2 Two dimensions

The target two-dimensional equations of interest are the Laplace and Poisson equations. As an example of physical utility, suppose we are given a distribution of charge $\rho(\mathbf{r})$, the charge-per-unit-volume at the point $\mathbf{r}$. Then Maxwell's equations tell us that (for this time-independent distribution) the associated potential solves:

$$\nabla^2 V(\mathbf{r}) = -\frac{\rho(\mathbf{r})}{\epsilon_0}. \tag{4.14}$$

In regions where there is no charge, where $\rho(\mathbf{r}) = 0$, we have a degenerate form of the Poisson problem called the Laplace problem, and in that case, we can find the potential $V(\mathbf{r})$ from

$$\nabla^2 V(\mathbf{r}) = 0. \tag{4.15}$$

The elephant in the room, for both of these problems, is the boundary conditions. In the Laplace problem, the need for boundaries is clear. The Laplacian operator $\nabla^2$ in two dimensions reads:

$$\nabla^2 V(\mathbf{r}) = \left(\frac{\partial^2}{\partial x^2} + \frac{\partial^2}{\partial y^2}\right) V(x, y), \tag{4.16}$$

and there are any number of solutions for $V(x, y)$ – for example,

$$V(x, y) = V_0 \tag{4.17}$$

a constant, satisfies (4.15). But then, so does:

$$V(x, y) = \frac{1}{2}A(x^2 - y^2) + Bx + Cy + D \tag{4.18}$$

for arbitrary constants $A$, $B$, $C$, and $D$ (this solution includes (4.17) as a special case).

---

**Multiplicative separation**

In fact, we can go pretty far down the Laplace road – take $V(x, y) = X(x)Y(y)$, then the Laplace equation becomes:

$$X''(x)Y(y) + X(x)Y''(y) = 0, \qquad (4.19)$$

and if we divide by $X(x)Y(y)$, then we get, upon rearrangement:

$$\frac{X''(x)}{X(x)} = -\frac{Y''(y)}{Y(y)}. \qquad (4.20)$$

Now, the left-hand side depends only on $x$, and the right-hand side depends only on $y$, so they must separately be constant, i.e.

$$\frac{X''(x)}{X(x)} = A = -\frac{Y''(y)}{Y(y)} \qquad (4.21)$$

where $A$, in general complex, is called the "separation constant." This equation admits the following continuous, infinite, set of solutions:

$$V(x, y) = \left(\alpha e^{\sqrt{A}x} + \beta e^{-\sqrt{A}x}\right)\left(\gamma e^{\sqrt{-A}y} + \delta e^{-\sqrt{-A}y}\right) \qquad (4.22)$$

where $\alpha$, $\beta$, $\gamma$, and $\delta$ are still more undetermined constants.

---

Which one of these infinite families of solutions, (4.18) or (4.22), is *correct*? After all, we can't have all of these different answers to a physically well-posed problem. Well, the problem, as a physical, or mathematical entity, is not yet well-posed – we have not provided the boundary conditions. There are, as with Newton's second law, a variety of ways to pin down the relevant solution given a physical configuration.

In our two-dimensional, Cartesian (coordinates $x$ and $y$) setting, one option is to specify the *functions*

$$V(x, y_0) = b_1(x) \quad V(x, y_f) = b_2(x) \quad V(x_0, y) = b_3(y) \quad V(x_f, y) = b_4(y) \tag{4.23}$$

where $y_0$, $y_f$, $x_0$ and $x_f$ define a rectangular boundary on which we know the potential. We might take the functions $b_1(x)$, $b_2(x)$, $b_3(y)$ and $b_4(y)$ to be constants. In this case, then, we are surrounding our region with no charge by perfect conductors (the potential on a conducting surface is constant).

As an example, let's take an infinite two-dimensional space with no charge inside it. Our implicit boundaries are at infinity, where we assume the potential $V$ must vanish. For this to be the case, we must have:

$$V(x, -\infty) = 0 \quad V(x, \infty) = 0 \quad V(-\infty, y) = 0 \quad V(\infty, y) = 0, \quad (4.24)$$

and if we insert this demand into any of our three solutions, we will find:

$$V(x, y) = 0, \quad (4.25)$$

as "the" solution.

We could also imagine surrounding our charge-free region with a square conducting wire (of side length $a$), so that:

$$V(x, 0) = V_0 \quad V(x, a) = V_0 \quad V(0, y) = V_0 \quad V(a, y) = V_0, \quad (4.26)$$

and this time, we get the solution:

$$V(x, y) = V_0. \quad (4.27)$$

An important property of solutions to the Laplace *problem* (that is, the Laplace differential equation (4.15), together with an appropriate set of boundary conditions like (4.23)) is that the solution is unique – so we know, once we have satisfied $\nabla^2 V(\mathbf{r}) = 0$ on the interior of some domain, and matched the values given on the boundary of that domain, we are done.

---

**Example**
Using the multiplicative separation we obtained in (4.22), let's see how the introduction of boundary conditions allows us to fix all the free constants. Suppose we take a square box of side length $a$, and we ground all sides except the bottom, there we set the potential to be $V_0 \sin(\pi x/a)$. Explicitly, we have the set:

$$V(x, 0) = V_0 \sin\left(\frac{\pi x}{a}\right) \quad V(x, a) = 0 \quad V(0, y) = 0 \quad V(a, y) = 0. \quad (4.28)$$

For reasons that will be clear in a moment, take the separation constant in (4.21) to be $A = -P^2$, with $P$ a positive real number. Then our solution reads:

$$V(x, y) = \left(Ae^{iPx} + Be^{-iPx}\right)\left(Ce^{Py} + De^{-Py}\right) \quad (4.29)$$

for constants $A$, $B$, $C$, and $D$. The condition at $x = 0$ tells us that $B = -A$, and then we note that:

$$A\left(e^{iPx} - e^{-iPx}\right) = 2i A \sin(Px) \quad (4.30)$$

and our solution so far is:

$$V(x, y) = 2i A \sin(Px)\left(Ce^{Py} + De^{-Py}\right). \quad (4.31)$$

Now for $x = a$, we again need $V(a, y) = 0$ (for all $y$). We could accomplish this by setting $A = 0$, but then we would not be able to impose any additional boundary conditions, since $V(x, y)$ would be zero everywhere. Instead, we place a constraint on $P$ by demanding that $\sin(Pa) = 0$, telling us that:

$$Pa = n\pi \quad n = 1, 2, \ldots \tag{4.32}$$

so we again have an infinite family of solutions, indexed now by the positive integer $n$. The solution at this stage looks like

$$V(x, y) = 2iA \sin\left(\frac{n\pi}{a}x\right)\left(Ce^{\frac{n\pi}{a}y} + De^{-\frac{n\pi}{a}y}\right). \tag{4.33}$$

Moving on to the $y$ boundaries – at $y = 0$, we must get $V_0 \sin(\pi x/a)$, and with our current form, we have:

$$V(x, 0) = 2iA \sin\left(\frac{n\pi}{a}x\right)(C + D) = V_0 \sin\left(\frac{\pi x}{a}\right), \tag{4.34}$$

from which we learn two things: First, that $2iA(C + D) = V_0$, and second, that $n = 1$, else we cannot satisfy the functional requirement on the right for all $x$.

At $y = a$, our solution is:

$$V(x, a) = 2iA \sin\left(\frac{\pi}{a}x\right)\left(Ce^{\pi} + De^{-\pi}\right), \tag{4.35}$$

and we have the simultaneous pair:

$$2iA(C + D) = V_0 \quad Ce^{\pi} + De^{-\pi} = 0 \longrightarrow \boxed{\begin{array}{l} C = \frac{V_0(1 - \coth(\pi))}{4iA} \\[2mm] D = \frac{V_0(1 + \coth(\pi))}{4iA}. \end{array}} \tag{4.36}$$

The final solution to our fully specified problem is then:

$$V(x, y) = \frac{V_0}{2} \sin\left(\frac{\pi}{a}x\right)\left((1 - \coth(\pi))e^{\frac{\pi}{a}y} + (1 + \coth(\pi))e^{-\frac{\pi}{a}y}\right). \tag{4.37}$$

The same boundary requirements hold for the Poisson problem – there, the general solution (the analogue of (4.22)) is not so easy to write down (without a specific source, $s$), so it is not as obvious how the boundary conditions are used to isolate the unique solution to the problem (except in specific cases). In either the Laplace or Poisson setup, it is clear that whatever numerical method we choose, boundary conditions should be easy to implement, and the method should be tailored towards this type of data, rather than, say, the value and first derivatives specified at some set of points. The matrix approach will easily allow us to solve either the Laplace or Poisson problem subject to the specification of boundary values.

## 4.2 Finite difference in one dimension

The matrix formulation of (4.3) can be carried out in one dimension, where it is easier to connect to our previous (ODE) discrete methods. We'll start in this simplified setting and focus on a generic second-order differential equation, since that requires two boundary conditions, and extends naturally to the second-order $\nabla^2$ operator.

Any linear second-order ODE can be written in the form:

$$\boxed{f''(x) + p(x)f'(x) + q(x)f(x) = s(x)} \tag{4.38}$$

for unknown $f(x)$, and given functions $p(x)$, $q(x)$, and $s(x)$ (the source). We also need boundary conditions at $x = x_0$ and $x = x_f$ (i.e. $f(x_0) = f_0$ and $f(x_f) = f_f$ are provided). We could solve this equation using, for example, Runge–Kutta methods with shooting, and those will work even when the ODE is not linear.

Instead, take a regularly spaced grid, $x_n = n\Delta x$ for $n = 1 \longrightarrow N$. For concreteness, we will set $x_0 = 0$ so that $n = 0$ represents a boundary point. In addition, take $(N+1)\Delta x = x_f$, so that the integers $1 \to N$ label interior points. Now, for any point $x_n$, we can form approximations to the first and second derivatives of $f(x)$ at $x$ that are accurate to second order in $\Delta x$:

$$\frac{f(x_n + \Delta x) - f(x_n - \Delta x)}{2\Delta x} = f'(x_n) + \frac{1}{6}f'''(x_n)\Delta x^2 + O(\Delta x^3)$$

$$\frac{f(x_n + \Delta x) - 2f(x_n) + f(x_n - \Delta x)}{\Delta x^2} = f''(x_n) + \frac{1}{12}f''''(x_n)\Delta x^2 + O(\Delta x^3). \tag{4.39}$$

Inputting the expressions on the left for the derivatives in (4.38) gives the following grid approximation – let $f_n \equiv f(x_n)$ etc.

$$f_n\left(q_n - \frac{2}{\Delta x^2}\right) + f_{n+1}\left(\frac{p_n}{2\Delta x} + \frac{1}{\Delta x^2}\right) + f_{n-1}\left(-\frac{p_n}{2\Delta x} + \frac{1}{\Delta x^2}\right) = s_n. \tag{4.40}$$

Remember that what we have here are equations for each value of $n = 1, 2, \ldots, N$. Let's write them out explicitly:

$$f_1\left(q_1 - \frac{2}{\Delta x^2}\right) + f_2\left(\frac{p_1}{2\Delta x} + \frac{1}{\Delta x^2}\right) + f_0\left(-\frac{p_1}{2\Delta x} + \frac{1}{\Delta x^2}\right) = s_1$$

$$f_2\left(q_2 - \frac{2}{\Delta x^2}\right) + f_3\left(\frac{p_2}{2\Delta x} + \frac{1}{\Delta x^2}\right) + f_1\left(-\frac{p_2}{2\Delta x} + \frac{1}{\Delta x^2}\right) = s_2$$

$$\vdots$$

$$f_{N-1}\left(q_{N-1} - \frac{2}{\Delta x^2}\right) + f_N\left(\frac{p_{N-1}}{2\Delta x} + \frac{1}{\Delta x^2}\right) + f_{N-2}\left(-\frac{p_{N-1}}{2\Delta x} + \frac{1}{\Delta x^2}\right) = s_{N-1}$$

$$f_N\left(q_N - \frac{2}{\Delta x^2}\right) + f_{N+1}\left(\frac{p_N}{2\Delta x} + \frac{1}{\Delta x^2}\right) + f_{N-1}\left(-\frac{p_N}{2\Delta x} + \frac{1}{\Delta x^2}\right) = s_N.$$

$$(4.41)$$

The left-hand side is linear in the (unknown) values $\{f_j\}_{j=1}^N$. If we define a vector $\mathbf{f} \in \mathbb{R}^N$ with entries that are the $f_j$ (the unknown approximations to $f(x_j)$), then we can encode the above equations as a matrix-vector multiplication on the left. Our vector, once we solve for it, is giving us approximations to the solution on the grid – that is, for $n = 1 \longrightarrow N$. What about the values, needed for the first and last equations, of $f_0 = f(0)$ and $f_{N+1} = f(x_{N+1})$? Those two values are provided by the boundary data, and we have to insert them by hand; they will eventually move over to the "source" side (the right).

In order to form the relevant matrix, let's write the generic (4.40), suggestively, as a dot product,

$$\left(\left(-\frac{p_n}{2\Delta x} + \frac{1}{\Delta x^2}\right) \ \left(q_n - \frac{2}{\Delta x^2}\right) \ \left(\frac{p_n}{2\Delta x} + \frac{1}{\Delta x^2}\right)\right) \cdot \begin{pmatrix} f_{n-1} \\ f_n \\ f_{n+1} \end{pmatrix} = s_n. \qquad (4.42)$$

Now, suppose we build a matrix, $\mathbb{D}$, that has precisely the row structure on the left (with zeroes for all entries except the diagonal and its nearest neighbors) for the appropriate $n = 2 \longrightarrow N - 1$ interior points – then we will be approximating the differential equation at almost all points on the interior of the domain – on the right, we just make the vector of values $s_n \equiv s(x_n)$, and we have the discrete setup we are after:

$$\mathbb{D}\mathbf{f} = \mathbf{s}. \qquad (4.43)$$

Almost – we still need to define the difference problem for the two grid points, $x_1$ and $x_N$ – these depend on $f_0$ and $f_f$ respectively, and those values are given. These two special cases are easy enough to handle; let's do the $n = 1$ case:

$$f_1\left(q_1 - \frac{2}{\Delta x^2}\right) + f_2\left(\frac{p_1}{2\Delta x} + \frac{1}{\Delta x^2}\right) = s_1 - f_0\left(-\frac{p_1}{2\Delta x} + \frac{1}{\Delta x^2}\right). \quad (4.44)$$

Notice that because $f_0$ is given, the term that cannot be defined in terms of the grid goes over to the right. This process gives us a modification of the discrete problem that has no analogue in the continuous problem. Nevertheless, it is a pretty useful modification – even in the case $s(x) = 0$, where we would normally have to find the null space of the matrix $\mathbb{D}$ in order to find solutions (or content ourselves with a null solution if $\mathbb{D}$ is invertible), we discover that the boundary terms are ensuring that the the problem is solvable. Incidentally, it is the ease with which we can handle this particular type of boundary condition that motivated our discretization – one could imagine enlarging the support points (e.g. by introducing $f(x \pm 2\Delta x)$ in (4.39)) for either the first or second derivative to improve the accuracy of the method, but then the boundary conditions become more difficult to impose.

We are ready to fully define $\mathbb{D}$:

$$\mathbb{D} \doteq$$

$$\begin{pmatrix}
\left(q_1 - \frac{2}{\Delta x^2}\right) & \left(\frac{p_1}{2\Delta x} + \frac{1}{\Delta x^2}\right) & 0 & 0 & \cdots & 0 \\
\left(-\frac{p_2}{2\Delta x} + \frac{1}{\Delta x^2}\right) & \left(q_2 - \frac{2}{\Delta x^2}\right) & \left(\frac{p_2}{2\Delta x} + \frac{1}{\Delta x^2}\right) & 0 & \cdots & 0 \\
0 & \left(-\frac{p_3}{2\Delta x} + \frac{1}{\Delta x^2}\right) & \left(q_3 - \frac{2}{\Delta x^2}\right) & \left(\frac{p_3}{2\Delta x} + \frac{1}{\Delta x^2}\right) & \cdots & 0 \\
0 & 0 & \ddots & \ddots & \ddots & 0 \\
0 & \cdots & 0 & \left(-\frac{p_{N-1}}{2\Delta x} + \frac{1}{\Delta x^2}\right) & \left(q_{N-1} - \frac{2}{\Delta x^2}\right) & \left(\frac{p_{N-1}}{2\Delta x} + \frac{1}{\Delta x^2}\right) \\
0 & \cdots & 0 & 0 & \left(-\frac{p_N}{2\Delta x} + \frac{1}{\Delta x^2}\right) & \left(q_N - \frac{2}{\Delta x^2}\right)
\end{pmatrix}$$

$$(4.45)$$

The full equation, replacing (4.43) with both source and boundary vectors on the right, is

$$\boxed{\mathbb{D}f = s - b,} \quad (4.46)$$

with

$$s \doteq \begin{pmatrix} s_1 \\ s_2 \\ s_3 \\ \vdots \\ s_N \end{pmatrix} \quad b \doteq \begin{pmatrix} f_0\left(-\frac{p_1}{2\Delta x} + \frac{1}{\Delta x^2}\right) \\ 0 \\ 0 \\ \vdots \\ f_f\left(\frac{p_N}{2\Delta x} + \frac{1}{\Delta x^2}\right) \end{pmatrix}. \quad (4.47)$$

A function that develops this matrix and the right-hand side of the difference equation (4.46) is shown in Implementation 4.1. The inputs to the function `MakeDifference` are the functions p, q and s from (4.38), the boundary values $f(0)$ and $f(x_f)$ (called `f0` and `fNp1`), the final point (called `xf`), and the number of steps to take, `Ns`. Note that in this setup, we take 0 and $x_f$ to be the boundary points, so that if we want $N$ interior points, i.e. $N$ grid points, we must have $\Delta x = x_f / (N + 1)$.

---

**Implementation 4.1** One-dimensional difference setup

```
MakeDifference[p_, q_, s_, f0_, fNp1_, xf_, Ns_] := Module[
   {retD, retsmb, index, dx, x},
   retD = Table[0.0, {i, 1, Ns}, {j, 1, Ns}];
   retsmb = Table[0.0, {i, 1, Ns}];
   dx = xf / (Ns + 1);
   x = dx;
   retD[[1, 1]] = (q[x] - 2 / dx^2);
   retD[[1, 2]] = (p[x] / (2 dx) + 1 / dx^2);
   retsmb[[1]] = s[x] - (-p[x] / (2 dx) + 1 / dx^2) f0;

   For[index = 2, index < Ns, index = index + 1,
    x = x + dx;
    retD[[index, index - 1]] = (-p[x] / (2 dx) + 1 / dx^2);
    retD[[index, index]] = (q[x] - 2 / dx^2);
    retD[[index, index + 1]] = (p[x] / (2 dx) + 1 / dx^2);
    retsmb[[index]] = s[x];
   ];
   x = x + dx;
   retD[[Ns, Ns - 1]] = (-p[x] / (2 dx) + 1 / dx^2);
   retD[[Ns, Ns]] = (q[x] - 2 / dx^2);
   retsmb[[Ns]] = s[x] - (p[x] / (2 dx) + 1 / dx^2) fNp1;
   Return[N[{retD, retsmb}]];
  ]
```

---

---

**Damped driven harmonic oscillator redux**

Let us start off simple to test the setup. We'll solve the dimensionless damped, driven harmonic oscillator:

$$q''(s) = -\omega^2 q(s) - q'(s) + \tilde{F}(s) \qquad (4.48)$$

from (4.13). The first observation to make is that our method will clearly fail to produce results in the case $q(0) = 0$, $q(1) = 0$, and $\tilde{F}(s) = 0$. For that choice of boundary condition and forcing, the right-hand side vector $\mathbf{s} - \mathbf{b}$ in (4.46) is zero, and the null space of the matrix $\mathbb{D}$ is empty (at least for some choices of $\omega^2$), so that the solution we get is $\mathbf{f} = 0$. If we instead chose $q(0) = 1$, $q(1) = -1$, we would correctly recover an approximation to $\cos(\omega s)$ (for $\omega = 2\pi p$, and integer $p$) even if $\tilde{F}(s) = 0$. Alternatively, we can correctly solve the equation of motion with $q(0) = 0 = q(1)$ provided $F(s)$ is not zero everywhere. The issue here is that $q''(s) + q'(s) = -\omega^2 q(s)$ with $q(0) = 0 = q(1)$ is really an eigenvalue problem – it can be solved for *some* choices of $\omega$. We saw how to solve a similar equation ($q''(s) = -\omega^2 q(s)$) with these boundary conditions in Chapter 3 where it came up in the context of the infinite square well. There, we found both the function and energy (analogous to $q(s)$ and $\omega$). We'll return to the eigenvalue problem explicitly in Chapter 10.

For now, let's take a non-trivial source function. Set $\tilde{F}(s) = \cos(2\pi s)\sin(10\pi s)$, and we'll take $\omega = (2\pi)$ – this choice corresponds to $p = 1$, $q = (2\pi)^2$ and source $\cos(2\pi s)\sin(10\pi s)$ in the setup (4.38). For boundary conditions, we are free to set $q(0) = 0 = q(1)$ here because of the non-zero source. The solution is analytically available, and we have plotted it together with the numerical result obtained by inverting the matrix output by Implementation 4.1 in Figure 4.1.

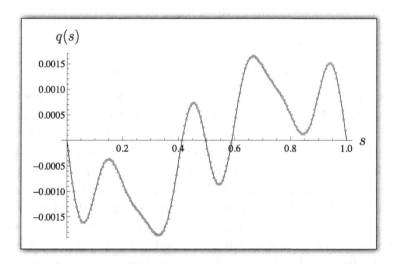

Figure 4.1 The numerical solution (dots), found via $\mathbf{f} = \mathbb{D}^{-1}(\mathbf{s} - \mathbf{b})$, plotted on top of the exact solution, for forcing $\tilde{F}(s) = \cos(2\pi s)\sin(10\pi s)$. Here, we used $N = 200$ points for $s = 0 \longrightarrow 1$.

We can also use the numerical solution to probe the decay of transients in a damped, driven harmonic oscillator. Take $\tilde{F}(s) = \cos(\frac{1}{2}\pi s)$, and we'll let $q(0) = -\frac{1}{2}$, $q(s_f) = 0$ where $s_f = 25.01167873217457$ (a dollar if you can figure out how the right-hand endpoint was determined). In Figure 4.2, we see the result; the numerical solution appears as dots, and the exact solution as a solid curve. You can clearly see the initial oscillation frequency changing as time goes on.

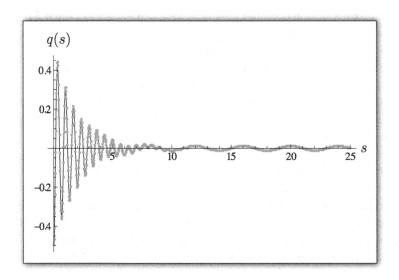

Figure 4.2 The numerical solution (dots), found via $\mathbf{f} = \mathbb{D}^{-1}(\mathbf{s} - \mathbf{b})$, plotted on top of the exact solution, for forcing $\tilde{F}(s) = \cos(\frac{1}{2}\pi s)$. There are $N = 1000$ points taken for $s = 0 \longrightarrow s_f$.

### 4.3 Finite difference in two dimensions

Our focus, in two dimensions, will be Poisson's problem on a rectangular grid, but the basic ideas hold for more complicated differential operators and different geometries. We will again end up with a matrix inverse problem of the form $\mathbb{D}\mathbf{f} = \mathbf{s} - \mathbf{b}$, and the only real difference moving to two dimensions introduces is in defining the correct vector $\mathbf{f}$ – there is an ordering issue that we must decide upon (and stick with). Consider the grid shown in Figure 4.3 – this has $N_x = 3$, $N_y = 3$ interior points (colored black), and we have labelled them by their integer indices, while the grid itself has constant spacing $\Delta x$ and $\Delta y$, not necessarily the same.

We need to form a vector out of the grid-function $f_{nm} = f(x_n, y_m)$. This vector will be the solution to Poisson's problem obtained by matrix inversion just as in the one-dimensional case. Define the embedding:

$$\mathbf{f} \doteq \begin{pmatrix} f_{11} \\ f_{21} \\ f_{31} \\ f_{12} \\ f_{22} \\ f_{32} \\ f_{13} \\ f_{23} \\ f_{33} \end{pmatrix}. \tag{4.49}$$

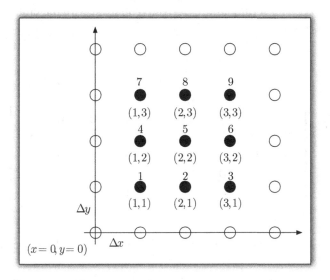

Figure 4.3 A two-dimensional grid for Cartesian coordinates. This one has three interior points in each direction.

Then the fourth entry, say, of **f** is: $f_4 = f(x_1, y_2)$. It is easy to keep track of the vector indices by introducing the function:

$$g(n, m) = (m - 1)N_x + n \tag{4.50}$$

returning the vector index associated with the grid point $(x_n, y_m)$. If we ask for the vector location of the point $(x_1, y_2)$, then, we get: $g(1, 2) = 3 + 1 = 4$, correct. We can use this function to generate the difference matrices, since now $f_{g(n,m)} = f_{nm}$.

Let's build separate matrices for the $\partial^2/\partial x^2$ and $\partial^2/\partial y^2$ approximations. We will take the differences to be

$$\frac{\partial^2 f(x_n, y_m)}{\partial x^2} \approx \frac{f(x_n - \Delta x, y_m) - 2f(x_n, y_m) + f(x_n + \Delta x, y_m)}{\Delta x^2} \tag{4.51}$$

and similarly for $\partial^2 f(x_n, y_m)/\partial y^2$. Once again, we have to watch the boundary conditions, but it's just a few extra copies of the one-dimensional case.

Using the function (4.50), we can write the difference (4.51) in terms of entries from **f** in (4.49),

$$\frac{\partial^2 f_{g(n,m)}}{\partial x^2} \approx \frac{f_{g(n+1,m)} - 2f_{g(n,m)} + f_{g(n-1,m)}}{\Delta x^2}. \tag{4.52}$$

If we define $i \equiv g(n - 1, m)$, $j \equiv g(n, m)$ and $k \equiv g(n + 1, m)$, then the three entries we need in the matrix $\mathbb{D}_x$ to capture (4.52) upon multiplication by **f** are:

$$(\mathbb{D}_x)_{ji} = \frac{1}{\Delta x^2} \quad (\mathbb{D}_x)_{jj} = -\frac{2}{\Delta x^2} \quad (\mathbb{D}_x)_{jk} = \frac{1}{\Delta x^2}. \tag{4.53}$$

In the $N_x = N_y = 3$ setting, the $x$-derivative matrix, with the embedding shown in (4.49), takes the form:

$$\mathbb{D}_x \doteq \frac{1}{\Delta x^2} \begin{pmatrix} -2 & 1 & 0 & 0 & 0 & 0 & 0 & 0 & 0 \\ 1 & -2 & 1 & 0 & 0 & 0 & 0 & 0 & 0 \\ 0 & 1 & -2 & 0 & 0 & 0 & 0 & 0 & 0 \\ 0 & 0 & 0 & -2 & 1 & 0 & 0 & 0 & 0 \\ 0 & 0 & 0 & 1 & -2 & 1 & 0 & 0 & 0 \\ 0 & 0 & 0 & 0 & 1 & -2 & 0 & 0 & 0 \\ 0 & 0 & 0 & 0 & 0 & 0 & -2 & 1 & 0 \\ 0 & 0 & 0 & 0 & 0 & 0 & 1 & -2 & 1 \\ 0 & 0 & 0 & 0 & 0 & 0 & 0 & 1 & -2 \end{pmatrix}. \tag{4.54}$$

The $\mathbb{D}_y$ form is similar (with changing $m$ instead of $n$) and can be found in the chapter notebook.

Adding these two, we have the full $\mathbb{D} = \mathbb{D}_x + \mathbb{D}_y$ operator. Once again, the boundary conditions must be specified – these are now functions, and we need four of them (functions of $x$ for the $y$ boundaries, and functions of $y$ for the $x$ boundaries). The source becomes a vector, embedded like **f** (we have to be consistent), so that the source vector is just the function $s(x_n, y_m)$ (for source $s(x, y)$ in $\nabla^2 f = s$) ordered appropriately (see the chapter notebook for the implementation details). The source and boundary terms appear on the right, and we again have a matrix inverse problem, just as in the one-dimensional case.

### Arbitrary grounded boundary

The ordering we have chosen in (4.49) relies on our grid being rectangular – but what if we want to change the shape of the boundary? Or suppose we want to introduce a conducting surface inside our rectangular domain, how would we do that? We'll generate an approach that is based on a square grid (so that the number of grid points in the $x$ and $y$ direction are the same, although the spacings need not be) and introduce grounded conductors inside. The idea generalizes and provides a simple way to construct the Laplacian finite difference matrix.

Start with a square matrix ($N \times N$) of ones, these will represent the points at which we want to calculate the potential. Introduce in that matrix zeroes where you want the grounded points to be (actually drawing on a piece of paper and scanning it in is one of the easiest ways to make such a matrix if the grounded points form a strangely shaped internal boundary). Now we make a list of all the non-zero points in the matrix, and for each element of that list, we include the matrix location (the $n$ and $m$ values) to the left ($n - 1, m$), right ($n + 1, m$), above ($n, m + 1$), and below ($n, m - 1$), unless those points have matrix entry zero, in which case we put a zero

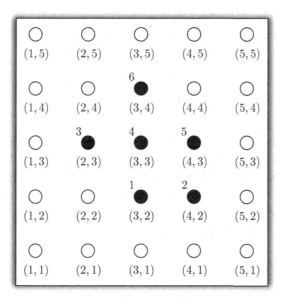

Figure 4.4 The black circles represent the region in which we want to know the solution to Poisson's problem – the rest of the points are grounded. Below each grid point is the $(n, m)$ index as in Figure 4.3, and at the upper left of the black circles is its vector index, assigned by sweeping through the grid and sequentially numbering the black circles.

in our list. We now have a vector of *just the points of interest*, i.e. the points where we do not know the value of the potential. And each of these points has its nearest neighbors (or ground) stored with it. In addition to this list, we make a matrix that has the vector index of each point at the correct location – just the original matrix, but with increasing numbers at the points of interest (rather than simply ones) ordered as in Figure 4.3. We can make this as we make our list of points and neighbors. Call the matrix hmat, then the function $g(n, m)$ replacing (4.50) here is: $g(n, m) =$ hmat[[N-(m-1),n]].

To be explicit, suppose someone gave us Figure 4.4 – there we have a $5 \times 5$ matrix (so $N = 5$) with black circles representing the points where the potential will be calculated, and the open circles are points where $V = 0$ (grounded). The grid index is shown beneath each point, with the vector index appearing at the upper left on the six relevant points. We make a vector, now, with left/right/top/bottom information – the first element of that vector, associated with the grid point located at $(3, 2)$, would be: $\{\{3, 2\}, 0, \{4, 2\}, \{3, 3\}, 0\}$, identifying the grid location, then a zero indicating that the point to the left is at ground, then $\{4, 2\}$ is the grid point to the right, and $\{3, 3\}$ is the point above, the point below is grounded, hence the final zero in the list. As another example, the fifth entry in our vector would be:

$\{\{4, 3\}, \{3, 3\}, 0, 0, \{4, 2\}\}$. The matrix hmat for this example would be:

$$\text{hmat} \doteq \begin{pmatrix} 0 & 0 & 0 & 0 & 0 \\ 0 & 0 & 6 & 0 & 0 \\ 0 & 3 & 4 & 5 & 0 \\ 0 & 0 & 1 & 2 & 0 \\ 0 & 0 & 0 & 0 & 0 \end{pmatrix}. \tag{4.55}$$

To generate the Laplacian matrix, we just go through our vector of points and neighbors, making the $x$ and $y$-directed finite differences out of the neighbors, and ignoring any zeroes (indicating a grounded point). The whole process is harder to describe than it is to implement, so take a look at the functions Hashmat and cMakeD2 from the chapter notebook that do the actual work. We'll use these below to do an example, and you'll make an appropriate grounded boundary matrix in Problem 4.15.

## 4.4 Examples

We'll set up and solve a few concrete cases, both with complete solution, and without. All of these examples come from E&M, and should be familiar to you.

### 4.4.1 Empty box

Returning to our box example from Section 4.1.2 for Laplace's equation, we want a numerical solution to:

$$\nabla^2 V(x, y) = 0 \tag{4.56}$$

on $x \in [0, 1]$, $y \in [0, 1]$ with:

$$V(x, 0) = \sin(\pi x) \quad V(x, a) = 0 \quad V(0, y) = 0 \quad V(a, y) = 0. \tag{4.57}$$

Here, the boundary conditions are the only contributors to the right-hand side of $\mathbb{D}\mathbf{f} = \mathbf{s} - \mathbf{b}$ – when we invert the matrix, we will get the solution vector $\mathbf{f}$ ordered according to (4.49), and it takes some unpacking to provide a reasonable graphical representation. In Figure 4.5, we see the contour plot for the numerical solution (using $N_x = N_y = 25$), matching the exact solution given by (4.37).

### 4.4.2 Parallel plate capacitor

We can also look at more physical (read "finite") distributions that are familiar from the idealized cases we think about in electrostatics. Consider the parallel

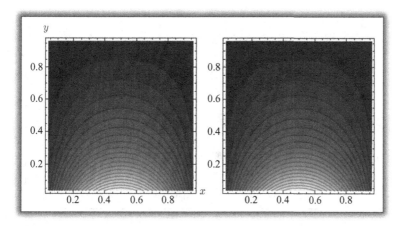

Figure 4.5 On the left, the numerical solution to the Laplace problem with boundaries given by (4.57) (with $a = 1$ m), and on the right, the exact solution from (4.37).

plate capacitor – the field in between the plates is roughly uniform, but there is fringing around the edges. We should be able to look at that fringing field by using a source for the Poisson equation.

Our grid here is necessarily finite, and so there are some finite edge effects to watch out for. If we put our parallel plate capacitor in the center of a box with grounded walls, then we are bringing infinity in to a finite location (required on a finite grid, of course). Nevertheless, the finite parallel plate capacitor gives a good, qualitative description of fringing fields. In this case, we have a source, and all we need to do is pick a row of our grid and give it non-zero (but constant) line charge, and mirror it with a negatively charged uniform line. The results are shown in Figure 4.6. On the left, we show the contour plot of the potential, and on the right, the associated electric field.

### 4.4.3 Constant surface charge box

Take a two-dimensional "box" with uniform line charge density $\lambda$ on all four sides. Each side has length $2a$, centered as shown in Figure 4.7. To find the potential at any point inside, consider a single line charge, oriented as shown in Figure 4.8. For this configuration, the potential at the point $\mathbf{r} = x\hat{\mathbf{x}} + y\hat{\mathbf{y}}$ is determined by the integral:

$$V(x, y) = -\frac{\lambda}{2\pi\epsilon_0} \int_{-a}^{a} \log \sqrt{(x - x')^2 + y^2}\, dx', \qquad (4.58)$$

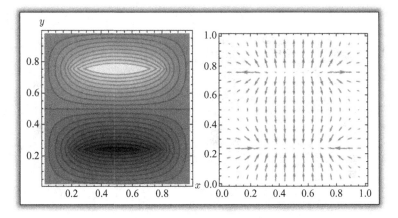

Figure 4.6 Parallel plate potential – this is for $N_x = N_y = 40$. The potential for all grid points is shown on the left. On the right is the (numerically computed) electric field from $\mathbf{E} = -\nabla V$.

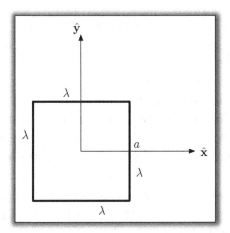

Figure 4.7 Two-dimensional version of the box problem.

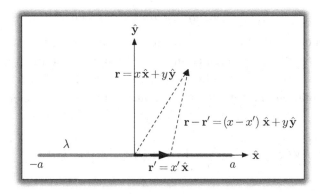

Figure 4.8 Uniform line charge, centered, with length $2a$.

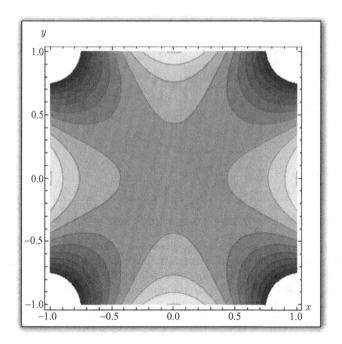

Figure 4.9 Potential for the superposition (4.60), using $a = 1$ m, and $\lambda/2\pi\epsilon_0 = 1$ N m/C.

the two-dimensional analogue of our usual Coulomb potential, summed over the line. The integral is

$$V(x, y) = \frac{\lambda}{2\pi\epsilon_0}\left[2a + y\tan^{-1}\left(\frac{x-a}{y}\right) - y\tan^{-1}\left(\frac{x+a}{y}\right)\right.$$

$$\left. + (x-a)\log\sqrt{(x-a)^2 + y^2} - (x+a)\log\sqrt{(x+a)^2 + y^2}\right]. \quad (4.59)$$

If we take the axes shown in Figure 4.7 and apply $V(x, y)$ to the four line charges using superposition, we get

$$V_\square(x, y) = V(x, y+a) + V(y, a-x) + V(-x, a-y) + V(-y, x+a) \quad (4.60)$$

where we are just shifting the axes for the four line charges (starting at the bottom and moving counter-clockwise). Using $a = 1$ and "natural" units, the potential is shown in Figure 4.9. The Laplacian of $V$ is zero inside the box (that's somewhat surprising given the form of (4.60), but of course, we created it with this property in place). We can compare the exact potential solution with the numerical $\mathbb{D}$ inversion – using a $50 \times 50$ grid in $x - y$, and a square box with $\lambda$ constant, we get the potential shown in Figure 4.10 by inversion.

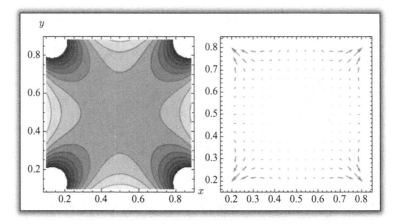

Figure 4.10 Numerical solution, both potential (left) and electric field (right), for a box with uniform λ (in two dimensions). The problem is set up with grounded outside walls, so there is an "exterior" potential outside the box itself. The numerical grid we use goes from $0 \to 1$, so the problem is solved in shifted and scaled coordinates here, compared to Figure 4.9.

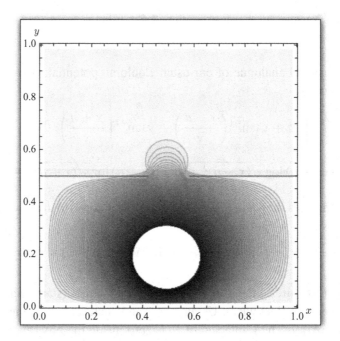

Figure 4.11 A charged particle sits below a grounded line with a slit in the middle (shown). The potential leaks through the slit. As usual, the four walls of the box are also grounded.

### 4.4.4 Grounded plate with hole

Finally, we'll use our internal grounded boundary approach. If we take a grounded box and put a grounded line through its vertical center, with a small horizontal opening in the middle, then placing a charge in the bottom half of the box gives field in the top half, a sort of single-slit experiment. We start with a matrix that has ones everywhere except along the grounded line, and then place a source at some appropriate location. The Laplacian matrix is then generated and the usual inverse problem solved. The result is shown in Figure 4.11.

### Further reading

1. Arfken, George B. & Hans J. Weber. *Mathematical Methods for Physicists*. Academic Press, 2001.
2. Baumgarte, Thomas W. & Stuart L. Shapiro. *Numerical Relativity: Solving Einstein's Equations on the Computer*. Cambridge University Press, 2010.
3. Bender, Carl M. & Steven A. Orszag. *Advanced Mathematical Methods for Scientists and Engineers*. McGraw-Hill Book Company, 1978.
4. Boas, Mary. *Mathematical Methods in the Physical Sciences*. Wiley, 2005.
5. Golub, Gene H. & James M. Ortega. *Scientific Computing and Differential Equations: An Introduction to Numerical Methods*. Academic Press, 1992.
6. Griffiths, David J. *Introduction to Electrodynamics*. Prentice Hall, 1999.
7. Stoer, J. & R. Bulirsch. *Introduction to Numerical Analysis*. Springer-Verlag, 1993.

### Problems

One of the goals of the lab problems is to generate the discrete form of the Laplacian for two-dimensional spherical coordinates (think of cylindrical coordinates with $z = 0$):

$$\nabla^2 V(s, \phi) = \frac{1}{s} \frac{\partial}{\partial s} \left( s \frac{\partial V}{\partial s} \right) + \frac{1}{s^2} \frac{\partial^2 V}{\partial \phi^2}. \tag{4.61}$$

The computational structure of the curvilinear coordinate system is virtually identical to the Cartesian case. Thinking about the form of the Laplacian in cylindrical coordinates, you should be able to directly modify the finite difference matrix-making routines in the chapter notebook to obtain the relevant matrices, vectors and appropriate boundary conditions. The first few problems below set the stage, and provide concrete target calculations to check your solver.

### Problem 4.1

In Section 4.3, we made a two-dimensional Cartesian grid and mapped the points in the grid to a vector in order to provide a discrete matrix representation of $\nabla^2$. In particular, we defined the function:

$$g(n, m) = (m - 1)N_x + n \tag{4.62}$$

that takes the grid index $(n, m)$ and turns it into the vector index given our ordering.

For the circular case, we can order as shown in Figure 4.12, and go all the way around in $\phi$ – the function:

$$g(n, m) = \text{Mod}(m - 1, N_\phi)N_s + n \tag{4.63}$$

gives us the vector index associated with the grid index $(n, m)$ for the circular grid (with $N_s$ interior grid points in the $s$ direction, and $N_\phi$ grid points going around in $\phi$). The Mod function is ensuring that the $\phi$ index is appropriately periodic. Here, we associate the first index with the $s$ grid location, the second index with $\phi$, and order our vector via (for $N_s = 2$ as in Figure 4.12):

$$\mathbf{f} \doteq \begin{pmatrix} f_{11} \\ f_{21} \\ f_{12} \\ f_{22} \\ f_{13} \\ f_{23} \\ \vdots \end{pmatrix}. \tag{4.64}$$

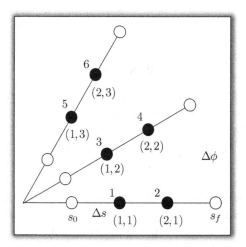

Figure 4.12 A circular grid with constant $\Delta s$ and $\Delta\phi$ spacing. The points are ordered as shown starting at $\phi = 0$. The radial boundaries are $s_0$ and $s_f$.

Write the discretized form of the Laplace operator for two-dimensional spherical coordinates (analogous to, for example (4.51)) at the grid point $(n, m)$ on the interior of the grid (i.e. for $2 \le n \le N_s - 1$). Do the same for the $n = 1$ and $n = N_s - 1$ boundary points using $V_0$ and $V_f$ as the known values of the solution at $s_0$ and $s_f$.

## Problem 4.2

In two dimensions, Poisson's equation for electrostatics is

$$\nabla^2 V(s, \phi) = -\frac{\rho_g}{\epsilon_0} \qquad (4.65)$$

where $\rho_g$ has dimension of charge per unit area (a two-dimensional density). Notice that if we imagine the Lorentz force law, $\mathbf{F} = q\mathbf{E} = -q\nabla V$, is unchanged in two dimensions, then $\epsilon_0$ cannot be the usual constant from E&M (think of its dimension).

(a) Suppose we have a configuration with $\rho_g$ confined to a ring of width $\Delta s$ at $R$. Find the electric field inside and outside the ring (use Gauss's law, appropriate to two dimensions). The electric field is discontinuous in this case, what is $E_{out}(R) - E_{in}(R)$?

(b) We will be checking this discontinuity, so we need a numerical expression for it – the electric field is the negative $s$-derivative of the potential (which is what we will be solving for). We want to check:

$$-\left(\frac{\partial V_{out}}{\partial s} - \frac{\partial V_{in}}{\partial s}\right)\Bigg|_{s=R} = E_{out}(R) - E_{in}(R), \qquad (4.66)$$

and in order to do this, we need the derivative of the potential with respect to $s$ for the "inner" ($s < R$) and "outer" ($s > R$) forms, evaluated at $R$. Write the appropriate numerical approximation for these using one-sided differences. If we take $R$ to lie on a grid point $s_k$, for example, then your answer should involve values like $V(s_{k+1}) - V(s_k)$ and $\Delta s$ (for the outer derivative approximation). For the right-hand side, replace the difference in electric fields by the value of the discontinuity you found in part (a).

## Problem 4.3

In two dimensions, find the potential in the region between two concentric circles of radii $s_0$ and $R$ held at potentials $V_0$ and zero, respectively (assume that there is no charge in this region) – that is, solve $\nabla^2 V = 0$ with $V(s_0) = V_0$ and $V(R) = 0$.

## Problem 4.4

There is an averaging feature to the solution of Laplace's equation, even in one dimension (and it holds for higher dimension as well). Show that the solution to $f''(x) = 0$ has the property that $f(x) = \frac{1}{2}(f(x + \epsilon) + f(x - \epsilon))$ for small $\epsilon$. So the value of $f(x)$ at any point is the average of the value of $f$ at nearby points. Using this observation, argue that on any finite domain, $x \in [x_\ell, x_r]$, $f(x)$ (satisfying $f''(x) = 0$) can have max/min values only at the boundaries of the domain. This is clearly true in one dimension, where the solutions represent straight lines.

## Problem 4.5

Here, we will use the fact, motivated by the previous problem, that the solution to Laplace's equation has minima/maxima only on boundaries, to establish the uniqueness of the solution to the Poisson problem. Take $\nabla^2 f = s$ in a volume $\Omega$, with the value of $f$ specified on the boundary of that volume, $f(\partial\Omega) = g$ given. Now suppose you had two solutions to the PDE and boundary conditions, subtract them and use the min/max idea to show that the two solutions must be identical everywhere in $\Omega$ and on its boundary, hence the two solutions are identical.

## Problem 4.6

For linear PDEs, like the Poisson problem, where superposition holds, we can build solutions for general sources out of the point-source solution. A point source at $\mathbf{r}'$ is represented by the delta distribution: $s = \delta^n(\mathbf{r} - \mathbf{r}')$ in $n$ dimensions, and the solution to the Poisson problem with this source,

$$\nabla^2 G(\mathbf{r}, \mathbf{r}') = \delta^n(\mathbf{r} - \mathbf{r}') \tag{4.67}$$

is called a "Green's function" (note that $\mathbf{r}'$ is the source location, $\mathbf{r}$ is a vector pointing to any point in $n$ dimensions, and $\nabla^2$ is the Laplacian w.r.t. the coordinates). Show that for a general source, $s(\mathbf{r})$, the solution to $\nabla^2 f(\mathbf{r}) = s(\mathbf{r})$ is given by the integral

$$f(\mathbf{r}) = \int G(\mathbf{r}, \mathbf{r}') s(\mathbf{r}') d\tau' \tag{4.68}$$

where the integral is over all space (an $n$-dimensional volume integral – note that the integral is over the source locations, $\mathbf{r}'$).

## Problem 4.7

In two dimensions the solution to $\nabla^2 G(x, y) = \delta^2(x, y)$ for a point source at the origin (in this case) is $G(x, y) = \frac{1}{2\pi} \log(s/s_0)$ where $s = \sqrt{x^2 + y^2}$ and $s_0$ is an arbitrary radial distance. Given that we have, in theory, an integral solution (4.68) for any source distribution using this Green's function, why do we need our finite difference approach to solving Poisson's problem?

## Problem 4.8

We have three infinite sheets of charge, they are parallel to each other and separated by a distance $d$. The top and bottom sheets have charge-per-unit area $\sigma$ and the middle sheet has $-2\sigma$, find the electric field everywhere.

## Problem 4.9

In Section 4.1.2, we used multiplicative separation of variables to solve for the potential inside a square box (side length $a$) with grounded sides and top (so that $V = 0$ there). On the bottom, we placed a non-zero potential: $V(x, 0) = V_0 \sin(\pi x/a)$. The solution to the Laplace problem for that setup is given in (4.37). Using the same boundary conditions, solve using the Helmholtz operator, so that inside the box,

$$(\nabla^2 - \mu^2) V = 0 \tag{4.69}$$

and you must match your solution to the boundaries.

## Lab problems

We'll start with problems defined in a Cartesian setting (on grids that extend from $0 \to 1$ in both $x$ and $y$), so that you can use the Cartesian solver provided in the notebook for this chapter. The final four problems will use your circular solver and checks from Problems 4.1, 4.2, and 4.3.

### Problem 4.10
Solve the ODE (an example of (4.38)):

$$f''(x) + \sin(x) f'(x) + f(x) = 0 \tag{4.70}$$

with $f(0) = 0$ and $f(4\pi) = 1$ by making the matrix $\mathbb{D}$ and vector $\mathbf{s} - \mathbf{b}$ using Implementation 4.1 with $N = 100$, and inverting. Solve the same problem using the shooting techniques from the previous chapter.

### Problem 4.11
We can re-create all the familiar point sources in approximation using the finite difference approach. In all of the cases below, use a grounded box (so that we have "moved" spatial infinity in to a finite location, the fields near the sources will then approximate the fields for sources far from spatial infinity) with length and height 1 – in this and subsequent problems, grid indices are given for source positions ($n$ gives the $x$-location $n\Delta x$, and $m$ the $y$, $m\Delta y$).
(a) On a grid with $N_x = N_y = 49$, put a single point charge of strength $\rho_g/\epsilon_0 = 1$ at $n = m = 25$, with zero at all other locations (you can use an If statement in the definition of your source function to accomplish this). Solve Poisson's equation and make a contour plot of this approximation to the monopole solution.
(b) Using a grid with $N_x = 49$, $N_y = 50$, put a point charge with $\rho_g/\epsilon_0 = 1$ at $n = 25$ and $m = 25$ and one with $\rho_g/\epsilon_0 = -1$ at $n = 25$, $m = 26$. Again, solve and make a contour plot of the potential (which should look like the dipole potential).
(c) For the quadrupole, take $N_x = N_y = 50$, and put point charges with $\rho_g/\epsilon_0 = 1$ at $n = 25$, $m = 25$, and $n = 26$, $m = 26$, and ones with $\rho_g/\epsilon_0 = -1$ at $n = 25$, $m = 26$ and $n = 26$, $m = 25$. Make a contour plot of the solution.

### Problem 4.12
In Figure 4.10, we see the potential inside a "box" of charge – but we can explore both the interior and exterior solutions numerically. Set up the box of line charge on a grid with $N_x = N_y = 60$, and put $\rho_g/\epsilon_0 = 1$ for $n = 15$ to 46 and the same range for $m$ (so the

source is zero except for $n = 15 \rightarrow 46$ at $m = 15$, and $m = 46$, and $m = 15 \rightarrow 46$ at $n = 15, 46$). Solve for the potential to see the interior solution connect to the exterior.

## Problem 4.13

For a grid with $N_x = N_y = 49$, set the source to zero for all points except $n = m = 25$, where you'll take $\rho_g/\epsilon_0 = 5000$. Ground the top, left, and right walls of your box, but set the bottom to potential $V(x, 0) = \sin(\pi x)$ (here, $\{x, y\} \in [0, 1]$). Solve for the potential and make a contour plot of the result.

## Problem 4.14

Set up three parallel lines of charge on a grid with $N_x = N_y = 49$ as follows: make a source that is zero everywhere except along a line at $m = 25$ extending from $n = 5$ to $n = 46$, where you'll take $\rho_g/\epsilon_0 = -2$, a line with the same width at $m = 15$ with $\rho_g/\epsilon_0 = 1$, and one at $m = 35$ with $\rho_g/\epsilon_0 = 1$. Solve for the potential, and then calculate the electric field using the GradV function from the chapter notebook. Does it look reasonable given your solution to Problem 4.8?

## Problem 4.15

We'll find the potential given a point charge that sits inside a grounded triangle. Make a $40 \times 40$ matrix that has zeroes everywhere except above (and on) the diagonal, where it will have ones. Using the functions Hashmat and cMakeD2, generate the discrete Laplacian matrix using $N_x = 40 = N_y \equiv N$ and $\Delta x = \Delta y = 1/(N + 1)$. Write a source function that puts a point charge of strength $\rho_g/\epsilon_0 = 1$ at grid location $n = 35, m = 20$. Solve for the potential and make a contour plot of your solution. Notice how the contours have to squeeze themselves so as to achieve the boundary conditions on the left and top in addition to the diagonal.

---

The remaining problems all involve solving the Poisson problem with circular boundaries. Using the Cartesian case as a template, implement a Poisson solver for circular coordinates. For the angular portion, set $\Delta\phi = 2\pi/N_\phi$ for $N_\phi$ given. The radial coordinate has finite extent given by $s_0$ and $s_f$ (boundary points) – take $\Delta s = \frac{(s_f - s_0)}{(N_s + 1)}$ and remember to calculate $s_j$ using

$$s_j = s_0 + j\Delta s. \tag{4.71}$$

## Problem 4.16

Compute, using your circular PDE solver, the potential for a disk with interior $s_0 = 0.1$, exterior $s_f = 2.0$ and $N_s = 20$, $N_\phi = 40$ points in the $s$ and $\phi$ directions – set the interior potential to $V_0 = 10$, and the exterior to ground. Plot the potential as a function of $s$ for $\phi = 0$, and compare with the theoretical prediction from Problem 4.3 (i.e. plot the theoretical curve on top of your data points).

**Problem 4.17**

Using the same grid as in the previous problem, put a non-uniform charge distribution at the center of your spherical grid: $V(s = s_0, \phi) = \sin \phi$, and ground the outer ring at $s_f$, so that $V(s_f) = 0$. Find the potential on your grid, and make a contour plot.

**Problem 4.18**

Ground the interior and exterior boundaries of a disk region with $s_0 = 0.1$, $s_f = 2.0$ and $N_s = N_\phi = 40$. Place a ring of charge at $R = s_0 + 10\Delta s$ with $\rho_g/\epsilon_0 = 1000$. Find the potential, and make a contour plot of your solution. The solution should be spherically symmetric, so take a slice at $\phi = 0$ and plot the potential as a function of the radial coordinate. There is a discontinuity in the derivative of $V$ at the ring, calculate the appropriate approximate derivative "inside" and "outside" the ring, and show that the difference is consistent with the discontinuity you computed in
Problem 4.2.

**Problem 4.19**

Use the same setup as in the previous problem, but this time give the ring angular dependence – place a ring of charge at $R = s_0 + 10\Delta s$ with $\rho_g/\epsilon_0 = 1000 \cos(\phi)$. Find the potential, and make a contour plot of your solution.

# 5

# Time-dependent problems

We have looked at the finite difference method applied to two-dimensional PDEs, and it is pretty clear how that approach would generalize to higher dimension. There are two fundamental difficulties when applying the class of finite difference methods in Chapter 4. One is the linear (or linearized) form that the differential equations must take to be accessible. The other is the size of the spatial grids – in three dimensions, if we had 100 grid points, we would need to construct and invert a matrix of size: $100^3 \times 100^3$, which can be difficult.[1]

When time and space are together in a PDE setting, there is a natural splitting (anathema to modern physics) between the spatial and temporal derivatives. Using that splitting, we can develop techniques that allow us to solve nonlinear PDEs, and capture some interesting physical behavior. It is the formulation of the "boundaries" in the spatial versus temporal directions that allows us to separate the treatment of time and space. The temporal part of the problem will typically have an initial value formulation (given an initial position and velocity, propagate a solution forward in time), while the spatial part is most naturally phrased in boundary condition language (given the potential on all four sides of a box, find the potential inside the box).

## 5.1 Physical motivation

Our model problems will be PDEs with both spatial and temporal derivatives. The most familiar PDE of this form is the wave equation:

$$-\frac{1}{v^2}\frac{\partial^2 u(x,t)}{\partial t^2} + \frac{\partial^2 u(x,t)}{\partial x^2} = 0, \tag{5.1}$$

---

[1] We will see ways to approximately invert these large, sparse matrices in Chapter 11.

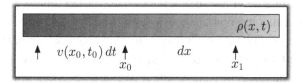

Figure 5.1 A distribution of charge $\rho(x, t)$. In the interval between $x_0$ and $x_1$, there is total charge $\rho(x_0, t_0)dx$ at time $t_0$. During a time $dt$, the charge located in the interval between $x_0 - v(x_0, t_0)dt$ and $x_0$ will enter the region, so we'll have an influx of $\rho(x_0, t_0)v(x_0, t_0)dt$ – similarly, charge will leave the region on the right.

which comes up in the study of fluids, continuous elastic bodies, sound, and of course, E&M, where it governs the electromagnetic potentials (in Lorentz gauge). The meaning of $u(x, t)$ depends on context, it could be a mass density, the height of a string, or the value of the electric potential at a particular place and time. The constant $v$ in the equation represents the speed with which the wave propagates. That speed is set by physical properties of the medium in which the wave motion is occurring, so that $v$ itself could depend on position and/or time.

Another example of a PDE governing a function of time and space is the time-dependent Schrödinger equation:

$$-\frac{\hbar^2}{2m}\frac{\partial^2}{\partial x^2}\Psi(x, t) + V(x)\Psi(x, t) = i\hbar\frac{\partial\Psi(x, t)}{\partial t}. \tag{5.2}$$

This precursor to the time-independent form allows us to determine the probability density, as a function of time ($\rho(x, t) = \Psi(x, t)^*\Psi(x, t)$) associated with the position of a particle of mass $m$ moving in a physical environment described by $V(x)$, the potential energy.

Both of these PDEs are linear, and that is a natural place to begin looking at properties of numerical methods meant to solve them. We'll start with the simplified class of conservative PDEs – these are first order in time and space (unlike the wave equation and the Schrödinger equation).

### 5.1.1 Conservation laws

Think of charge conservation in electricity and magnetism. We are given a charge density (charge per unit volume) $\rho(x, t)$ (in one spatial dimension, $\rho$ has the interpretation of charge per unit length), and the velocity of the flow of charge at each point $x$ for all times $t$: $v(x, t)$. Take an interval extending from $x_0$ to $x_0 + dx = x_1$ as shown in Figure 5.1.

The amount of charge in the spatial interval $[x_0, x_1]$ at time $t_0$ is: $\rho(x_0, t_0)dx$. If the total charge in this interval changes over a time $dt$, it must be because charge entered or left the spatial region. The amount of charge coming in from the left is: $\rho(x_0, t_0)v(x_0, t_0)dt$, and the charge leaving on the right is: $-\rho(x_0 + dx, t_0)v(x_0 + dx, t_0)dt$, so the change in the total amount of charge over the time interval $dt$ is:

$$\rho(x_0, t_0 + dt)dx - \rho(x_0, t_0)dx$$
$$= (\rho(x_0, t_0)v(x_0, t_0) - \rho(x_0 + dx, t_0)v(x_0 + dx, t_0))dt \quad (5.3)$$

or

$$\frac{\partial \rho(x_0, t_0)}{\partial t}dtdx \approx -\frac{\partial}{\partial x}(\rho(x_0, t_0)v(x_0, t_0))\,dxdt \quad (5.4)$$

using Taylor expansion. For any point $x$, and any time $t$, then, we have:

$$\frac{\partial \rho(x, t)}{\partial t} + \frac{\partial}{\partial x}(\rho(x, t)v(x, t)) = 0. \quad (5.5)$$

Such a relation between the temporal and spatial evolution of a density is called a "conservation law." There are many examples, and you can generate them by replacing "charge" with whatever it is you like to think about.[2]

There is an associated integral form for conservation laws, also familiar from E&M – if we integrate (5.5) over an interval $[x_0, x_f]$, then:

$$\frac{d}{dt}\int_{x_0}^{x_f} \rho(x, t)dx = -\int_{x_0}^{x_f} \frac{\partial}{\partial x}(\rho(x, t)v(x, t))\,dx = \big(\rho(x_0, t) - \rho(x_f, t)\big)$$

$$(5.6)$$

which generalizes to three dimensions: $\frac{dQ}{dt} = -\oint_{\partial\Omega} \mathbf{J} \cdot d\mathbf{a}$ where $\Omega$ is a volume of interest (with boundary $\partial\Omega$), and $Q$ is the charge enclosed in that volume.

### 5.1.2 The wave equation

The wave equation

$$-\frac{1}{v^2}\frac{\partial^2 u(x, t)}{\partial t^2} + \frac{\partial^2 u(x, t)}{\partial x^2} = 0 \quad (5.7)$$

is not clearly a conservative equation.

---

[2] There are physical examples, like charge, or fluid density. But there are also gauge choices that can be expressed as conservation laws – Lorentz gauge, after all, reads $\frac{1}{c^2}\frac{\partial V}{\partial t} = -\nabla \cdot \mathbf{A}$, making $V$ a conserved quantity with "current" $\mathbf{A}$.

We can make (5.7) look more like (5.5) by multiplying by $v^2$ and factoring:

$$\left(-\frac{1}{v^2}\frac{\partial^2}{\partial t^2} + \frac{\partial^2}{\partial x^2}\right)u(x,t) = 0 \longrightarrow \boxed{\left(\frac{\partial}{\partial t} - v\frac{\partial}{\partial x}\right)\left(\frac{\partial}{\partial t} + v\frac{\partial}{\partial x}\right)u(x,t) = 0.}$$

(5.8)

Now one way to satisfy the wave equation is to have:

$$\frac{\partial u(x,t)}{\partial t} + \frac{\partial}{\partial x}(vu(x,t)) = 0$$

(5.9)

(so that the second operator in (5.8) kills $u(x,t)$ by itself), and this *is* of the form (5.5), with a constant speed $v$ associated with the wave motion.

The general solutions to (5.9) are well known – we have: $u(x,t) = f(x - vt)$ for any function $f(p)$, and this corresponds to picking up the function $f(p)$ at time $t = 0$, and sliding it along with speed $v$ as time goes on. To see this interpretation, take the value of the function at $p = 0$, $u(0,0) = f(0)$, and ask where this value occurs at time $t$ – that question is answered by solving $u(x,t) = f(x - vt) = f(0)$, so that $x - vt = 0$ giving $x = vt$. The value of $u(0,0)$ occurs at $x = vt$ at time $t$ (constant speed, to the right of the starting point).

Aside from that argument, there is a useful way to visualize the form of these right-traveling solutions. In the two-dimensional $x - t$ plane (where $x$ is the horizontal axis, $t$ the vertical one), curves defined by $x(t) = x_0 + vt$ have the same value of $u(x(t), t)$ for all time $t$ since the derivative of $u(x(t), t)$ vanishes along them:

$$\frac{du(x(t),t)}{dt} = \frac{\partial u}{\partial x}\dot{x} + \frac{\partial u}{\partial t} = \frac{\partial u}{\partial x}v + \frac{\partial u}{\partial t} = 0$$

(5.10)

using $x = x_0 + vt$ with $\dot{x} = v$, and the PDE (5.9) itself. That means if you know the value of $u(x_0, 0)$ (which you must, given the initial condition), and you draw a line from $x_0$ with slope $1/v$ in the $x - t$ plane, then all along that line the solution $u(x(t), t)$ takes on the value $u(x_0, 0)$.

These curves, along which the solution $u(x,t)$ is constant, are an example of "characteristic curves," and drawing them can tell you a lot about the evolution of $u(x,t)$ in more complicated settings. In the case of a fixed $v$, and initial function $u(x,0) = f(x)$, we can see the family of curves, and the movement of $u(x,t)$ along them, in Figure 5.2.

What happens if we think about a medium, like air, in which the speed of sound changes over space? Sound speed, physically, is governed by factors like temperature, pressure, humidity, and these vary over some distance. The speed with which sound moves through the medium then depends on where the sound is, so

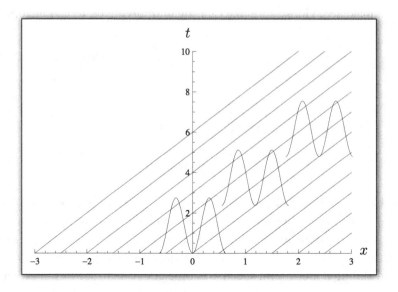

Figure 5.2 The characteristic curves for a solution to (5.9). The curves are lines, $x = x_0 + vt$, so have slope $1/v$. A few different values of $x_0$ are displayed. On top of the lines is an example initial waveform $u(x, 0)$ that takes on the same value along each line.

that:

$$\frac{\partial u(x, t)}{\partial t} + \frac{\partial}{\partial x}(v(x)u(x, t)) = 0, \tag{5.11}$$

or

$$\frac{\partial u(x, t)}{\partial t} + v(x)\frac{\partial u(x, t)}{\partial x} = -u(x, t)\frac{dv(x)}{dx}, \tag{5.12}$$

and this is not as easily solvable. For this more general case, the characteristic curves are defined as those for which (see [27], for example):

$$\dot{x}(t) = v(x(t)) \quad x(0) = x_0, \tag{5.13}$$

which had simple linear solution in the constant $v$ case. The characteristic curves can be used to turn the PDE governing $u(x, t)$ into an ODE by evaluating $u$ along $x(t)$:

$$\frac{d}{dt}u(x(t), t) = \frac{\partial u}{\partial x}v(x(t)) + \frac{\partial u}{\partial t} = -u(x(t), t)v'(x(t)), \tag{5.14}$$

where the final equality comes from inserting $\frac{\partial u}{\partial t}$ from (5.12). Notice that only in special cases will the solution $u(x, t)$ be constant along the characteristics (that happened in the wave equation with constant speed because $v'(x(t)) = 0$).

### 5.1.3 Traffic flow

To take a specific example (following [36]) that demonstrates some of the general difficulty, and also provides a model for discussion, let's consider the conservation equation:

$$\frac{\partial \rho(x, t)}{\partial t} + \frac{\partial}{\partial x}(\rho(x, t)v(x, t)) = 0 \tag{5.15}$$

where $\rho(x, t)$ is the density of cars on a road. In this case, the speed with which the cars move is itself a function of $\rho(x, t)$ – if there are no cars on the road, you travel at $v_m = 65$ mph, but if there are a lot of cars around, so that the density at a particular point (and time) is large, you cannot go as fast. Barring a fundamental law of nature, we'll insert the simplest possible relation (i.e. linear) satisfying these two basic assumptions:

$$v(x, t) = v_m \left(1 - \frac{\rho(x, t)}{\rho_m}\right) \tag{5.16}$$

where $\rho_m$ is the maximum allowed density, and at this density, the cars are stopped. Notice that even this linear choice for $v$ renders the conservation law nonlinear. If we insert our form for $v(x, t)$ in (5.15), then:

$$\frac{\partial \rho(x, t)}{\partial t} + \left[v_m \left(1 - \frac{2\rho(x, t)}{\rho_m}\right)\right] \frac{\partial \rho(x, t)}{\partial x} = 0. \tag{5.17}$$

In this case, the characteristic curves along which solutions are constant satisfy:

$$\dot{x} = v_m \left(1 - \frac{2\rho(x, t)}{\rho_m}\right) \tag{5.18}$$

since then:

$$\frac{d\rho(x(t), t)}{dt} = \frac{\partial \rho}{\partial t} + \frac{\partial \rho}{\partial x}\left[v_m \left(1 - \frac{2\rho(x, t)}{\rho_m}\right)\right] = 0. \tag{5.19}$$

This observation gives us a way to get a qualitative picture of the actual solution to (5.17) for short times. Suppose we are given $\rho(x, 0)$, a smooth function, the characteristic equation for $t$ near zero is

$$\dot{x} = v_m \left(1 - \frac{2\rho(x_0, 0)}{\rho_m}\right). \tag{5.20}$$

The right-hand side is constant for $t \approx 0$, and then the characteristic curves are approximately

$$x(t) = v_m \left(1 - \frac{2\rho(x_0, 0)}{\rho_m}\right) t + x_0, \tag{5.21}$$

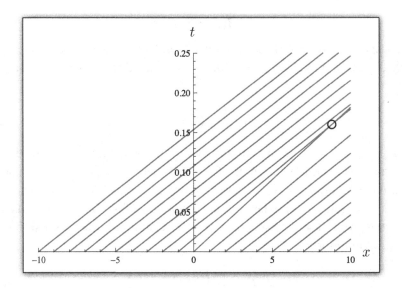

Figure 5.3 Characteristic curves for an initial density given by $\rho_0(x) = 20e^{-x^2}$ cars/mile, with $v_m = 65$ mph and $\rho_m = 264$ cars/mile (assuming a car takes up about 20 ft of room on the road). The initial value of $\rho$ is preserved along each line – at some time (circled), the characteristic lines begin to cross, indicating that our solution is double-valued. This situation suggests that we have over-extended the region of validity of our solution (that is no surprise, since (5.21) is only valid for short times).

so given an initial profile, $\rho(x, 0)$, we draw curves (lines in this case) emanating from $x_0$ with slope determined by $\rho(x_0, 0)$. Along these curves, the solution $\rho(x, t) = \rho(x(t), 0)$. The problem is that these curves can cross, and the solution becomes double-valued – the solution at this point is not physically valid.

In Figure 5.3, we see a typical set of characteristic curves for the traffic equation given an initial Gaussian density of cars centered at $x = 0$. The crossing of characteristic curves is typical of nonlinear PDEs (the wave equation's characteristic curves do not cross) – a new type of solution is formed (one that avoids the double-valued catastrophe).

### 5.1.4 Shock solutions

Nonlinear PDEs support discontinuous solutions known as "shocks" – the hallmark of which is characteristic crossing. Shocks can arise when the speed of a particle, say, travels faster than the characteristic (local) speed set by the PDE. The name is reminiscent of shocks in fluid dynamics, another nonlinear set of PDEs. When, for example, an airplane travels faster than the speed of sound, a shock can form.

Working in the context of our traffic equation, we can study shocks via the "Riemann problem" (a "shock tube").

In the Riemann problem, we take a solution that consists of a moving boundary separating constant values, so that we start with an ansatz:

$$\rho(x, t) = \begin{cases} \rho_l & x < F(t) \\ \rho_r & x > F(t) \end{cases}.$$ (5.22)

Now we use the conservation PDE itself to determine the form of $F(t)$. It is clear that the constant solutions to the left and right of the boundary each solve the traffic flow PDE, so now we need to use the PDE to constrain $F(t)$.

Suppose we integrate equation (5.5) with respect to $x$ as we did in the charge example at the end of Section 5.1.1. Keeping in mind that the solutions are constant on either side of $x = F(t)$, we can write

$$\int_{-\infty}^{\infty} \frac{\partial \rho(x, t)}{\partial t} dx = \rho_\ell v(\rho_\ell) - \rho_r v(\rho_r).$$ (5.23)

With the ansatz (5.22), we can evaluate the left-hand side of (5.23) by writing the full solution in terms of the step function: $\theta(x) = 0$ if $x < 0$, and 1 for $x > 0$, then the content of (5.22) is encapsulated by the function

$$\rho(x, t) = (\rho_\ell - \rho_r) \theta(F(t) - x) + \rho_r$$ (5.24)

with $F(t)$ to be determined. The derivative of the step function is the Dirac delta function: $\theta'(x) = \delta(x)$ which has the property that $\delta(x) = 0$ for $x \neq 0$, and $\delta(0) = \infty$, but a very special sort of infinity – an integrable one: $\int_{x_\ell}^{x_r} \delta(x) dx = 1$ as long as $x_\ell < 0$ and $x_r > 0$.

The time derivative of $\rho(x, t)$ from (5.24) is then:

$$\frac{\partial \rho(x, t)}{\partial t} = (\rho_\ell - \rho_r) F'(t) \delta(F(t) - x),$$ (5.25)

and inserting this in the left-hand side of (5.23), we have (the delta distribution gets rid of the integral):

$$(\rho_\ell - \rho_r) F'(t) = \rho_\ell \left( 1 - \frac{\rho_\ell}{\rho_m} \right) v_m - \rho_r \left( 1 - \frac{\rho_r}{\rho_m} \right) v_m$$ (5.26)

(where we have input the expressions for speed on the right using (5.16)), and this equation can be solved, in terms of the constants, for $F(t)$:

$$F(t) = \left( 1 - \frac{\rho_\ell + \rho_r}{\rho_m} \right) v_m t.$$ (5.27)

The shock itself, meaning the discontinuity, travels at a specific speed, and to the left and right of this shock location, the solution has constant density. As an example,

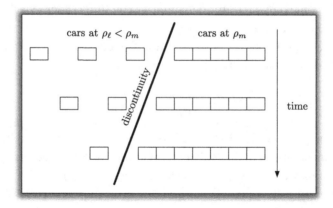

Figure 5.4 Cars on the right are at maximum density (and so have $v_r = 0$). Cars coming in from the left travel to the right until they hit the line of stopped cars, at which point they, too, stop. The discontinuity travels to the *left*.

suppose $\rho_r = \rho_m$, so that we have a line of cars that is at maximum density (and hence, zero velocity). If $\rho_\ell < \rho_m$, then in this case, the location of the shock in time is given by $F(t) = -\frac{\rho_\ell}{\rho_m} v_m t$. The shock travels in a direction opposite the flow of traffic. This makes sense: as cars reach the region of maximum density, they must stop, and so the line of cars with maximum density grows to the *left* as in Figure 5.4.

### 5.1.5 Fluids

A fluid in one dimension is described by a density $\rho(x, t)$ (the mass per unit volume at a point $x$ at time $t$) and a velocity $v(x, t)$ at all points and times. The density is related to the mass by the now-familiar equation

$$\frac{\partial \rho}{\partial t} + \frac{\partial(\rho v)}{\partial x} = 0, \tag{5.28}$$

but we need another equation governing $v$. The additional physics is provided by Newton's second law; there are external forces acting on the fluid, so we need to write $m\frac{dv}{dt} = -\frac{dU}{dx}$ in appropriate form.

It is natural to replace the particle mass, $m$, appearing in the second law, with the density of the fluid, $\rho$. The potential energy must become an energy density, but then a pretty literal transcription gives

$$\rho \frac{dv}{dt} = -\frac{\partial U}{\partial x}. \tag{5.29}$$

The total time derivative of a function of position and time can be written generically – for a function $f(x, t)$, we have

$$\frac{df(x, t)}{dt} = \frac{\partial f}{\partial x}\dot{x} + \frac{\partial f}{\partial t} \tag{5.30}$$

and for our fluid, $\dot{x} = v$ – the change in $x$ is given by the velocity $v$ at $x$ and $t$. Then (5.29) can be written as

$$\rho\frac{\partial v}{\partial t} + \rho v\frac{\partial v}{\partial x} = -\frac{\partial U}{\partial x}. \tag{5.31}$$

Finally, we can bring this field "equation of motion" into conservative form by collecting $t$ and $x$ derivatives (and using (5.28)):

$$\boxed{\frac{\partial(\rho v)}{\partial t} + \frac{\partial}{\partial x}(\rho v^2 + U) = 0.} \tag{5.32}$$

In this setting, what we have is called "conservation of momentum" for the fluid (see [6, 36, 37]). The potential energy density $U$ is typically a "pressure," but could include additional forcing. Our development of the equations here can be extended to their relativistic forms by defining the modified relativistic momentum (and handling the transformation of densities properly), just as we did for particles in Section 2.1.2 – that process is nicely laid out in [7, 46].

There is clearly nonlinearity even in the absence of an interesting $U$ – since the derivative with respect to $x$ acts on $\rho v^2$, and $\rho$ and $v$ are the target solution quantities,[3] but we expect additional dependence on $\rho$ from the pressure itself, that is, $U \equiv p(\rho) \sim \rho^\gamma$ (for some $\gamma$) is a typical relation between pressure and density. Note finally that in some cases, fluids also have conserved energy, and this could be added to our mass and momentum conservation laws ((5.28) and (5.32) respectively). Many fluid setups do *not* conserve energy, because heat is allowed to flow in or out of the fluid (if we hold the fluid at constant temperature, for example). When conservation of energy holds, and is added to (5.28) and (5.32), the set is known as the "Euler equations."

### 5.1.6 Quantum mechanics

Schrödinger's equation does not need a direct conservative form, as we shall see, for a numerical solution to be useful. The basic structure is:

$$i\hbar\frac{\partial\Psi(x, t)}{dt} - \left[-\frac{\hbar^2}{2m}\frac{\partial^2\Psi(x, t)}{\partial x^2} + V(x)\Psi(x, t)\right] = 0, \tag{5.33}$$

---

[3] It would be natural to consider $\rho$ and $\rho v$ as the variables of interest, so that the product $\rho v$, appearing in (5.32), is not by itself a "nonlinear" term.

so this is first order in the temporal derivative, second order in spatial ones, and linear in $\Psi(x, t)$.

We'll look at quantum mechanical scattering dynamics – a particle travels to the right, with $V(x) = 0$, and encounters a potential of some sort as in Section 3.1.3 – this time, though, we will use the full dynamics – an initial wave function $\Psi(x, 0)$ must be provided, and we'll take the Gaussian:

$$\Psi(x, 0) = \left(\frac{2}{\pi}\right)^{\frac{1}{4}} \left(\frac{2Em - p_0^2}{\hbar^2}\right)^{\frac{1}{4}} e^{-\left(\frac{2Em - p_0^2}{\hbar^2}\right)x^2} e^{i\frac{p_0}{\hbar}x} \tag{5.34}$$

which corresponds to a particle localized initially at the origin (but with probability density that spreads out like a Gaussian), initial momentum expectation value: $\langle p \rangle = p_0$, and $\langle H \rangle = E$.[4] Given this initial condition, and some boundary values (like $\psi(\pm\infty) = 0$), we can "look at" the time evolution of the probability density as the wave packet interacts with an arbitrary (but localized) potential.

## 5.2 Exactly solvable cases

There are times when a PDE can be split up naturally into a temporal part and a spatial part. Such cases provide good checks for our methods, and we should be aware of them as limits of nonlinear, "unsolvable," PDEs. For the linear wave equation with constant speed $v$, $\frac{\partial u}{\partial t} + v\frac{\partial u}{\partial x} = 0$, we know that:

$$u(x, t) = f(x - vt) \tag{5.36}$$

for any function $f(p)$ is a solution. We could also take a separation of variables approach as in Section 4.1.2: $u(x, t) = \phi(t)X(x)$, then the conservative form of the wave equation tells us that

$$\frac{\dot{\phi}(t)}{\phi(t)} + v\frac{X'(x)}{X(x)} = 0 \tag{5.37}$$

and we see that the first term depends only on $t$, the second only on $x$, so these terms must each be equal to a constant, $S$, for equality to hold at all times

$$\dot{\phi}(t) = -S\phi(t) \quad X'(x) = \frac{S}{v}X(x). \tag{5.38}$$

---

[4] The averages here are with respect to the density defined by $\Psi(x, t)$, namely $\rho(x, t) = \Psi(x, t)^*\Psi(x, t)$, so that

$$\langle H \rangle = \int_{-\infty}^{\infty} \Psi(x, t)^* H\Psi(x, t)dx, \tag{5.35}$$

for example. We are being given an initial density with the specification of $\Psi(x, 0)$, so we can compute the averages using that particular case at $t = 0$.

The solution here is $\phi(t) = e^{-St}$ and $X(x) = Ae^{\frac{S}{v}x}$, or, put together,

$$u(x, t) = Ae^{\frac{S}{v}(x-vt)}.$$  (5.39)

Taking $S$ to be imaginary, we have a single Fourier mode, and since the wave equation is linear, we can invoke superposition to build up any general function.

Similar arguments hold for the Schrödinger equation for particular potentials (time-independent potentials imply that Schrödinger's equation is separable). When an exact form is available, even if only as the limit of a problem of more immediate interest, use it.

## 5.3 Discretization and methods

We'll solve PDEs of the form:

$$\frac{\partial u(x, t)}{\partial t} + \frac{\partial}{\partial x}(v(x, t)u(x, t)) = 0$$  (5.40)

numerically to set up the methods. We need to provide the initial data: $u(x, 0) = u_0(x)$, and some boundary condition(s) (to be discussed in a moment), in addition to the function $v(x, t)$ (either written in terms of $u(x, t)$ as with the traffic flow example, or provided as a function based on external considerations, as we might do for charge conservation).

Our approach will be to discretize in time and space separately. We'll work on fixed, uniform grids, with $t_n = n\Delta t$ for some constant $\Delta t$, and $x_j = j\Delta x$ for some (other constant) spacing $\Delta x$. We'll refer to the discrete solution indexed as follows:

$$u_j^n = u(x_j, t_n).$$  (5.41)

Technically, the description and analysis of the methods we develop should follow the conventions of Section 2.3, and they will but with relaxed notational reminders (there is already enough decoration on objects like $u_j^n$ without the addition of bars and hats).

We'll start with the simplified (and exactly solvable) linear problem from (5.9):

$$\frac{\partial u(x, t)}{\partial t} + \frac{\partial}{\partial x}(vu(x, t)) = 0$$  (5.42)

for constant $v$ (the speed, appearing in (5.9)). Consider the simplest possible temporal discretization, a one-sided approach in time, with a centered spatial difference:

$$\frac{u_j^{n+1} - u_j^n}{\Delta t} = -v\frac{u_{j+1}^n - u_{j-1}^n}{2\Delta x},$$  (5.43)

so that

$$u_j^{n+1} = u_j^n - v\frac{\Delta t}{2\Delta x}\left(u_{j+1}^n - u_{j-1}^n\right). \tag{5.44}$$

This is an example of the "forward Euler method" (which we first encountered in the ODE setting) allowing us to take a solution at $n = 0$ and propagate it forward in time for all grid points $x_j$. We can find the local truncation error order by inserting $u(x, t)$ in (5.44) and using Taylor expansions (so that, as in Section 2.3, we apply the method to the projection of the exact solution):

$$u_j^{n+1} = u_j^n + \frac{\partial u(x_j, t_n)}{\partial t}\Delta t + O(\Delta t^2)$$

$$u_{j+1}^n - u_{j-1}^n = 2\frac{\partial u(x_j, t_n)}{\partial x}\Delta x + O(\Delta x^3) \tag{5.45}$$

and then (5.44), with the projection of the solution inserted, is

$$u_j^n + \frac{\partial u(x_j, t_n)}{\partial t}\Delta t + O(\Delta t^2) = u_j^n - v\frac{\partial u(x_j, t_n)}{\partial x}\Delta t + O(\Delta x^2\Delta t). \tag{5.46}$$

The PDE (5.42) is satisfied (locally, meaning step-to-step) up to errors of size $O(\Delta t^2)$.

We can think about the stability of the method – if we define the vector $\mathbf{u}^n$ to have entries: $u_i^n$ (a vector made up of the spatial solution at time level $n$, in other words), then stability means[5] that $\|\mathbf{u}^n\|$ does not grow with $n$. Let's see what happens to the norm of $\mathbf{u}^n$ using the forward Euler method. Start by writing the method as a matrix operating on $\mathbf{u}^n$,

$$\mathbf{u}^{n+1} = \left(\mathbb{I} - v\frac{\Delta t}{\Delta x}\mathbb{Q}\right)\mathbf{u}^n \equiv \mathbb{P}\mathbf{u}^n \tag{5.47}$$

where the entries of $\mathbb{Q}$ are defined by (5.44) (each row has at most two entries, $\pm\frac{1}{2}$). As in Section 2.5, we know that the solution is $\mathbf{u}^n = \mathbb{P}^n\mathbf{u}^0$, and growth occurs if the maximum eigenvalue (in absolute value) of $\mathbb{P}$ is greater than one. The eigenvalues of $\mathbb{P}$ depend on the size of the spatial grid, but they end up being greater than one for all values of $v\Delta t/\Delta x$, and for all grid sizes. So the story is the same as in Section 2.5, the method is unstable, and sequences generated by the method will inevitably grow.

One way around the stability issue is to define the "implicit" Euler method via:

$$\frac{u_j^{n+1} - u_j^n}{\Delta t} = -v\frac{u_{j+1}^{n+1} - u_{j-1}^{n+1}}{2\Delta x}, \tag{5.48}$$

---

[5] This is a very conservative and informal definition of stability, but it captures the sentiment – for a careful discussion, see [27].

which cannot be solved explicitly for $u_j^{n+1}$ (since it couples the $j$ and $n$ indices) – in terms of $\mathbf{u}^n$, the above implicit method is equivalent to:

$$\mathbf{u}^n = \mathbb{T}\mathbf{u}^{n+1} \tag{5.49}$$

where $\mathbb{T}$ is a matrix that holds all the details of the method inside it – then the update for $\mathbf{u}^n$ can be solved by matrix inversion – a slow, but safe way to proceed (we'll see this type of approach applied to the quantum mechanical scattering problem).

We can also add stability by modifying the "stencil" (the set of points required to make a temporal update) even in the explicit setting. For example, the "Leapfrog" method is related to the original (5.43) by replacing the one-sided temporal derivative approximation with a centered one:

$$\frac{u_j^{n+1} - u_j^{n-1}}{2\Delta t} = -v\frac{u_{j+1}^n - u_{j-1}^n}{2\Delta x}, \tag{5.50}$$

which can be solved for $u_j^{n+1}$:

$$\boxed{u_j^{n+1} = u_j^{n-1} - v\frac{\Delta t}{\Delta x}\left(u_{j+1}^n - u_{j-1}^n\right).} \tag{5.51}$$

Many methods come from this general pattern, some finite difference approximation for derivatives applied to the temporal and spatial sides separately – another popular variant is known as Lax–Friedrichs:

$$\boxed{u_j^{n+1} = \frac{1}{2}\left(u_{j-1}^n + u_{j+1}^n\right) - v\frac{\Delta t}{2\Delta x}\left(u_{j+1}^n - u_{j-1}^n\right).} \tag{5.52}$$

We have replaced, on the right, the point $u_j^n$ from the Euler method recursion (5.44) with an average of the points to the left and right. This is enough to ensure stability (with appropriate relation between $\Delta t$ and $\Delta x$, namely: $|v|\Delta t \leq \Delta x$) for the method applied to the linear problem.

### 5.3.1 Nonlinear modification

Our general conservation law (5.40) involves a varying $v(x, t)$. There are a number of ways we can take a method defined for constant $v$, and update it to include a changing $v(x, t)$. For example, we could take the Lax–Friedrichs scheme applied to the full problem:

$$u_j^{n+1} = \frac{1}{2}\left(u_{j+1}^n + u_{j-1}^n\right) - \frac{\Delta t}{2\Delta x}\left(u_{j+1}^n v_{j+1}^n - u_{j-1}^n v_{j-1}^n\right). \qquad (5.53)$$

Let's think about this method in the context of the traffic flow example – we set $u_j^n \equiv \rho_j^n$, and $v_j^n = v_m(1 - \rho_j^n/\rho_m)$, then we just need to provide the discretized initial data: $\rho_j^0$ and the method will update all points for any time interval of interest.

Now we come, finally, to the question of boundary conditions. It is clear that we will need $\rho_{N+1}^n$ for any time level $n$, and in addition, we will need the $\rho_0^n$ value. If we think of a spatial grid with $x_j = j\Delta x$ for $j = 1 \longrightarrow N$ so that $j = 0$ is the left-hand endpoint, and $j = N + 1$ gives the location of the right-hand endpoint, then we need to provide boundary values for the grid points at $x_0$ and $x_{N+1}$ in order to fully specify the numerical solution. There are a few common choices, and they depend on the problem of interest.

One simple boundary condition is to set the values $\rho_0^n = \rho_1^n$ and $\rho_{N+1}^n = \rho_N^n$ – this is a good choice when the "action" of the problem is occurring far from the boundaries, as it basically assumes that the function $\rho(x, t)$ has zero derivatives on the left and right (so that $\rho$ is constant near the boundaries). Setting derivatives to zero like this allows values to pass out of the region in which the numerical solution is taking place. Another, similar, choice is to take $\rho_0^n = 0$ and $\rho_{N+1}^n = 0$ – this time, we are setting the value of $\rho(x, t)$ to zero at the edges of the numerical domain. Physically, this can be justified by, for example, considering functions that must vanish at spatial infinity, and bringing those points in to finite values (there may also be mechanical considerations – perhaps we have a string, and we pin the ends down). Finally, periodic boundary conditions can be used to focus on a single "cell" of a computation – we assume $\rho(x + a, t) = \rho(x, t)$ for some spacing $a$ (the numerical domain of the solution, for example). Then we can set $\rho_0^n = \rho_N^n$ and $\rho_{N+1}^n = \rho_1^n$. This results in a wrapping – solution values leave on the right, and re-enter from the left.

The code that implements nonlinear Lax–Friedrichs is shown in Implementation 5.1. We take $u$ to be the variable (like density), and $v$ the velocity (a vector), so we need to provide the spatial values for $u$ at time $t = 0$, called u0, we need to give the function, v (taking arguments $x$ and $u$, although we could easily add $t$ dependence) that calculates velocity, and then a bunch of numerical details: the time step dt, the number of steps (in time) to take, Nt, and the spatial step size dx. Finally, the implementation allows for a variety of boundary conditions, so there is an input variable that allows a user to specify which boundary condition to use. Note that the initial u0 must satisfy the boundary condition and include the boundary points as its first and last entries, then the value of $u$ at all other time steps will satisfy the boundary conditions by explicit enforcement.

---

**Implementation 5.1** Lax–Friedrichs solver

```
LFPDE[u0_, v_, dt_, Nt_, dx_, bndtype_] := Module[{retu, u, uup, Nx, i, x, j},
  u = u0;
  Nx = Length[u0];
  uup = Table[0, {k, 1, Nx}];
  retu = Table[0, {k, 1, Nt}];
  For[i = 1, i ≤ Nt, i = i + 1,
    x = dx;
    For[j = 2, j < Nx, j = j + 1,
      uup[[j]] = (1 / 2) (u[[j + 1]] + u[[j - 1]])
        - dt / (2 dx) (u[[j + 1]] v[x + dx, u[[j + 1]]] - u[[j - 1]] v[x - dx, u[[j - 1]]]);
      x = x + dx;
    ];
    If[bndtype == "periodic",
      uup[[1]] = uup[[Nx - 1]];
      uup[[Nx]] = uup[[2]];
    ];
    If[bndtype == "diffuse",
      uup[[1]] = uup[[2]];
      uup[[Nx]] = uup[[Nx - 1]];
    ];
    If[bndtype == "zero",
      uup[[1]] = 0.0;
      uup[[Nx]] = 0.0;
    ];
    u = uup;
    retu[[i]] = uup;
  ];
  Return[retu];
]
```

---

### *Traffic flow shock solution*

We can apply the Lax–Friedrichs method to our traffic flow problem; the movie is shown in the chapter notebook. We start with a smooth Gaussian, and the solution rapidly develops a shock when the characteristic curves cross – that shock then travels to the right with speed determined by the values of density to its left and right as in (5.27) – an example of this behavior is shown in Figure 5.5.

### *5.3.2 Lax–Wendroff*

The Lax–Wendroff method takes the analytic treatment one step further – consider the Taylor expansion of $u(x, t + \Delta t)$:

$$u(x, t + \Delta t) \approx u(x, t) + \frac{\partial u}{\partial t}\Delta t + \frac{1}{2}\frac{\partial^2 u}{\partial t^2}\Delta t^2. \qquad (5.54)$$

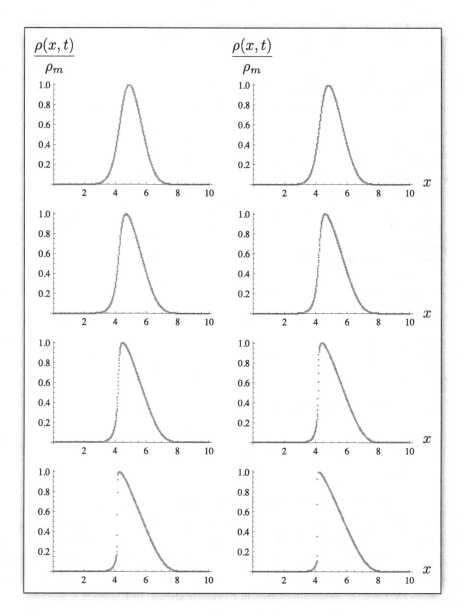

Figure 5.5 An initial Gaussian density profile steepens into a discontinuity (a shock forms). The numerical solution is shown at $10\Delta t$ intervals starting at $10\Delta t$.

We'll use the wave equation (5.42) (for example) to replace $\frac{\partial u}{\partial t} = -v\frac{\partial u}{\partial x}$ (for constant $v$), so we have:

$$\frac{\partial^2 u}{\partial t^2} = \frac{\partial}{\partial t}\left(-v\frac{\partial u}{\partial x}\right) = -v\frac{\partial}{\partial x}\left(\frac{\partial u}{\partial t}\right) = +v^2\frac{\partial^2 u}{\partial x^2}, \qquad (5.55)$$

where we have used cross derivative equality: $\frac{\partial^2 u}{\partial t \partial x} = \frac{\partial^2 u}{\partial x \partial t}$. But then the Taylor series reads:

$$u(x, t + \Delta t) \approx u(x, t) - v\frac{\partial u}{\partial x}\Delta t + \frac{1}{2}v^2\frac{\partial^2 u}{\partial x^2}\Delta t^2, \tag{5.56}$$

and we can discretize this directly:

$$u_j^{n+1} = u_j^n - \frac{v\Delta t}{2\Delta x}\left(u_{j+1}^n - u_{j-1}^n\right) + \frac{1}{2}\frac{\Delta t^2}{\Delta x^2}v^2\left(u_{j+1}^n - 2u_j^n + u_{j-1}^n\right). \tag{5.57}$$

The nonlinearization of this method takes a variety of forms. One way to generate a good nonlinear version is to define an intermediate solution at shifted grid points – this is the "Richtmyer two-step Lax–Wendroff method":

$$\boxed{\begin{aligned} u_{j+\frac{1}{2}}^{n+\frac{1}{2}} &= \frac{1}{2}\left(u_j^n + u_{j+1}^n\right) - \frac{\Delta t}{2\Delta x}\left(u_{j+1}^n v_{j+1}^n - u_j^n v_j^n\right) \\ u_j^{n+1} &= u_j^n - \frac{\Delta t}{\Delta x}\left(u_{j+\frac{1}{2}}^{n+\frac{1}{2}} v_{j+\frac{1}{2}}^{n+\frac{1}{2}} - u_{j-\frac{1}{2}}^{n+\frac{1}{2}} v_{j-\frac{1}{2}}^{n+\frac{1}{2}}\right) \end{aligned}} \tag{5.58}$$

where $v$ is a function of $x$ and $t$, potentially through a dependence on $u$. The intermediate points make a direct comparison with the linear form (5.57) difficult, but allow for easy implementation.

## 5.4 Crank–Nicolson for the Schrödinger equation

Schrödinger's equation is an example of a PDE that is not in conservative form. It is also a "linear" PDE in the sense that only $\Psi(x, t)$ and its derivatives show up. Schrödinger's equation in one spatial dimension is

$$i\hbar\frac{\partial \Psi(x, t)}{\partial t} = -\frac{\hbar^2}{2m}\frac{\partial^2 \Psi(x, t)}{\partial x^2} + V(x)\Psi(x, t). \tag{5.59}$$

Remember that this equation tells us about the (quantum) dynamics of a particle of mass $m$ under the influence of a potential $V(x)$, and plays a role in quantum mechanics analogous to Newton's second law in classical mechanics.

In preparation for our numerical work, let's perform the usual scalings that allow us to write the equation in dimensionless form. Let $x = aq$ for some length scale $a$ set by the problem, and $t = t_0 s$. We can simplify by letting $t_0 = \frac{2ma^2}{\hbar}$, a natural time scale given $a$. If we set $V(x) = \frac{\hbar^2}{2ma^2}\tilde{V}(x)$ to get dimensionless $\tilde{V}$, then

$$\frac{\partial \Psi(q, s)}{\partial s} = -i\left(-\frac{\partial^2 \Psi(q, s)}{\partial q^2} + \tilde{V}(q)\Psi(q, s)\right). \tag{5.60}$$

With the constants out of the way, we are ready to discretize.

It is tempting to take, as the discretization of (5.60) (again using $\Psi_j^n = \Psi(q_j, s_n)$ to represent the discrete solution, for a regularly spaced grid $q_j = j\Delta q, s_n = n\Delta s$):

$$\frac{\Psi_j^{n+1} - \Psi_j^n}{\Delta s} = -i\left(-\frac{1}{\Delta q^2}\left(\Psi_{j+1}^n - 2\Psi_j^n + \Psi_{j-1}^n\right) + \tilde{V}(q_j)\Psi_j^n\right). \quad (5.61)$$

This is the forward Euler method once again, and has the unfortunate stability issues discussed in Section 5.3.

So we move to the implicit Euler method:

$$\frac{\Psi_j^{n+1} - \Psi_j^n}{\Delta s} = -i\left(-\frac{1}{\Delta q^2}\left(\Psi_{j+1}^{n+1} - 2\Psi_j^{n+1} + \Psi_{j-1}^{n+1}\right) + \tilde{V}(q_j)\Psi_j^{n+1}\right). \quad (5.62)$$

The method is stable (for some values of $\Delta s/\Delta q^2$), and it is worth working out its details. All we have done is evaluate the right-hand side of (5.60) at time level $n + 1$, rather than $n$. Suppose our discrete spatial grid has $q_j = j\Delta x$ for $j = 1 \longrightarrow N$. Then we can define the vector:

$$\Psi^n \doteq \begin{pmatrix} \Psi_1^n \\ \Psi_2^n \\ \vdots \\ \Psi_N^n \end{pmatrix} \quad (5.63)$$

and write (5.62) in terms of the matrix-vector multiplication (implied by the dot product):

$$\left(-i\frac{\Delta s}{\Delta q^2} \quad \left(1 + i\Delta s\tilde{V}(q_j) + 2i\frac{\Delta s}{\Delta q^2}\right) \quad -i\frac{\Delta s}{\Delta q^2}\right) \cdot \begin{pmatrix} \Psi_{j-1}^{n+1} \\ \Psi_j^{n+1} \\ \Psi_{j+1}^{n+1} \end{pmatrix} = \Psi_j^n. \quad (5.64)$$

This should be reminiscent of the matrix-vector formulation of ODEs on a finite grid from Section 4.1.1– by constructing the matrix $\mathbb{P}$ made up of rows of appropriately shifted entries from (5.64) (and all other entries in the row zero), we have reduced the problem to one of matrix inversion. The implicit Euler method is

$$\mathbb{P}\Psi^{n+1} = \Psi^n \longrightarrow \Psi^{n+1} = \mathbb{P}^{-1}\Psi^n. \quad (5.65)$$

For simplicity, we take the boundaries in $\Psi$ to be zero. That is, we'll work in a one-dimensional box defined for $q = 0 \longrightarrow 1$, with $V(0) = V(1) = \infty$, so that the wave function must vanish at 0 and 1 for all times $t$. Our spatial grid extends from $q_1$ to $q_N$, with $q_0$ and $q_{N+1}$ as boundary points. Since $q_{N+1} = 1$, we can define the step size $\Delta q = 1/(N + 1)$. Then we construct our matrix $\mathbb{P}$ with these boundary conditions, and invert it to find the update of the wave function. Note that the matrix does not change in time, so we only need to invert it once.

The problem with this method is that it does not preserve the normalization of the wave function – that is, if $\int_0^a \Psi(x, 0)^* \Psi(x, 0) dx = 1$ initially, we do not necessarily have $\int_0^a \Psi(x, t)^* \Psi(x, t) dx = 1$ for all time $t$. On the continuum side, Schrödinger's equation itself ensures that the norm of its solution is constant, as it must be for a probabilistic interpretation. What we have is a numerical lack of norm conservation. Remember that the implicit Euler method, defined for ODEs in Section 2.5, had the property that for a simple harmonic oscillator, energy was lost over time. That decay is associated with stable methods, and can be beneficial. In the current quantum mechanical context, the decay manifests itself as a loss of probability – if a particle starts inside the infinite square well at time $t = 0$, then it is in the well with probability one. Over time, the numerical solution will have probability (of finding the particle in the box) less than one, with no physical mechanism by which the particle could leave. So the method is stable, but difficult to interpret.[6] It turns out that if you take a combination of the implicit and explicit forms of Euler's method, you recover a norm-preserving numerical method, called "Crank–Nicolson" (laid out for quantum mechanical scattering in [17]), and you will explore its properties in the lab problems.[7]

Finally, if we think back to the finite difference operators we set up in Chapter 4 for two-dimensional PDEs, it is relatively easy to move our problem from one spatial dimension to two. Remember, from Section 4.3, that discretization turns the Laplace operator into a matrix,

$$\nabla^2 f = s \longrightarrow \mathbb{D}\mathbf{f} = \mathbf{s} - \mathbf{b} \tag{5.66}$$

with boundary conditions imposed on the right combined with the modified source (appropriately vectorized). For two spatial dimensions, if we introduce a dimensionless coordinate $r$ via: $y = ar$ (with $a$ the same fundamental length as for the $x$ coordinate, so we are working on a square), then (5.60) becomes:

$$\frac{\partial \Psi(q, r, s)}{\partial s} = -i \left[ -\left( \frac{\partial^2 \Psi(q, r, s)}{\partial q^2} + \frac{\partial^2 \Psi(q, r, s)}{\partial r^2} \right) + \tilde{V}(r, s)\Psi(q, r, s) \right],$$

$$\tag{5.67}$$

and we can define a new matrix with spatial discretization, $\nabla^2 \longrightarrow \mathbb{D}$, that gives Euler-type methods (both implicit and explicit) identical in form to the one-dimensional case.

As an example – take the explicit Euler method and suppose we discretize on a grid with $q_j = j\Delta q$, and $r_j = j\Delta q$ (again, a square grid, with the same spacing in both the horizontal and vertical directions), and $\mathbf{\Psi}^n \in \mathbb{R}^{N^2}$ for $N$ grid points

---

[6] It's better than the explicit case, where the total probability starts at one and then *grows*.
[7] The Crank–Nicolson method can also be developed for the heat equation – that can be found in [20], for example.

in each direction (using an embedding like (4.49)). The (Euler) finite difference approximation to (5.67) becomes:

$$\frac{\Psi^{n+1} - \Psi^n}{\Delta s} = -i\left(-\mathbb{D} + \tilde{\mathbb{V}}\right)\Psi^n \tag{5.68}$$

where $\tilde{\mathbb{V}}$ is the matrix representing the potential $\tilde{V}$ projected appropriately onto the grid (in one dimension, a diagonal matrix with $\tilde{\mathbb{V}}_{jj} = \tilde{V}(q_j)$). We can again write the update as:

$$\Psi^{n+1} = \left[\mathbb{I} - i\,\Delta s\left(-\mathbb{D} + \tilde{\mathbb{V}}\right)\right]\Psi^n \longrightarrow \Psi^{n+1} = \mathbb{Q}\Psi^n. \tag{5.69}$$

## Further reading

1. Batchelor, G. K. *An Introduction to Fluid Dynamics*. Cambridge University Press, 1967.
2. Baumgarte, Thomas W. & Stuart L. Shapiro. *Numerical Relativity: Solving Einstein's Equations on the Computer*. Cambridge University Press, 2010.
3. Griffiths, David J. *Introduction to Quantum Mechanics*. Pearson Prentice Hall, 2005.
4. Gustaffson, Bertil, Heinz-Otto Kreizz, & Joseph Oliger. *Time Dependent Problems and Difference Methods*. John Wiley & Sons, 1995.
5. Infeld, Eryk & George Rowlands. *Nonlinear Waves, Solitons and Chaos*. Cambridge University Press, 2000.
6. LeVeque, Randal J. *Numerical Methods for Conservation Laws*. Birkhäuser, 1992.
7. Wilson, James R. & Grant J. Matthews. *Relativistic Numerical Hydrodynamics*. Cambridge University Press, 2003.

## Problems

### Problem 5.1

**(a)** Prove that a function $u(x, t)$ that satisfies the wave equation also satisfies:

$$\left(\frac{\partial}{\partial t} + v\frac{\partial}{\partial x}\right)\left(\frac{\partial}{\partial t} - v\frac{\partial}{\partial x}\right)u = 0 \tag{5.70}$$

for constant $v$ (this is just a check of the factorization of the wave equation).

**(b)** Sketch the wave solution: $u(x, t) = U(x + vt)$ with $U = u_0 e^{-\alpha x^2}$ at time $t = 0$ and at time $t$ (by sketch, I mean identify the "center" and height at the center).

### Problem 5.2

One of the defining properties of the delta function is its behavior under the integral:

$$\int_{-1}^{1} \delta(x)f(x)dx = f(0). \tag{5.71}$$

Show that the derivative of the step function:

$$\theta(x) \equiv \begin{cases} 0 & x < 0 \\ 1 & x \geq 0 \end{cases} \tag{5.72}$$

behaves just like a delta function – that is, show that:

$$\int_{-1}^{1} \frac{d\theta(x)}{dx} f(x)dx = f(0) \tag{5.73}$$

(Hint: use integration by parts).

## Problem 5.3

Generate (by hand) a solution to the PDE:

$$\frac{\partial u(x,t)}{\partial t} + \frac{\partial}{\partial x}(\alpha x u(x,t)) = 0 \tag{5.74}$$

that has $u(x, 0) = x$.

## Problem 5.4

Take the Gaussian initial wave function, from (5.34), and rewrite $\Psi(x,0) = 1/\sqrt{a}\chi(x,0)$, for dimensionless $\chi$. Now using the dimensionless parameters from Section 5.4, i.e. $x = aq$, $t = \frac{2ma^2}{\hbar}s$, $E = \frac{\hbar^2}{2ma^2}\tilde{E}$, find the constant in $p_0 = \alpha\tilde{p}_0$ for dimensionless $\tilde{p}_0$, and show that $\chi(q, 0)$, in terms of $\tilde{E}$, $\tilde{p}_0$, and $q$, is

$$\chi(q, 0) = \left(\frac{2}{\pi}\right)^{1/4}\left(\tilde{E} - \frac{1}{4}\tilde{p}_0^2\right)^{1/4} e^{-(\tilde{E} - \frac{1}{4}\tilde{p}_0^2)q^2} e^{i\frac{\tilde{p}_0}{2}q}. \tag{5.75}$$

## Problem 5.5

Schrödinger's equation supports wave function normalization – if an initial wave function is normalized, so that $\int_{-\infty}^{\infty} \Psi(x,0)^*\Psi(x,0)dx = 1$, then prove that the solution to Schrödinger's equation that has this initial wave function also has, for all time, $\int_{-\infty}^{\infty} \Psi^*(x,t)\Psi(x,t)dx = 1$. Use the one-dimensional form of Schrödinger's equation:

$$-\frac{\hbar^2}{2m}\frac{\partial^2}{\partial x^2}\Psi(x,t) + V(x)\Psi(x,t) = i\hbar\frac{\partial}{\partial t}\Psi(x,t), \tag{5.76}$$

and assume that the wave function vanishes as $x \longrightarrow \pm\infty$ for all times.

## Problem 5.6

Run the Lax–Friedrichs solver for the wave equation with $v = 50$ and an initial waveform $u_0(x) = \sin(2\pi x)$ on a spatial grid extending from $x = 0$ to $1 + \Delta x$ (including boundary points) in steps of $\Delta x = 0.01$. Set $\Delta t = \Delta x/v_m$ as the time step, and run the solver for one hundred steps using periodic boundary conditions. Compare your numerical solution with the true solution by taking a few temporal snapshots from your numerical result and overlaying the exact solution.

**Problem 5.7**

Show that the Richtmyer two-step *is* Lax–Wendroff when $v(x, t) = v$, a constant.

**Problem 5.8**

There is a simple relationship between the matrix form of the implicit and explicit Euler updates for Schrödinger's equation. Call $\mathbb{Q}$ the update found in the explicit version, $\mathbf{\Psi}^{n+1} = \mathbb{Q}\mathbf{\Psi}^n$, and let $\mathbb{P}$ be the implicit matrix: $\mathbb{P}\mathbf{\Psi}^{n+1} = \mathbf{\Psi}^n$. Show that:

$$\mathbb{P} = \mathbb{Q}^*, \tag{5.77}$$

then if you just make the $\mathbb{Q}$ matrix for explicit Euler, you obtain the $\mathbb{P}$ matrix by simple conjugation (assume zero boundary condition for all times).

**Problem 5.9**

To see how a mixed method, like Crank–Nicolson, preserves the norm of the wave function, we'll work out a simple one-dimensional case. Take the ODE $f'(s) = if(s)$, for which the solution is $f(s) = f_0 e^{is}$, with constant magnitude $f_0$.

**(a)** Using the explicit Euler update:

$$\frac{f_{j+1} - f_j}{\Delta s} = i f_j \tag{5.78}$$

find the "propagator," the factor $\alpha_e$ in $f_{j+1} = \alpha_e f_j$. The forward Euler recursion relation is solved by $f_j = \alpha_e^j f_0$, so if $|\alpha_e| > 1$, the solution grows. Show that $|\alpha_e| > 1$ for all $\Delta s > 0$.

**(b)** Now consider the implicit Euler update:

$$\frac{f_{j+1} - f_j}{\Delta s} = i f_{j+1}. \tag{5.79}$$

Again, find the propagator $\alpha_i$ such that $f_{j+1} = \alpha_i f_j$. Show that this constant has $|\alpha_i| < 1$ for all $\Delta s > 0$.

**(c)** Finally, for the "Crank–Nicolson" propagator, start from the "average" evaluation:

$$\frac{f_{j+1} - f_j}{\Delta s} = i\left(\frac{f_{j+1} + f_j}{2}\right), \tag{5.80}$$

or

$$\left(1 - \frac{1}{2}i\Delta s\right) f_{j+1} = \left(1 + \frac{1}{2}i\Delta s\right) f_j. \tag{5.81}$$

Solving for $f_{j+1}$ will give a new propagator $\alpha_{CN}$ – what is the magnitude of this?

**Problem 5.10**

The stability of a method is associated with a solution's growth as a function of time. We can get a handle on this growth by considering test solutions of the form:

$$u(x_j, t^n) = s(k)^n e^{ikx_j} \longrightarrow u_j^n = s(k)^n e^{ikj\Delta x}. \tag{5.82}$$

This ansatz is reasonable, since we are looking at a single Fourier mode, with wave number $k$ – the growth of the solution is determined by the size of $s$ – if $|s| < 1$ (for all $k$), then the method is stable (the formal mathematics of this type of stability analysis can be found in [27, 31]). The idea is that any function, including the solution, can be decomposed into a sum of Fourier modes:

$$u(x, t) = \sum_{k=-\infty}^{\infty} a_k(t)e^{ikx}, \tag{5.83}$$

and we are making sure that for no $k$ does the associated coefficient $a_k(t)$ grow with $n$.

(a) Find $s(k)$ for the explicit Euler update: $u_j^{n+1} = u_j^n - \frac{v\Delta t}{2\Delta x}(u_{j+1}^n - u_{j-1}^n)$. Once again, it should be clear that $|s(k)| > 1$ for all $k$, meaning the method is unstable.

(b) Find $s(k)$ for the Leapfrog method (using constant speed $v$ as in part (a)). What condition must you place on the relation between $\Delta t$ and $\Delta x$ to obtain a stable method?

(c) Find $s(k)$ for the Lax–Friedrichs method. What relation between $\Delta t$ and $\Delta x$ ensures stability here?

This approach to defining and evaluating the stability of a method is called a von Neumann stability analysis (a nice discussion is in [7], and includes the examples in this problem).

## Lab problems

**Problem 5.11**

Implement the Richtmyer two-step (Lax–Wendroff) method in full nonlinear form. We'll test its behavior on the linear problem, where $v(x, 0) = v_m$.

(a) Run the method for $\Delta x = 0.01$, $v_m = 50$, $\Delta t = \Delta x/v_m$, and periodic boundary conditions using $u(x, 0) = \sin(2\pi x)$ with $x \in [0, 1 + \Delta x]$ (you have to go one step beyond the target endpoint to employ periodic boundary conditions). Run for two hundred steps and make a movie of the resulting traveling wave.

(b) Try using all the same values, but set $\Delta t = 1.1\Delta x/v_m$. Can you see the lack of stability in the solution?

**Problem 5.12**

Take the wave equation, $\frac{\partial u}{\partial t} + \frac{\partial}{\partial x}(v_m u) = 0$ with $v_m = 1.0$, and set

$$u(x, 0) = \begin{cases} 1 & x \le 0.5 \\ 0 & x > 0.5 \end{cases}, \tag{5.84}$$

the initial conditions appropriate to the Riemann problem. For both the Lax–Friedrichs and Lax–Wendroff methods, let $\Delta x = 0.01$, $\Delta t = \Delta x/2$, and use 101 spatial grid points (i.e. $x \in [0, 1]$) with "diffuse" boundary conditions. Run for 100 time steps and make a movie of the output.

We know that the solution $u(x, t)$ is just the initial waveform $u(x, 0)$ moving to the right with speed $v_m$. But both numerical solvers fail to resolve the discontinuity perfectly, and they do so in ways that are typical of their application to more complicated (nonlinear) problems (one of the solvers smooths out the discontinuity, the other exhibits an oscillatory overshoot).

## Problem 5.13

Imagine a medium made of two different materials placed next to each other in a repeating pattern (glass, for example, with different index of refraction). The characteristic speed in each material is different, so we have an alternating pattern of speeds. Such a configuration could be modeled by letting:

$$v(x) = v_m \sin^2(2\pi f x) \tag{5.85}$$

where $v(x)$ goes from $v_m$ to zero in an oscillatory manner.

Solve $\frac{\partial u}{\partial t} + \frac{\partial}{\partial x}(vu) = 0$ using Lax–Friedrichs with zero boundary conditions given the above $v(x)$. Work on $x \in [0, 1]$ with $\Delta x = 0.005$, $v_m = 50$, $\Delta t = \Delta x / v_m$, and $f = 5$. Take a sharply peaked Gaussian as the initial wave function: $u(x, 0) = e^{-100(x-1/2)^2}$ and run two hundred steps of the solver using "zeroed" boundaries.

## Problem 5.14

Solve the Riemann problem for traffic flow using the Lax–Wendroff solver (with "diffuse" boundary conditions) – use, as in Section 5.1.4, $v_m = 65$, $\rho_\ell = 0$, $\rho_r = \frac{1}{2}\rho_m$ with $\rho_m = 264$. Run on a grid with $\Delta x = 0.01$ for $x \in [0, 10]$ – switch from $\rho_\ell$ to $\rho_r$ at $x = 5$ initially. Take 200 steps in time and calculate the speed of the discontinuity. Does it match the prediction in (5.27)? Try it with the Lax–Friedrichs solver and $\rho_r = \rho_m$, $\rho_\ell = \frac{1}{2}\rho_m$, and again check the "shock speed."

---

The remaining five problems deal with various numerical solutions of the Schrödinger equation (including its two-dimensional form), and provide experience with implicit vs. explicit methods – in each of the first three problems below, the PDE solution can be implemented by generating a matrix such that $\mathbf{\Psi}^{n+1} = \mathbb{M}\mathbf{\Psi}^n$ (in the implicit case, $\mathbb{M}$ will be the inverse of another matrix). Make movies of your solutions that clearly show their behavior as they move to the right and encounter the potential step defined below in (5.86).

## Problem 5.15

Implement the explicit Euler method to solve the Schrödinger equation on $q = -\frac{1}{2}$ to $\frac{1}{2}$ (note, then, that $q_j = j\Delta x - \frac{1}{2}$, we'll assume that the wave function vanishes at the endpoints $-\frac{1}{2}$ and $\frac{1}{2}$) with

$$V(q) = \begin{cases} 0 & q \leq \frac{1}{8} \\ 5000 & q > \frac{1}{8} \end{cases} \tag{5.86}$$

Use, for $\chi(q, 0)$, the Gaussian in (5.34) rendered dimensionless, as in (5.75), with constants $\tilde{E} = 4500$, $\tilde{p} = 100$, $\Delta q = 1/(201)$ (so $N = 200$ grid points extending from $-\frac{1}{2} + \Delta q$ to $\frac{1}{2} - \Delta q$) and $\Delta s = \Delta q / 2000$. You should check that in the zero potential case, you get a wave function that is expanding and moving to the right (take, say, 500 steps).

### Problem 5.16
Implement the implicit Euler method, and solve the same problem as above.

### Problem 5.17
The Crank–Nicolson method is a kind of average of the implicit and explicit Euler methods. Referring to (5.61), we could write the explicit Euler update as:

$$\frac{\boldsymbol{\Psi}^{n+1} - \boldsymbol{\Psi}^n}{\Delta s} = -i\mathbb{H}\boldsymbol{\Psi}^n, \tag{5.87}$$

for $\mathbb{H}$ the discretized Hamiltonian operator. The implicit update comes from:

$$\frac{\boldsymbol{\Psi}^{n+1} - \boldsymbol{\Psi}^n}{\Delta s} = -i\mathbb{H}\boldsymbol{\Psi}^{n+1}. \tag{5.88}$$

To get Crank–Nicolson, we take:

$$\frac{\boldsymbol{\Psi}^{n+1} - \boldsymbol{\Psi}^n}{\Delta s} = -i\mathbb{H}\frac{\boldsymbol{\Psi}^{n+1} + \boldsymbol{\Psi}^n}{2}. \tag{5.89}$$

Separating the dependence on $\boldsymbol{\Psi}^{n+1}$ and $\boldsymbol{\Psi}^n$, we have:

$$\underbrace{\left(\mathbb{I} + \frac{1}{2}i\,\Delta s\mathbb{H}\right)}_{\equiv\bar{\mathbb{P}}} \boldsymbol{\Psi}^{n+1} = \left(\mathbb{I} - \frac{1}{2}i\,\Delta s\mathbb{H}\right) \boldsymbol{\Psi}^n. \tag{5.90}$$

Implement the method, and solve the step problem again.

### Problem 5.18
Your wave function has some initial normalization (from our current point of view, it doesn't much matter what its value is, although ideally, it should be 1):

$$\int_0^1 \Psi(s, q)^* \Psi(s, q) dq = A. \tag{5.91}$$

Schrödinger's equation respects this normalization (so that $\Psi(x, t)$ solving it has constant normalization). Your numerical methods will not, in general. Take the following, plausible (our next subject is precisely integration), approximation to the integral in (5.91):

$$\sum_{j=1}^{200} \|\psi_j^n\|^2 \Delta q \tag{5.92}$$

and plot the norm of your numerical solution for $n = 1 \longrightarrow 500$ for each of the methods (explicit Euler, implicit Euler, and Crank–Nicolson). Which one preserves the initial normalization the best?

## Problem 5.19

Implement the Crank–Nicolson solution for Schrödinger's equation in two dimensions – the update will look like (5.90), but this time the matrix $\mathbb{H}$ includes the Laplacian matrix approximation within it. In terms of the matrices $\mathbb{D}$ and $\tilde{\mathbb{V}}$ from (5.68), we have:

$$\underbrace{\left[ \mathbb{I} + \frac{1}{2} i \Delta s \left( -\mathbb{D} + \tilde{\mathbb{V}} \right) \right]}_{= \bar{\mathbb{P}}} \boldsymbol{\Psi}^{n+1} = \left[ \mathbb{I} - \frac{1}{2} i \Delta s \left( -\mathbb{D} + \tilde{\mathbb{V}} \right) \right] \boldsymbol{\Psi}^n. \tag{5.93}$$

Now write a function that performs the update $\boldsymbol{\Psi}^{n+1} = \bar{\mathbb{P}}^{-1} \bar{\mathbb{P}}^* \boldsymbol{\Psi}^n$ by first forming the product $\bar{\mathbb{P}}^* \boldsymbol{\Psi}^n$, and then doing the matrix inversion using the built-in function LinearSolve. Your function should take in the initial wave function and the number of temporal steps to take (along with $\Delta s$) in addition to the matrix $\bar{\mathbb{P}}$. The solutions will be embedded in the usual rectangular manner from Section 4.3 and you can make contour or three-dimensional plots of the probability densities to generate movies. As a problem for testing, take a $25 \times 25$ grid with $\Delta q = \Delta r = \frac{1}{26}$ ($q$ and $r$ are the scaled spatial variables), a finite circular well potential:

$$V(q, r) = \begin{cases} -1000 & \sqrt{(q - 1/2)^2 + (r - 1/2)^2} \leq 0.1 \\ 0 & \text{else} \end{cases} \tag{5.94}$$

and initial wave function:

$$\psi^0(q, r) = \left( \frac{2}{\pi} \right)^{1/4} \left( \tilde{E} - \frac{\tilde{p}^2}{4} \right)^{1/4} e^{-\left( \tilde{E} - \frac{\tilde{p}^2}{4} \right)(q - 1/4)^2} e^{i \frac{\tilde{p}}{2}(q - 1/2)}$$

$$\times \left( \frac{2}{\pi} \right)^{1/4} \left( \tilde{E} - \frac{\tilde{p}^2}{4} \right)^{1/4} e^{-\left( \tilde{E} - \frac{\tilde{p}^2}{4} \right)(r - 1/2)^2}, \tag{5.95}$$

a Gaussian centered at $q = 1/4, r = 1/2$ and with momentum in the $q$ direction initially – take $\tilde{E} = 450$, $\tilde{p} = 40$ and $\Delta s = 0.001$. Run the solution for 10 steps and make a movie of the result – check that the norm of the probability density is relatively constant over this range (use simple rectangular sums to approximate the integrals as in (5.92)).

# 6

# Integration

We have been focused on solving differential equations, and to the extent that solving, for example:

$$m\ddot{x} = F \tag{6.1}$$

for $x(t)$ involves, in some sense, integrating, we have really *been* integrating all along.

That's an indirect use of integration, though – when $F$ depends only on time, and not on spatial locations, we could indeed integrate directly to find, say, the velocity:

$$v(t) = \frac{1}{m} \int_0^t F(\bar{t})d\bar{t}, \tag{6.2}$$

and it is this sort of "area under the curve" integration that we will be looking at numerically[1] in this chapter, using two very different approaches.

## 6.1 Physical motivation

Integrals, in physics, represent continuous sums, generally over three-dimensional spatial volumes. Sometimes, we want the quantity represented by the integral directly: We are given a local description, the charge density for example, and we want to know how much total charge we have in a particular volume. Or we use the integral solution to a differential equation to obtain information, the period of a pendulum, for example. These are "direct" integrals, and we'll set up some examples of these first. Then we'll move on and look at integrals that solve partial differential equations. These generally come from linear PDEs where a Green's

---

[1] The numerical approximation to integrals is sometimes called "quadrature."

function (the point source solution) is known, and solutions for arbitrary sources represent summing up the contributions from a continuum of point sources.

### 6.1.1 Direct integrals

There are a variety of densities of interest: mass density, energy density, probability density. All these densities end up integrated when we want to consider large regions. As a simple one-dimensional example, suppose we have a rod of length $L$ that has mass per unit length $\lambda(x)$. The amount of mass at any point is zero. But small intervals $dx$ do have mass: $dM(x) = \lambda(x)dx$, and we can find the total mass by adding up all the individual contributions along the rod:

$$M = \int_0^M dM(x) = \int_0^L \lambda(x)dx. \tag{6.3}$$

The same idea holds in two and three dimensions. Given a volume charge density, $\rho(\mathbf{r})$, the charge at a point is zero, but the charge in an infinitesimal volume surrounding a point is: $dQ(\mathbf{r}) = \rho(\mathbf{r})dxdydz$, and so again we take:

$$Q = \int_\Omega \rho(\mathbf{r})dxdydz \tag{6.4}$$

to find the total charge $Q$ in a volume $\Omega$.

The story continues – the goal of solving Schrödinger's equation:

$$i\hbar \frac{\partial \Psi(x,t)}{\partial t} = -\frac{\hbar^2}{2m} \frac{\partial^2 \Psi(x,t)}{\partial x^2} + V(x)\Psi(x,t) \tag{6.5}$$

is a functional description of $\Psi(x,t)$ – and the interpretation of the solution is that the position probability density $\rho(x,t)$ associated with a particle of mass $m$ moving under the influence of $V(x)$ is given by $\Psi(x,t)^*\Psi(x,t)$. Even at this stage, an integral is important. From (6.5), it is clear that if $\Psi(x,t)$ is a solution, then so is $A\Psi(x,t)$, for arbitrary constant $A$. We require that $\Psi(x,t)$ be "normalized,"

$$A^2 \int_{-\infty}^{\infty} \Psi(x,t)^*\Psi(x,t)dx = 1, \tag{6.6}$$

and use this requirement to set the value of $A$. But in order to do that, we need to be able to integrate the normalization condition. Once we have the wave function appropriately normalized, we can compute the probability that the particle is between $a$ and $b$ (two spatial locations) as a function of time:

$$P(a,b,t) = \int_a^b \Psi^*(x,t)\Psi(x,t)dx, \tag{6.7}$$

again, a question answered by an integral. We can also compute the expectation value of position,

$$\langle x \rangle = \int_{-\infty}^{\infty} \Psi^*(x,t) x \Psi(x,t) dx \tag{6.8}$$

or other quantities of interest. In general, given a probability density, we can determine the probability of finding a particle in an interval $dx$ at time $t$: $dP = \rho(x,t)dx$, and construct averages:

$$\langle f \rangle = \int_{-\infty}^{\infty} \rho(x,t) f(x) dx \tag{6.9}$$

for any function of position, $f(x)$, with similar expressions in three dimensions.

---

**Probability density**

A specification of the probability density gives us all the information we need to compute statistically relevant quantities (in the corner of statistical information that we inhabit). For example, the density associated with a Gaussian distribution is:

$$\rho(x) = \frac{1}{\sqrt{2\pi}\sigma} e^{-\frac{(x-a)^2}{2\sigma^2}} \tag{6.10}$$

and has:

$$\int_{-\infty}^{\infty} \rho(x) = 1 \tag{6.11}$$

so the probability of getting a value $x \in [-\infty, \infty]$ is one. The average value is:

$$\langle x \rangle \equiv \int_{-\infty}^{\infty} \rho(x) x dx = a, \tag{6.12}$$

and the average squared value is:

$$\langle x^2 \rangle \equiv \int_{-\infty}^{\infty} \rho(x) x^2 dx = \sigma^2 + a^2. \tag{6.13}$$

From $\langle x \rangle$ and $\langle x^2 \rangle$, we can construct the variance:

$$\langle x^2 \rangle - \langle x \rangle^2 = \sigma^2. \tag{6.14}$$

---

*Pendulum*

As another case in which the numerical result of an integral is of immediate interest, we can compute the period of a realistic pendulum (the particulars of this setup

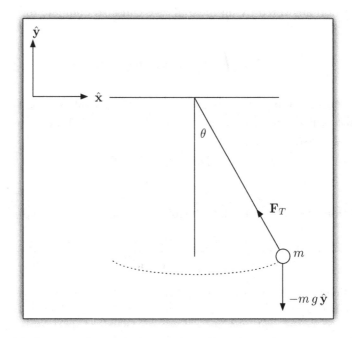

Figure 6.1 A mass $m$ is attached to a string of length $L$ – it moves up and back.

come from [3]). For a pendulum, a "bob" of mass $m$ is suspended by a string of length $L$, as in Figure 6.1. We know the forces in the $x$ and $y$ direction, so Newton's second law reads:

$$m\ddot{x} = -F_T \sin\theta \quad m\ddot{y} = F_T \cos\theta - mg. \tag{6.15}$$

In addition, we know that the pendulum bob is constrained to travel along a path given by:

$$x(t) = L\sin\theta(t) \quad y(t) = -L\cos\theta(t). \tag{6.16}$$

Taking the second derivatives of $x(t)$ and $y(t)$ and inputting them into (6.15), we get the angular acceleration associated with $\theta(t)$:

$$\ddot{\theta} = -\frac{g}{L}\sin\theta. \tag{6.17}$$

If we multiply this equation by $\dot{\theta}$, then the left-hand side can be written as $\frac{d}{dt}\left(\frac{1}{2}\dot{\theta}^2\right)$, and the right-hand side is $\frac{g}{L}\frac{d}{dt}\cos\theta$, both total derivatives in time. Integrate once to get

$$\dot{\theta}^2 = \frac{2g}{L}\cos\theta + \alpha. \tag{6.18}$$

We can set the constant $\alpha$ using the initial conditions, suppose the pendulum bob starts from rest at an angle $\theta_M$. Then $\alpha = -\frac{2g}{L}\cos\theta_M$, and the equation we want to solve is

$$\dot\theta = \pm\left[\frac{2g}{L}(\cos\theta - \cos\theta_M)\right]^{1/2}. \qquad (6.19)$$

Instead of solving for $\theta(t)$ explicitly (using RK, for example), let's find the period of oscillation, a quantity that should be obtainable by integrating (6.19). In order to avoid the choice of sign for $\dot\theta$, we'll take a well-defined quarter of the period – as the bob swings from $\theta_M$ to 0, the minus sign is relevant, and the quarter period is:

$$t = -\sqrt{\frac{L}{2g}}\int_{\theta_M}^{0} \frac{d\theta}{\sqrt{\cos\theta - \cos\theta_M}}. \qquad (6.20)$$

The period is $T = 4t$, so:

$$T = 4\sqrt{\frac{L}{2g}}\int_{0}^{\theta_M} [\cos\theta - \cos\theta_M]^{-1/2}\, d\theta. \qquad (6.21)$$

The integral here can be turned into a familiar expression, from a certain point of view. Note that $\frac{1}{2}(1 - \cos\theta) = \sin^2(\theta/2)$, then we have:

$$\begin{aligned}
\cos\theta - \cos\theta_M &= (1 - 2\sin^2(\theta/2)) - (1 - 2\sin^2(\theta_M/2)) \\
&= 2(\sin^2(\theta_M/2) - \sin^2(\theta/2)). 
\end{aligned} \qquad (6.22)$$

Take a substitution of the form $\sin(\theta/2) = A\sin(\phi)$, motivated by the above, and since $\sin(\theta_M/2) = A\sin(\phi_M)$ is just a constant, we set $A \equiv \sin(\theta_M/2)$ so that $\phi_M = \frac{1}{2}\pi$. Now the difference of cosines in (6.22) can be written:

$$\cos\theta - \cos\theta_M = 2A^2(1 - \sin^2\phi) = 2A^2\cos^2\phi. \qquad (6.23)$$

To finish the change of variables, note that:

$$\frac{1}{2}\cos(\theta/2)\, d\theta = A\cos\phi\, d\phi \qquad (6.24)$$

so that the period integral in (6.21) is

$$T = 4\sqrt{\frac{L}{2g}}\int_{0}^{\frac{1}{2}\pi} \frac{2A\cos\phi\, d\phi}{\cos(\theta/2)\sqrt{2}A\cos\phi}. \qquad (6.25)$$

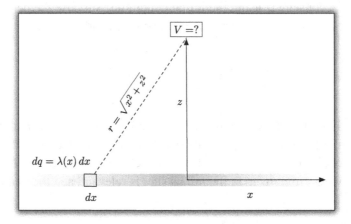

Figure 6.2 An infinite line of charge carries a known distribution $\lambda(x)$ (charge per unit length) – what is the electrostatic potential a height $z$ above the middle of the line?

and use the trigonometric identity: $\cos^2(\theta/2) + \sin^2(\theta/2) = 1$ to replace the $\cos(\theta/2)$ on the right with $\sqrt{1 - A^2 \sin^2 \phi}$. Our final expression is

$$T = 4\sqrt{\frac{L}{g}} \int_0^{\frac{1}{2}\pi} \frac{d\phi}{\sqrt{1 - A^2 \sin^2 \phi}} \qquad A \equiv \sin(\theta_M/2). \qquad (6.26)$$

This integral has a name, it is called an "elliptic integral," and numerical values exist in tables. But those tables had to be made somehow ...

### 6.1.2 Integrals that solve PDEs

Aside from finding total charge, mass, probability, the various moments of a probability distribution, and direct integration of Newton's second law, many PDEs have integral formulations for their solution – these are manifestations of the superposition principle (itself a manifestation of the linearity of the PDEs). In E&M, the electrostatic potential satisfies $\nabla^2 V = -\rho/\epsilon_0$. We know that the solution for a general distribution of charge can be built up out of point-contributions, since the Poisson PDE is linear (so a sum of two solutions is a solution). That build-up has an integral as its endpoint.

#### One-dimensional line of charge

As an example, take an infinite line of charge, carrying charge-per-unit-length $\lambda(x) = \frac{Q}{a} e^{-x^2/a^2}$ for constants $Q$ and $a$. We want to know the electrostatic potential a height $z$ above the midpoint of the line (at $x = 0$); the setup is shown in Figure 6.2.

We know that each point along the line contributes to the potential as a point source – that is, if a little box of width $dx$ centered at $x$ contains charge $dq$, then its contribution to the potential at the point of interest is:

$$dV = \frac{dq}{4\pi\epsilon_0\sqrt{x^2 + z^2}}. \tag{6.27}$$

We also know that the potential satisfies superposition, so that all we have to do is "add up" all of the contributions, they don't interact with one another. We can perform that addition by noting that $dq = \lambda(x)dx$, from the definition of $\lambda(x)$, so that:

$$dV = \frac{\lambda(x)dx}{4\pi\epsilon_0\sqrt{x^2 + z^2}} \longrightarrow V = \int_{-\infty}^{\infty} \frac{\lambda(x)dx}{4\pi\epsilon_0\sqrt{x^2 + z^2}}, \tag{6.28}$$

and there it is, an integral that we have to compute. We can clean up the actual integral a bit by removing uninteresting constants, and noting that the integrand is even,

$$V(z) = \frac{2Q}{4\pi\epsilon_0 za} \int_0^{\infty} \frac{e^{-x^2/a^2}}{\sqrt{1 + x^2/z^2}}dx. \tag{6.29}$$

To make the units clear, we'll nondimensionalize the integrand – the constant $a$ sets the scale for the Gaussian decay, so let $x \equiv aq$, then:

$$V(z) = \frac{2Q}{4\pi\epsilon_0 z} \int_0^{\infty} \frac{e^{-q^2}}{\sqrt{1 + a^2q^2/z^2}}dq. \tag{6.30}$$

The constants out front are just right for a potential, and the integral is ready for numerical approximation. In this particular case, the answer is a known function of $z$, we end up with:

$$V(z) = \left[\frac{Q}{4\pi\epsilon_0 a}\right] e^{y^2} K_0(y^2) \quad y \equiv \frac{z}{\sqrt{2}a} \tag{6.31}$$

where $K_0$ is a modified Bessel function. Regardless of its name, in the general case, what we want is a numerical function that approximates the integral and returns that approximation for a given value of $z$.

### Electric field for a box of uniform charge

Maxwell's equation for a static distribution of charge in three dimensions is linear, so that:

$$\nabla \cdot \mathbf{E} = \frac{\rho}{\epsilon_0}, \tag{6.32}$$

has integral solution:

$$\mathbf{E} = \frac{1}{4\pi\epsilon_0} \int \frac{\rho(\mathbf{r}')}{R^2}\hat{\mathbf{R}}dx'dy'dz' \tag{6.33}$$

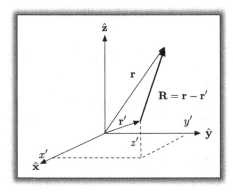

Figure 6.3 A vector $\mathbf{r}'$ points to an infinitesimal charge segment within the distribution of charge: $dQ = \rho(\mathbf{r}')d\tau'$. The vector $\mathbf{r}$ points to the "field point," the location of interest. The contribution that makes up the integral is: $\frac{dQ}{4\pi\epsilon_0 R^2}\hat{\mathbf{R}}$ with $\mathbf{R} \equiv \mathbf{r} - \mathbf{r}'$.

with $\mathbf{R} \equiv \mathbf{r} - \mathbf{r}'$ and $\hat{\mathbf{R}} = \mathbf{R}/R$. The integral form is important theoretically, it represents the continuous sum of point source contributions, available as a solution by superposition. In Figure 6.3, the vector $\mathbf{r}'$ points from the origin to a small volume $dx'dy'dz'$ containing charge $dQ = \rho(\mathbf{r}')dx'dy'dz'$. The electric field at $\mathbf{r}$ due to the charge at $\mathbf{r}'$ is:

$$dE = \frac{dQ}{4\pi\epsilon_0 R^2}\hat{\mathbf{R}} \tag{6.34}$$

and the integral (6.33) represents the sum of all the infinitesimal volume charge contributions to the electric field at $\mathbf{r}$.

Except in special cases, the integral (6.33) does not have nice, closed-form solutions. Numerically, however, we can carry out the full three-dimensional integration once we're given the distribution $\rho(\mathbf{r})$. As an example setup, let's think of computing the electric field outside a cube of side length $a$ centered at the origin.

Inside the cube, take $\rho = \rho_0 = q/a^3$, a constant charge density. Then the vector statement (6.33) is really three integrals:

$$E_x = \frac{\rho_0}{4\pi\epsilon_0} \int_{-a/2}^{a/2}\int_{-a/2}^{a/2}\int_{-a/2}^{a/2} \frac{(x-x')dx'dy'dz'}{\left((x-x')^2 + (y-y')^2 + (z-z')^2\right)^{3/2}}$$

$$E_y = \frac{\rho_0}{4\pi\epsilon_0} \int_{-a/2}^{a/2}\int_{-a/2}^{a/2}\int_{-a/2}^{a/2} \frac{(y-y')dx'dy'dz'}{\left((x-x')^2 + (y-y')^2 + (z-z')^2\right)^{3/2}} \tag{6.35}$$

$$E_z = \frac{\rho_0}{4\pi\epsilon_0} \int_{-a/2}^{a/2}\int_{-a/2}^{a/2}\int_{-a/2}^{a/2} \frac{(z-z')dx'dy'dz'}{\left((x-x')^2 + (y-y')^2 + (z-z')^2\right)^{3/2}}.$$

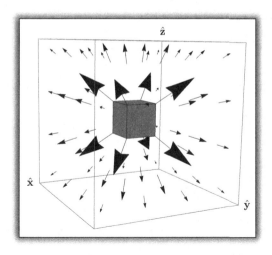

Figure 6.4 The vector field $\mathbf{E}(x, y, z)$ (evaluated at a grid of points $(x, y, z)$) outside a box with uniform charge density $\rho_0$.

The solution can be found numerically for arbitrary locations given by $x$, $y$, and $z$. An example of $\mathbf{E}$ for a box of length $a = 1.0$ is shown in Figure 6.4.

## 6.2 One-dimensional quadrature

We now turn to the method(s) for integrating a function numerically. One way to view the ODE solving methods developed in Chapter 2 is as a truncation of the definition of derivative:

$$\frac{df(x)}{dx} = \lim_{dx \to 0} \frac{f(x + dx) - f(x)}{dx}, \tag{6.36}$$

i.e. we erased the "limit" and took $dx \sim \Delta x$ to be a finite quantity. We can begin our study of integral methods with the same sort of bowdlerization of Calculus.

Given a function $f(x)$ and a grid of values $x_j = a + j\Delta x$ for $j = 0 \longrightarrow N$, the simplest method for approximating the area under the curve $f(x)$ is to replace the continuous summation with a discrete summation. That is, we'll approximate:

$$I = \int_a^b f(x)dx \tag{6.37}$$

with (take $a + N\Delta x = b$)

$$I \approx \sum_{j=0}^{N-1} f(x_j)\Delta x. \tag{6.38}$$

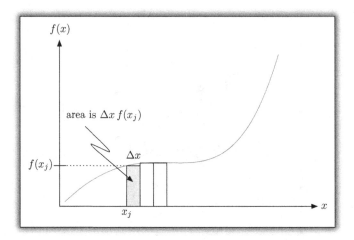

Figure 6.5 A rectangle area – the approximation (6.38) comes from adding up all of the rectangles from $x = a$ to $x = b$.

What we are doing, shown in Figure 6.5, is chopping the curve up into small rectangles, and summing the area of each rectangle.

As an example, let's integrate $f(x) = x^2$ from $x = 0$ to $b$. Take $N + 1$ points for our grid, given by $x_j = j\Delta x$ with $\Delta x \equiv b/N$. Then the approximation gives:

$$
\begin{aligned}
I &\approx \sum_{j=0}^{N-1} (j\Delta x)^2 \, \Delta x = \Delta x^3 \sum_{j=0}^{N-1} j^2 \\
&= \frac{1}{6} N(N-1)(2N-1)\,\Delta x^3 \\
&= \frac{1}{3}b^3 - \frac{1}{2}\Delta x b^2 + \frac{1}{6}\Delta x^2 b.
\end{aligned}
\tag{6.39}
$$

The actual value here is $I = \frac{1}{3}b^3$, so that our approximation is correct to leading order, with $O(\Delta x)$ corrections. In the limit $\Delta x \longrightarrow 0$, we of course recover the correct result. But we should be able to do better.

We can improve the method by taking a more relevant shape for our area approximation. In Figure 6.5, we made a rectangle because it has area that is easy to calculate – but the rectangle has height defined only by the *left* (in that figure, we could have used the right, of course) value of the curve. In Figure 6.6, we see a refined trapezoid that includes both the left and right-hand values to define a triangular top. Note that for the box shown in Figure 6.6, a simple rectangle would not capture the area under the curve very well. The area for the rectangle plus the triangle is:

$$
\Delta x f(x_k) + \frac{1}{2}\Delta x (f(x_{k+1}) - f(x_k)).
\tag{6.40}
$$

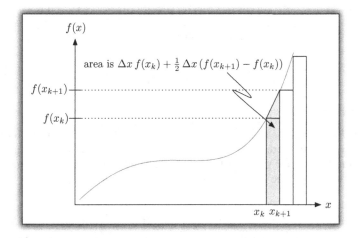

Figure 6.6 Using a rectangle plus a triangle to form a better approximation to the area under the curve.

This observation implies an approximation:

$$I \approx \sum_{j=0}^{N-1} \left[ \Delta x f(x_j) + \frac{1}{2} \Delta x \big( f(x_{j+1}) - f(x_j) \big) \right] = \sum_{j=0}^{N-1} \frac{1}{2} \Delta x \big( f(x_j) + f(x_{j+1}) \big).$$

(6.41)

On the far right, we can interpret our new approximation in terms of the average value of the function $f(x)$ on the interval between $x_j$ and $x_{j+1}$. This method is known as the "trapezoidal rule."

How does this new approach work for our test case? Again using $f(x) = x^2$, we have:

$$\begin{aligned}
I &\approx \sum_{j=0}^{N-1} \frac{1}{2} \Delta x^3 \big( j^2 + (j+1)^2 \big) = \frac{1}{2} \Delta x^3 \left[ \sum_{j=0}^{N-1} j^2 + \sum_{j=0}^{N-1} (j+1)^2 \right] \\
&= \frac{1}{2} \Delta x^3 \left[ \frac{1}{3} \big( N + 2N^3 \big) \right] \\
&= \frac{1}{3} b^3 + \frac{1}{6} b \Delta x^2.
\end{aligned}$$

(6.42)

The accuracy has improved – there is no $\Delta x$ term above, and the approximation to $I = \frac{1}{3} b^3$ goes like $O(\Delta x^2)$. We could keep going (and you will, in Problem 6.6), but we can generalize the approach without resorting to more and more complicated pictures.

## 6.3 Interpolation

We now look at how to generate methods using exact integration of polynomials that approximate a given function on some interval. Both the methods presented

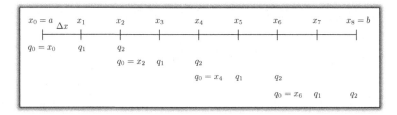

Figure 6.7 An example of a grid with eight points. We can subdivide it into four grids of two points. Here, $N = 8$, and $n = 2$ – we would approximate integrals on each of the four $n = 2$ grids, and then add those approximations together to get an approximation to the integral from $a$ to $b$.

above are special cases of interpolating integration techniques. What we want is an approximation to the function $f(x)$ over a small interval. This interval will be a subset of our larger discretization – in the previous section, we took $N + 1$ total points for our grid, $\{x_j \equiv a + j\Delta x\}_{j=0}^{N}$, and used the value of $f(x_j)$ at the grid points to approximate the integral.

What we're going to do now is consider a smaller grid, with just a few points, but with the same spacing $\Delta x$, and generate an approximation to the integral on this "local" grid, then build up the larger interval by piecing together the local approximation. Define our local grid via: $q_j = \alpha + j\Delta x$ (the same $\Delta x$ as above) for $j = 0$ to $n$. The value of $\alpha$ will be $x_k$ for some $k$, and we'll let $q_n \equiv \beta = \alpha + n\Delta x = x_{k+n}$. We will now generate methods by integrating (approximately) the function $f(x)$ on the grid $\{q_j\}_{j=0}^{n}$, and then in order to extend those methods to the entire grid (i.e. from $a$ to $b$, the integration region of interest), we must have $N$ divisible by $n$ (don't say I didn't warn you). An example of the grids for $N = 8$, $n = 2$ is shown in Figure 6.7.

Given a set of values $\{f_j = f(q_j)\}_{j=0}^{n}$ on the grid $q_j = \alpha + j\Delta x$ so that $q_0 = \alpha$ and $q_n = \beta$, there is a polynomial of degree (at most) $n$, $P_n(q)$ such that:

$$P_n(q_j) = f_j \quad \text{for } j = 0 \longrightarrow n, \tag{6.43}$$

since a polynomial of degree $n$ has $n + 1$ arbitrary constants to set. Polynomials are easy to integrate, so the idea here is to find a polynomial that goes through the $n + 1$ data points $\{f(q_j)\}_{j=0}^{n}$, for a function $f(q)$, and integrate that polynomial exactly.

Theoretically, we can construct $P_n(q)$ given the data $\{f_j \equiv f(q_j)\}_{j=0}^{n}$ – the formula is known as "Lagrange's interpolating formula," and reads:

$$P_n(q) = \sum_{j=0}^{n} f_j L_j(q) \tag{6.44}$$

where $L_j(q)$ is a polynomial of maximum degree $n$ that has $L_j(q_k) = \delta_{jk}$. We can construct such a polynomial explicitly:

$$L_j(q) = \prod_{k=0 k \neq j}^{n} \frac{q - q_k}{q_j - q_k}. \tag{6.45}$$

---

**Example**

Suppose we have $n = 3$, defining a grid with $q_0 = 0, q_1 = \Delta x, q_2 = 2\Delta x, q_3 = 3\Delta x$ (setting $\alpha = 0$). There are then four separate $L_j(q)$ polynomials of interest:

$$L_0(q) = -\frac{1}{6\Delta x^3}(q - \Delta x)(q - 2\Delta x)(q - 3\Delta x)$$

$$L_1(q) = \frac{1}{2\Delta x^3}q(q - 2\Delta x)(q - 3\Delta x)$$

$$L_2(q) = -\frac{1}{2\Delta x^3}q(q - \Delta x)(q - 3\Delta x) \tag{6.46}$$

$$L_3(q) = \frac{1}{6\Delta x^3}q(q - \Delta x)(q - 2\Delta x).$$

From these four, we can form a polynomial that takes on four given values, one at each grid point:

$$P_3(q) = f_0 L_0(q) + f_1 L_1(q) + f_2 L_2(q) + f_3 L_3(q). \tag{6.47}$$

An example function, taking four values at these four points and the associated interpolating polynomial are shown in Figure 6.8. The function here is $f(q) = \cos(2\pi q)$, and we're taking $\alpha = 0, \beta = 1$, and then $\Delta x = 1/3$.

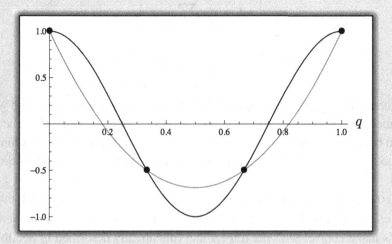

Figure 6.8 A function $f(q) = \cos(2\pi q)$ (black) with four points on a grid with $\Delta x = 1/3$. The interpolating polynomial from (6.47) is shown in gray.

We'll construct an approximation to the integral:

$$\int_{\alpha}^{\beta} f(q)dq \tag{6.48}$$

by integrating our polynomial interpolating function (6.44) – that is, we'll form:

$$J \equiv \int_{\alpha}^{\beta} P_n(q)dq \approx \int_{\alpha}^{\beta} f(q)dq. \tag{6.49}$$

A change of variables (following [42]) makes the integration a little easier – define $t$ by $q = \alpha + t\Delta x$, then:

$$L_j(q) = \bar{L}_j(t) \equiv \prod_{k=0 k \neq j}^{n} \frac{t-k}{j-k}, \tag{6.50}$$

and the integral of these polynomials, from $\alpha$ to $\beta = \alpha + n\Delta x$ is:

$$\int_{\alpha}^{\beta} L_j(q)dq = \Delta x \int_{0}^{n} \bar{L}_j(t)dt. \tag{6.51}$$

Now think of integrating the polynomial $P_n(q)$ term-by-term:

$$J = \int_{\alpha}^{\beta} P_n(q)dq = \int_{\alpha}^{\beta} \sum_{j=0}^{n} f_j L_j(q)dq = \Delta x \sum_{j=0}^{n} f_j \int_{0}^{n} \bar{L}_j(t)dt. \tag{6.52}$$

The integral of the $\bar{L}_j(t)$ is a definite integral, and so our integration approximation reduces to:

$$\boxed{\begin{array}{c} J = \Delta x \sum_{j=0}^{n} w_j f_j \\[2mm] w_j \equiv \int_{0}^{n} \bar{L}_j(t)dt. \end{array}} \tag{6.53}$$

where the set $\{w_j\}_{j=0}^{n}$ are the "weights" of the integral approximation, and will be determined once we choose $n$ – the polynomials in (6.45) make no reference to the function being integrated. The choice of $n$, the number of points in the subinterval (and maximum degree of the approximating polynomial) is enough to define a method.

$J$ is our approximation to $\int_{\alpha}^{\beta} f(q)dq$, and we build up the total integral, $\int_{a}^{b} f(q)dq$ by repeated application. We'll see how that works in a moment, but first let's make contact with the trapezoidal rule already described in Section 6.2. For the $n = 1$ approximation to the integral for the interval $\alpha = x_j$, $\beta = x_{j+1}$, we have

$$J = \Delta x \left( w_0 f(x_j) + w_1 f(x_{j+1}) \right) \tag{6.54}$$

with

$$w_0 = \int_0^1 \bar{L}_0(t)dt = \int_0^1 \frac{t-1}{0-1} dt = \frac{1}{2}$$

$$\tag{6.55}$$

$$w_1 = \int_0^1 \bar{L}_1(t)dt = \int_0^1 t\, dt = \frac{1}{2}.$$

This says that if we have two points, we can approximate the integral over a region from $x_j$ to $x_{j+1}$ by:

$$\frac{1}{2} \left( f(x_j) + f(x_{j+1}) \right) \Delta x \tag{6.56}$$

which is precisely what we got in developing the trapezoidal rule.

We can go further – suppose we take the integral $J$ over the range from $\alpha = x_j$ to $\beta = x_{j+2} = x_j + 2\Delta x$. We use the $n = 2$ polynomial:

$$J = \int_{x_j}^{x_{j+2}} P_2(x)dx = \Delta x \left( w_0 f(x_j) + w_1 f(x_{j+1}) + w_2 f(x_{j+2}) \right) \tag{6.57}$$

so we need the three definite integrals for the $w_0$, $w_1$, and $w_2$:

$$w_0 = \int_0^2 \frac{1}{2}(1-t)(2-t)\, dt = \frac{1}{3}$$

$$w_1 = \int_0^2 t(2-t)\, dt = \frac{4}{3} \tag{6.58}$$

$$w_2 = \int_0^2 \frac{1}{2}t(t-1)\, dt = \frac{1}{3}.$$

Given three "support" points, we can approximate the integral over the interval:

$$J = \frac{1}{3}\Delta x \left( f(x_j) + 4f(x_{j+1}) + f(x_{j+2}) \right) \approx \int_{x_j}^{x_{j+2}} f(x)dx. \tag{6.59}$$

This is a new method – when we use this ($n = 2$) $J$ to approximate $I$ by adding up over the interval $a$ to $b$, the method is called "Simpson's rule."

### *6.3.1 Arbitrary interval*

Suppose we now have an interval $[a, b]$ and we want to approximate the integral of the function $f(x)$ over this interval:

$$I = \int_a^b f(x)dx. \tag{6.60}$$

What we'll do is cut the interval up into segments of length $\Delta x$, with $x_j = a + j\Delta x$ for $j = 0 \longrightarrow N$ and $\Delta x = \frac{b-a}{N}$, then apply the appropriate form from above.

Using the trapezoidal rule, we have:

$$I \approx \sum_{j=0}^{N-1} \frac{1}{2}\Delta x\big(f(x_j) + f(x_{j+1})\big) \tag{6.61}$$

and using "Simpson's rule,"

$$I \approx \sum_{j=0,2,4,\ldots}^{N-2} \frac{1}{3}\Delta x\big(f(x_j) + 4f(x_{j+1}) + f(x_{j+2})\big). \tag{6.62}$$

Notice that in this case, we require that $N$ is divisible by 2 in order to fit the $n = 2$ grids evenly inside the larger one.

As an example, the code to integrate an arbitrary function $f(x)$ using Simpson's method is shown in Implementation 6.1. We provide the function to integrate, f, and the lower and upper limits, a and b. Numerically, we also specify the total number of grid points to use, Ns. For clarity, we have omitted the explicit check that Ns is divisible by 2, necessary for the correct functioning of the method.

---

**Implementation 6.1** Simpson's method

```
Simpson[f_, a_, b_, Ns_] := Module[{Ia, j, dx},
  dx = (b - a) / Ns;
  Ia = 0.0;
  For[j = 0, j ≤ Ns - 2, j = j + 2,
   Ia = Ia + (1 / 3) dx (f[a + j dx] + 4 f[a + (j + 1) dx] + f[a + (j + 2) dx]);
  ];
  Return[Ia];
  ]
```

---

### *Accuracy*

Using the interpolation approach, the accuracy of any method is easy to describe – we are always performing an exact polynomial integral, so a method with $n$ grid points in the interpolation will integrate polynomials of degree $n$ (or less) with no error. Using the Taylor expansion, we can connect those perfect polynomial integrations to errors in the integral approximation for any function.

For the Simpson's rule method, we are approximating general functions with polynomials of degree 2 – then we know that if we had a polynomial of degree 2 as the function to integrate, we could get the exact integral with only the points $x_0 = a$, $x_1$, and $x_2 = b$. So the method will accurately integrate any polynomial up to degree two perfectly. That is:

$$I = \int_a^b \left(Ax^2 + Bx + C\right) dx = J \tag{6.63}$$

in this case. We know that any function, in the vicinity of zero (say), can be expanded as a polynomial

$$f(x) = f(0) + f'(0)x + \frac{1}{2}f''(0)x^2 + \frac{1}{6}f'''(0)x^3 + \frac{1}{24}f''''(0)x^4 + O(x^5). \tag{6.64}$$

The first term not captured by our method goes like the third derivative of the function to be integrated. And, for our grid, we expect that the method itself will be dominated by errors of order $\Delta x^3$ (as it turns out, this is not the case – due to a "lucky" cancellation, the Simpson's rule integration routine is actually more accurate, as you will see in Problem 6.8).

### Controlling error

Our ability to control error comes from our choice of method, and the number of grid points, $N$, to use with that method. Once we have chosen a method, and we'll take Simpson's rule as an example, we know that the overall error will go like some power of $\Delta x \equiv \frac{b-a}{N}$. We can reduce the error by increasing $N$. Suppose someone gave us a target tolerance $\epsilon$, how could we choose $N$ such that $|I^{\text{num}} - I| \le \epsilon$ (i.e. the difference between the numerical approximation and the actual value of the integral is less than $\epsilon$)? The idea is to take two different values of $N$ and use those to extrapolate, given the error of the method, to find a value of $N$ that will lead to $|I^{\text{num}} - I| \le \epsilon$ (similar to determining adaptive step sizes from Section 2.4.1). For Simpson's method, we know that $I^{\text{num}} = I + O(\Delta x^4)$, so that if we take $N$ and $N'$ giving $\Delta x$ and $\Delta x'$, then:

$$I^{\text{num}}_{\Delta x} = I + A\Delta x^4 \qquad I^{\text{num}}_{\Delta x'} = I + A\left(\Delta x'\right)^4 \tag{6.65}$$

and we can estimate $A$, the constant (related to the fourth derivative of the function $f(x)$, if we are using Simpson's rule):

$$|A| = \left| \frac{I^{\text{num}}_{\Delta x'} - I^{\text{num}}_{\Delta x}}{(\Delta x')^4 - \Delta x^4} \right|. \tag{6.66}$$

Using this value, we can compute a third approximation to $I$, call it $I^{\text{num}}_{\Delta \bar{x}}$ with a desired accuracy $\epsilon$ – what we want is the step size, $\Delta \bar{x}$, that gives $\epsilon$ as its target accuracy,

$$|I_{\Delta\bar{x}}^{\text{num}} - I| = \epsilon = |A||\Delta\bar{x}^4 \longrightarrow \Delta\bar{x} = \left[\epsilon\left|\left(\frac{(\Delta x')^4 - \Delta x^4}{I_{\Delta x'}^{\text{num}} - I_{\Delta x}^{\text{num}}}\right)\right|\right]^{\frac{1}{4}}. \tag{6.67}$$

From $\Delta\bar{x}$, finally, we can estimate $\bar{N} = \frac{b-a}{\Delta\bar{x}}$ (we need to perform some massaging to make sure that this value is both an integer, and divisible by 2).

This $\Delta\bar{x}$ is not unique – it depends on the original choices of $\Delta x$ and $\Delta x'$. It is also not optimal, there is no guarantee that it is the largest grid step that achieves $\epsilon$. One needs to be careful in using (6.67) to keep $\Delta x$ and $\Delta x'$ coarse, otherwise the denominator may be very close to zero, $|I_{\Delta x'}^{\text{num}} - I_{\Delta x}^{\text{num}}| \approx 0$ (a problem, also, if you are integrating "perfectly," as can happen with polynomials, for example).

## 6.4 Higher-dimensional quadrature

Once we have a one-dimensional method, we can use it to calculate two- (and three-) dimensional integrals. Take a general two-dimensional integral:

$$I = \int_{a_x}^{b_x}\left[\int_{a_y}^{b_y} f(x, y)dy\right]dx \tag{6.68}$$

and define the one-dimensional integral

$$I_y(x) = \int_{a_y}^{b_y} f(x, y)dy, \tag{6.69}$$

where the integral is defined for each input value of $x$. To recover $I$, we just integrate the function $I_y(x)$ in $x$:

$$I = \int_{a_x}^{b_x} I_y(x)dx. \tag{6.70}$$

If we imagine discretizing in two dimensions, via $\Delta x = \frac{b_x - a_x}{N_x}$ and $\Delta y = \frac{b_y - a_y}{N_y}$, then we can construct an approximation to $I_y(x)$ using our one-dimensional integrator, and then integrate that function of $x$, again using a one-dimensional integral approximation. The code to implement this two-dimensional integration approximation using Simpson's rule is shown in Implementation 6.2 – as arguments, we need the function $f$, the limits in $x$, $ax$, and $bx$, and the number of steps to take in the $x$ direction, $Nx$, and we need the same set for the $y$ portion of the integration. The only trick in the function $Simpson2D$ is the local definition of $Iy$; this is a function of $x$ that gives precisely the approximation to (6.69), but it must be defined as a single-variable function so that we can integrate it, giving the final output.

**Implementation 6.2** Two-dimensional Simpson integration

```
Simpson2D[f_, ax_, bx_, Nx_, ay_, by_, Ny_] := Module[{Iy},
  Iy[x_] := Module[{Igrandy},
    Igrandy[y_] := f[x, y];
    Return[Simpson[Igrandy, ay, by, Ny]];
  ];
  Return[Simpson[Iy, ax, bx, Nx]];
]
```

*Modifying integration endpoints*

As another example of two-dimensional integration, suppose we want to compute the average value of a function $f(x, y)$ over a disk of radius $R$ – we'll use a Cartesian parametrization of the circle, so our integral of interest is:

$$I = \frac{1}{\pi R^2} \int_{-R}^{R} \left[ \int_{-\sqrt{R^2-x^2}}^{\sqrt{R^2-x^2}} f(x, y) dy \right] dx. \tag{6.71}$$

This problem introduces $x$ dependence in the integration limits for $y$, but is otherwise straightforward. In Implementation 6.3, we see the code that takes a function f, the radius of the circle, R, and the integration step size Ns, and returns the average value of $f$ over the circle.

**Implementation 6.3** Average over a circle

```
CircleAvg[f_, R_, Ns_] := Module[{Iy},
  Iy[x_] := Module[{Igrandy},
    Igrandy[y_] := f[x, y];
    Return[Simpson[Igrandy, -Sqrt[R^2 - x^2], Sqrt[R^2 - x^2], Ns]];
  ];
  Return[Simpson[Iy, -R, R, Ns] / (Pi R^2)];
]
```

*Three-dimensional integration*

The pattern continues. We can split a three-dimensional integral:

$$I = \int_{a_z}^{b_z} \left[ \int_{a_x}^{b_x} \left( \int_{a_y}^{b_y} f(x, y, z) dy \right) dx \right] dz \tag{6.72}$$

into a two-dimensional integral (which we could approximate using Implementation 6.2) and a one-dimensional integral.[2] The square brackets in (6.72) define a two-dimensional integral, given a value of $z$, and then we integrate over $z$. In Implementation 6.4, we see the generalization that takes a function f, and all relevant limits and steps sizes, and returns an approximation to $I$ from (6.72). This

---

[2] We could also view a three-dimensional integral as three one-dimensional integrals, but why throw out what we've done so far?

---

**Implementation 6.4** Three-dimensional Simpson integration

```
Simpson3D[f_, ax_, bx_, Nx_, ay_, by_, Ny_, az_, bz_, Nz_] := Module[{Ixy},
  Ixy[z_] := Module[{Igrandxy},
    Igrandxy[x_, y_] := f[x, y, z];
    Return[Simpson2D[Igrandxy, ax, bx, Nx, ay, by, Ny]];
  ];
  Return[Simpson[Ixy, az, bz, Nz]];
]
```

---

three-dimensional integral approximation can be used directly to find, for example, the electric field outside of a cube of charge, the problem set up in Section 6.1.2.

## 6.5 Monte Carlo integration

As the three-dimensional charge distribution example (6.35) shows (try running Implementation 6.4 using the integrands in (6.35) for a large number of field points), retaining accuracy in higher dimensions can be time consuming. In order to compute the average of a function over a circle, for example, we need to perform $N^2$ calculations ($N$ integrations, each requiring $N$ computations). As the dimension grows, the time it takes to integrate grows as well. There are two primary cases in which the multiple application of a one-dimensional integral will be unwieldy, and we will look at an alternate method of approximating these integrals. First, if the dimension of integration is large, the timing of an iterated one-dimensional algorithm will suffer. How big can a multi-dimensional integral be? There are, after all, only four dimensions (that we know about) – but think of statistical mechanics, where we might be interested in the volume of phase space for $N$ particles – that is a $6N$-dimensional integral. The second stumbling block is for three-dimensional volumes with boundaries that are difficult to parametrize. Our circle example above was easy to implement because the lower and upper endpoint for the $y$ integration is a simple function of $x$ – but if we had a crazy shape, the limits on the integrals would become unwieldy.

Enter Monte Carlo. The idea here is to *randomly sample* an integral, and approximate the result using a few sampling points, and our knowledge of the random numbers that generated them (see [16, 21, 40]). Think of a dartboard – if we throw darts at the board (with our eyes closed, so that there is no natural target), some will hit, some will miss, and the ratio of hits to misses can be used to give an estimate of the area. Suppose we count up all the darts that fall within a box of side length $2R$, enclosing the circular dartboard. The dart board takes up a fraction of the square area given by: $(\pi R^2)/(2R)^2$, and we expect the proportion of randomly thrown darts falling in the circle versus outside, to equal this ratio – in general, a shape with area $A$ will take up a fraction $A/(2R)^2$ of the square's area, and if we throw

$N$ darts, with $h$ hits inside the shape, we can estimate the area of the shape

$$\frac{A}{(2R)^2} \approx \frac{h}{N} \longrightarrow A \approx \frac{h}{N}(2R)^2 . \tag{6.73}$$

The code to implement the random throwing of darts at a circular target is shown in Implementation 6.5.

---

**Implementation 6.5** Monte Carlo for circle area

```
MCArea[R_, Ns_] := Module[{hit, index, x, y},
  hit = 0;
  For[index = 1, index ≤ Ns, index = index + 1,
   x = RandomReal[{-R, R}];
   y = RandomReal[{-R, R}];
   If[Sqrt[x^2 + y^2] ≤ R,
    hit = hit + 1;
   ];
  ];
  Return[(hit / Ns) (2 R) ^2];
 ]
```

---

Clearly, the random sampling idea that forms the heart of Monte Carlo depends on using an appropriate setup. For example, we want our square box enclosing the dart board to actually enclose the dart board. If we aim at a square box that is sitting next to the dart board, we will get, predictably, zero darts hitting the board. Just as important, for sampling, is the idea that the distribution from which we draw our random numbers matters. In Implementation 6.5, we used Mathematica's built-in Random function, which defaults to a uniform distribution. We characterize distributions by their probability density, $\rho(x)$ is the probability per unit length associated with a random number. A uniform distribution means that $\rho(x) = \rho_0$ (a constant) on any domain of interest.

We can use our interpretation of integrals as "area under a curve" to employ Monte Carlo as an integral approximator. Returning to the one-dimensional case, we want to evaluate

$$I = \int_a^b f(x)dx \tag{6.74}$$

and we will generate a Monte Carlo approximation by taking a box of width $b - a$ and height $\max(f(x) : x \in [a, b])$ (assuming the minimum value of the function is $\geq 0$ on the interval); this box has area $A_\square$. Then we'll generate uniformly distributed random numbers and calculate the approximate area:

$$A \approx \frac{h}{N}A_\square, \tag{6.75}$$

where $h$ is again the number of hits (the number of points that lie below the curve). We're using the function $f(x)$ here to define a "strange shape," extending from the

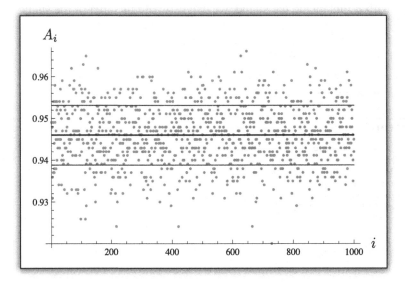

Figure 6.9 The various $A_i$ associated with $M = 1000$ separate runs of the Monte Carlo area approximation to the integral $\int_0^1 \frac{\sin(x)}{x} dx$. The gray lines display the average and one standard deviation on either side. The actual value is shown in black (and lies virtually on top of the Monte Carlo mean).

$x$-axis to the height of the function at $x$, and approximating this area just as we did for the dartboard.

The result $A$ will vary every time we run the Monte Carlo integration and we can calculate the statistics of that variation. If we take a Monte Carlo approximation with $N$ steps, and we repeat the calculation $M$ times, then the average $\langle A \rangle$ and (sample) variance $\sigma^2$ are given by:

$$\langle A \rangle = \frac{1}{M} \sum_{i=1}^{M} A_i \qquad \sigma^2 = \frac{1}{M-1} \sum_{i=1}^{M} (A_i - \langle A \rangle)^2 \qquad (6.76)$$

where $A_i$ is the $i$th area calculation, performed via (6.75).

As an example of this type of calculation, take the function $f(x) = \frac{\sin(x)}{x}$, we'll approximate the integral of this function from $x = 0 \longrightarrow 1$. If we run our Monte Carlo integrator with $N = 1000$, and $M = 1000$, we get an average value of $\langle A \rangle \approx$ .945 and $\sigma \approx 0.03$. We can compute the exact integral in this case, it ends up being: $I \approx 0.946083$. To get a feel for the scatter in the integration, the $M = 1000$ individual area calculations, along with the mean, standard deviation, and exact values, are shown in Figure 6.9.

### 6.5.1 Statistical connection

How did the notion of randomness enter a discussion of deterministic integrals? We can connect the Monte Carlo integration procedure to one-dimensional quadrature (without reference to areas), and it is this connection that we will develop in this section. Thinking of the integral:

$$I = \int_a^b f(x)dx, \tag{6.77}$$

we can view $I$ as an average – given a one-dimensional density $\rho(x)$, the integral can be expressed as:

$$I = \int_a^b \rho(x)\left(\frac{f(x)}{\rho(x)}\right) dx \equiv \left\langle \frac{f(x)}{\rho(x)} \right\rangle \tag{6.78}$$

so that this integral is itself an average of the function $f(x)/\rho(x)$. That introduces the random distribution. Now think of approximating the average by sampling at random points $\{x_j\}_{j=1}^{N_x}$. We would write:

$$I \approx \frac{1}{N_x} \sum_{j=1}^{N_x} \frac{f(x_j)}{\rho(x_j)}. \tag{6.79}$$

We have been using "flat" distributions for our Monte Carle integrations, so that for a function defined on $[a, b]$, the distribution governing the choice of $x_j \in [a, b]$ is just $\rho(x) = 1/(b - a)$. Inserting this into the sum,

$$\boxed{I \approx \frac{1}{N_x} \sum_{j=1}^{N_x} f(x_j)(b - a) = \sum_{j=1}^{N_x} f(x_j)\Delta x} \tag{6.80}$$

where $\Delta x \equiv (b - a)/N_x$. This final form provides some insight – it's just the rectangular approximation discussed in Section 6.2, only now the points $\{x_j\}_{j=1}^{N_x}$ are randomly chosen, not just the grid points we used there.

We need to connect (6.80) to the approach described by (6.75), where we take $I \approx \frac{h}{N}A_\Box$. In that setting, $A_\Box = (b - a)f_{max}$, where $f_{max}$ was the maximum value of $f(x)$ for $x \in [a, b]$. Suppose we take a set of $x$ values $\{x_j\}_{j=1}^{N_x}$ and for each value of $x_j$, we generate a random set of values $\{y_k\}_{k=1}^{N_y}$, then we can write the approximation as:

$$I \approx \frac{h}{N}A_\Box = \frac{(b - a)f_{max}}{N_x N_y} \sum_{j=1}^{N_x}\left(\sum_{k=1}^{N_y} H(x_j, y_k)\right) \tag{6.81}$$

where

$$H(x, y) = \begin{cases} 1 & y \le f(x) \\ 0 & y > f(x) \end{cases} \tag{6.82}$$

so that the sum over $j$ and $k$ gives us $h$, the total number of "hits" inside the area under the curve $f(x)$.

Rewriting the approximation one last time, we can make contact with (6.80):

$$I \approx \frac{1}{N_x} \sum_{j=1}^{N_x} (b - a) \left( \frac{f_{max}}{N_y} \sum_{k=1}^{N_y} H(x_j, y_k) \right), \tag{6.83}$$

and the term in parenthesis can be understood as an approximation to $f(x_j)$, the relevant ratio is:

$$\frac{f(x_j)}{f_{max}} \approx \frac{\sum_{k=1}^{N_y} H(x_j, y_k)}{N_y}, \tag{6.84}$$

where on the left, we have the ratio of the height $f(x_j)$ to the maximum height, and that should equal the quantity on the right (in the large $N_y$ limit), which is the ratio of the number of hits below $f(x_j)$ to the total number of "throws." From (6.84), we learn that

$$f(x_j) \approx \frac{f_{max}}{N_y} \sum_{k=1}^{N_y} H(x_j, y_k) \tag{6.85}$$

precisely the term appearing in (6.83). If we replace the term in parenthesis there with $f(x_j)$, we recover (6.80), and the two views (the statistical and geometrical ones) return the same integral approximation.

### 6.5.2 Extensions

Take the "integral-as-average" expression:

$$I = \int_a^b f(x)dx = \left\langle \frac{f(x)}{\rho(x)} \right\rangle. \tag{6.86}$$

The first question to ask is: What probability density $\rho(x)$ should we use to choose the random values $\{x_j\}_{j=1}^N$? We have been using a flat distribution, but this is not necessary, nor is it even desirable in most cases. Suppose our function $f(x)$ was sharply peaked, then a flat distribution of points would give $f(x_j) = 0$ for many values of $x_j$, rendering our sampling useless.

We can tailor our distribution to the function we are integrating – for a highly localized function, we might use a Gaussian distribution (6.10) for $\rho(x)$, where the mean was tuned appropriately. That way, we sample the portions of the function relevant to the area under the curve. Most numerical packages will include random numbers generated according to a Gaussian distribution, but how would we pick numbers according to some general $\rho(x)$ given only the ability to generate random numbers evenly on some interval?[3]

---

**Transformation**

We will work on $x \in [0, 1]$ with a probability density, $\rho(x)$, so that the probability of getting a number near $x$ is $\rho(x)dx$ and $\int_0^1 \rho(x)dx = 1$. Suppose we make a transformation to the variable $y(x)$, an invertible function of $x$. What happens to the density? The probability of choosing a number near $x$ can't change, since all we are doing is relabelling the point, so:

$$\rho(x)dx = \rho(x(y))dx \tag{6.87}$$

where $x(y)$ is the inverse of $y(x)$. But let's rewrite the whole expression in terms of the new variable $y$:

$$\rho(x)dx = \rho(x(y))dx = \rho(x(y))\frac{dx}{dy}dy \tag{6.88}$$

and we can think of the expression: $\rho(x(y))\frac{dx}{dy}$ as the probability density associated with $y$. That is, given $\rho(x)$, we can generate a new density for $y$ via:

$$\bar{\rho}(y) \equiv \rho(x(y))\frac{dx}{dy}. \tag{6.89}$$

That gives us a way to generate numbers drawn from any density we like: Start with a uniform density $\rho(x) = 1$ (for $x \in [0, 1]$), then given a target density $f(y)$, we solve:

$$f(y) = \frac{dx}{dy} \longrightarrow x = \int f(y)dy + \beta \tag{6.90}$$

($\beta$ is an integration constant) for $y(x)$ – if we generate random values for $x \in [0, 1]$ using $\rho(x) = 1$, then take each of these and run it through $y(x)$, we will have random values $y$ distributed according to the density $f(y)$.

As an example, suppose we have, as our target,

$$f(y) = -6y(y - 1), \tag{6.91}$$

a normalized density for $y \in [0, 1]$. Then the transformation $y(x)$ is defined by (6.90):

$$x = 3y^2 - 2y^3 + \beta, \tag{6.92}$$

---

[3] Generating truly random numbers, evenly or not, is already a tall order, but algorithms exist to do this approximately (see [34]), and we'll take such a flat distribution as our starting point.

where $\beta$ is chosen to be zero in this case, that way $x = 0$ and $x = 1$ go to $y = 0$ and $y = 1$. Now we have to invert the cubic, only one of the solutions satisfies our requirement that $y \in [0, 1]$:

$$y = \frac{1}{4}\left[2 + \frac{1 - i\sqrt{3}}{P} + (1 + i\sqrt{3})P\right]$$

$$P \equiv \left(-1 + 2x + 2\sqrt{x(-1 + x)}\right)^{1/3}.$$

(6.93)

We can take $N$ points $\{x_j\}_{j=1}^{N}$, uniformly distributed on $[0, 1]$, run them through (6.93) to make $\{y_j\}_{j=1}^{N}$ points distributed according to (6.91). A histogram of these values, for $N = 50\,000$, is shown in Figure 6.10; there you can clearly see the quadratic distribution taking shape.

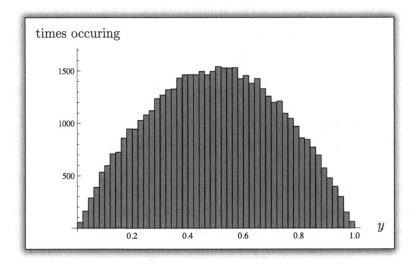

Figure 6.10 A histogram of $N = 50\,000$ points, starting from a flat distribution $x \in [0, 1]$; we generate the corresponding $\{y_j\}_{j=1}^{N}$ using (6.93) to get a quadratic distribution.

### *Further reading*

1. Gould, Harvey, Jan Tobochnik, & Wolfgang Christian. *Computer Simulation Methods: Applications to Physical Systems*. Pearson, 2007.
2. Pang, Tao. *An Introduction to Computational Physics*. Cambridge University Press, 2006.

3. Press, William H., Saul A. Teukolsky, William T. Vetterling, & Brian P. Flannery. *Numerical Recipes in C.* Cambridge University Press, 1996.
4. Stoer, J. & R. Bulirsch. *Introduction to Numerical Analysis.* Springer-Verlag, 1993.

## Problems

### Problem 6.1

Set up the electric field integral(s) for a flat disk (of radius $R$) carrying uniform charge $\sigma$ (i.e. a constant charge-per-unit-area) – the source points are given by:

$$\mathbf{r}' = r' \cos \phi' \hat{\mathbf{x}} + r' \sin \phi' \hat{\mathbf{y}} \tag{6.94}$$

for $\phi = 0 \longrightarrow 2\pi$. Call the field point $\mathbf{r} = x\hat{\mathbf{x}} + y\hat{\mathbf{y}} + z\hat{\mathbf{z}}$, and write your integral as a function of the Cartesian field point locations.

### Problem 6.2

Find the electric field a height $z$ above a disk of radius $R$ carrying constant surface charge $\sigma$, centered at the origin and lying in the $x - y$ plane. You can perform the integral from the previous problem, with $x = y = 0$ and arbitrary $z$ in this case. Make sure you recover the field of a point charge for $z \longrightarrow \infty$.

### Problem 6.3

An ellipse is the set of points, in the $x - y$ plane, satisfying: $x^2/a^2 + y^2/b^2 = 1$ for horizontal extent $2a$ and vertical extent $2b$ ($a$ and $b$ are the semi-major and semi-minor axes). This relation can be expressed in terms of the single parameter $\theta$ by writing:

$$x = a \sin \theta \quad y = b \cos \theta, \tag{6.95}$$

where $\theta \in [0, 2\pi]$ (note that it is *not* the angle $\phi \equiv \tan^{-1}(y/x)$).
(a) Express the infinitesimal length along the curve, $d\ell \equiv \sqrt{dx^2 + dy^2}$ in terms of $\theta$ and $d\theta$ (along with $a$ and $b$).
(b) The perimeter, $P$, of the ellipse comes from integrating this length for $\theta = 0 \longrightarrow 2\pi$. Set up the expression, and show that it can be written as:

$$P = 4a \int_0^{\pi/2} \sqrt{1 - e^2 \sin^2 \theta} \, d\theta \tag{6.96}$$

where $e \equiv \sqrt{a^2 - b^2}/a$ is the "eccentricity" of the ellipse. The integral here is a "complete elliptic integral of the second kind" (complete here refers to the upper integration limit, $\pi/2$, meaning that we will find the full circumference, and not just the arc length of a segment).
(c) Show that the perimeter reduces correctly in the circular case, $a = b \equiv R$.

**Problem 6.4**

For an ellipse carrying uniform line charge density $\lambda$, use source points:

$$\mathbf{r}' = a\sin(\theta')\hat{\mathbf{x}} + b\cos(\theta')\hat{\mathbf{y}} \tag{6.97}$$

to write the integral representing the electric field at a point $\mathbf{r} = x\hat{\mathbf{x}} + y\hat{\mathbf{y}} + z\hat{\mathbf{z}}$.

When $b = 0$, we should recover the integral expression for a uniformly charged finite line,

$$\mathbf{E}(x, y, z) = \frac{\lambda}{4\pi\epsilon_0} \int_{-a}^{a} \frac{\left((x - x')\hat{\mathbf{x}} + y\hat{\mathbf{y}} + z\hat{\mathbf{z}}\right)}{\left((x - x')^2 + y^2 + z^2\right)^{3/2}} dx'. \tag{6.98}$$

Check this limiting case for your full expression (be careful, your integral for the ellipse will go from $a$ to $-a$ and back, so you'll need to integrate from $\theta = -\frac{1}{2}\pi \longrightarrow \frac{1}{2}\pi$ to recover this limit).

**Problem 6.5**

Show that (6.38) and (6.41) are related: If you call the approximation (6.38) $I_b$, and (6.41) $I_t$, prove that

$$I_t = \frac{1}{2}\Delta x(-f(x_0) + f(x_N)) + I_b. \tag{6.99}$$

**Problem 6.6**

We can develop Simpson's rule as the quadratic extension of the discussion in Section 6.2. Given a function $f(x)$ (to integrate), and a grid with $x_j = j\Delta x$, write a quadratic function $G(x)$ such that $G(x_j) = f(x_j)$, $G'(x_j) = f'(x_j)$, $G''(x_j) = f''(x_j)$. Integrate this function from $x_{j-1}$ to $x_{j+1}$ and replace $f''(x)$ with $\frac{f_{j+1} - 2f_j + f_{j-1}}{\Delta x^2}$ to get the approximate integral over this region written in terms of $f_j$ and $f_{j\pm1}$. You should recover (6.59) (with $j \to j - 1$ there to match the indexing in this problem).

**Problem 6.7**

In Section 6.3, we obtained the weights and integral approximations for $n = 1$ and $n = 2$. Work out the next ($n = 3$) interpolating polynomial approximation. You have the polynomials, in (6.46); use them to compute the weights and $J$ in (6.53).

**Problem 6.8**

Show that Simpson's rule is more accurate than expected. To do this, take $f(x) = ax^3 + bx^2 + cx + d$, and evaluate the integral:

$$I = \int_0^{2\Delta x} f(x)dx, \tag{6.100}$$

then show that $J$, obtained using the Simpson rule for the points $x_0 = 0$, $x_1 = \Delta x$ and $x_2 = 2\Delta x$ gives the same result. Evidently, then, the Simpson method will integrate polynomials of degree 3 perfectly, and this is better than its $n = 2$ derivation would imply.

## Problem 6.9

We can define "spherical" coordinates in four dimensions – take $\{x, y, z, u\}$ as the four Cartesian coordinates in four dimensions, and then the natural generalization of spherical coordinates introduces an additional angle $\psi \in [0, \pi]$ with:

$$x = r \sin \psi \sin \theta \cos \phi$$
$$y = r \sin \psi \sin \theta \sin \phi$$
$$z = r \sin \psi \cos \theta \quad\quad\quad (6.101)$$
$$u = r \cos \psi$$

so that $x^2 + y^2 + z^2 + u^2 = r^2$. Using the volume element, $d\tau = r^3 \sin^2 \psi \sin \theta \, d\psi \, d\theta \, d\phi \, dr$, find the volume of a four-dimensional sphere of radius $R$ (work by analogy with the three-dimensional volume of a sphere: $V \equiv \int_0^R \int_0^{2\pi} \int_0^\pi \int_0^\pi d\tau$).

## Lab problems

## Problem 6.10

Implement an integrator based on your $n = 3$ expansion. Use it to compute the integral $\int_0^1 \frac{\sin \theta}{\theta} d\theta$. Compare with the Simpson's rule method to make sure your integrator works.

## Problem 6.11

For the Gaussian density: $\rho(x) = Ae^{-x^2}$, find $A$ (by normalizing the density), and the first two moments of the distribution $\langle x \rangle$ and $\langle x^2 \rangle$ – use $x_{-\infty} = -10$ and $x_\infty = 10$ as the limits of integration with your integrator from Problem 6.10.

## Problem 6.12

A pendulum of length $L = 1$ m that hangs 1 m above the ground (so that the bob just misses the ground when the pendulum is hanging vertically) starts a height 0.5 m above the ground – what is the period of oscillation for this pendulum (use your integrator from Problem 6.10 applied to (6.26))? What would the "linearized" period be (i.e. the one you use in introductory physics, assuming simple harmonic motion)?

## Problem 6.13

Apply your integrator to (6.30) with $a = 2$, $Q/(4\pi\epsilon_0 a) = 1$ V – first choose a value for numerical infinity, and justify your choice. Find the numerical value of the potential at $z = 2a$, and check your result using the built-in function `BesselK`. Plot the curve of $V(z)$ for $z = 1 \longrightarrow 10$.

## Problem 6.14

Write a function to calculate the perimeter of an ellipse (with semi-major axis $a$, semi-minor axis $b$) – use the error control scheme described in Section 6.3.1 with Simpson's method to ensure that your result is no more than $\epsilon = 10^{-9}$ from the true value. Test your function on the limiting circle ($a = b \equiv R$) and line ($b = 0$) cases. What is the perimeter if $a = 2, b = 1$?

## Problem 6.15

Using your integral from Problem 6.1, find the electric field for a grid of points above the flat circular disk with radius $R = 2$ m. Take $x = -5 \longrightarrow 5$ with $\Delta x = 1$, $y = -5 \longrightarrow 5$ with $\Delta y = 1$ and $z = 1 \longrightarrow 5$ with $\Delta z = 1$ to define the grid, and plot the resulting vectors (make a function that returns the electric field vector for a given $\{x, y, z\}$, then use `Table` to generate the list, and `ListVectorPlot3D` to plot the list). Finally, extract the $z$ component's magnitude for $z = 1 \longrightarrow 10$ and plot this, together with the exact result from Problem 6.2. In all cases, take $\sigma/(4\pi \epsilon_0) = 1$ V/m. You can use Implementation 6.2 to perform the two-dimensional integration with $r$ and $\phi$ as your "$x$" and "$y$" integration variables. Start with few points in the $r$ and $\phi$ directions to make sure everything is working.

## Problem 6.16

Find the electric field above the ellipse (set up in Problem 6.4) with $a = 2, b = 1$ m, and $\lambda/(4\pi \epsilon_0) = 1$ V. Display your result by making a table of the electric field vectors at all locations $x = -2.5 \longrightarrow 2.5$ in steps of $\Delta x = 1$, $y = -2.5 \longrightarrow 2.5$ in steps of $\Delta y = 1$, and $z = 0.5 \longrightarrow 2$ with $\Delta z = 0.5$. Check your result in the limiting case of a circle, $a = b = 2$ m, using the known electric field component in the $\hat{z}$ direction – plot your numerical result on top of the correct analytical one for $z = 1 \longrightarrow 10$. Use your integrator from Problem 6.10, but don't worry about the accuracy condition.

## Problem 6.17

The one-dimensional Monte Carlo integration approximates integrals via $I \approx \sum_{i=1}^{N} f(x_i)\Delta x$ for random locations $\{x_i\}_{i=1}^{N}$ as in (6.80). Take the integral of $f(x) = x^5$ from $\frac{1}{2}$ to 1; we know that

$$I = \int_{\frac{1}{2}}^{1} x^5 dx = \frac{1}{6}x^6 \Big|_{x=\frac{1}{2}}^{1} = \frac{21}{128}. \tag{6.102}$$

Write a function that approximates $I$ using (6.80) as follows: For a given $N$, set $\Delta x = \frac{1}{2}/N$, and then select a random set of $N$ points and evaluate the sum in (6.80). Using your function, make a table of the approximation to $I$ for $N = 1 \longrightarrow 500$ (in steps of 1). Plot this data together with the actual value of $I$ to see the "convergence" of one-dimensional Monte Carlo.

## Problem 6.18

Find the volume of a two-, three-, and four-dimensional sphere of radius 2 using Monte Carlo integration (6.75). A sphere, in each of these cases, is defined to be the set of points such that:

$$\sum_{i=1}^{D} x_i^2 \leq R^2 \tag{6.103}$$

where $D$ is the dimension of the space, $R$ is the radius of the sphere, and the set $\{x_i\}_{i=1}^{D}$ are the coordinates (so, for example, in $D = 2$, we have $x_1 = x$, $x_2 = y$). You must find the "area" of the enclosing cube, $A_\square$ in (6.75), and then use the condition above to determine if a dart has hit or missed. How many darts do you need to get each volume correct to (just) one decimal place after the zero (with reasonable consistency)?

# 7

# Fourier transform

The Fourier transform plays a major role in the solution of ordinary and partial differential equations, and so is of theoretical utility in physics. Just as important is its applications to data analysis, and signal processing, making the Fourier transform an indispensible tool in experimental physics as well. The reason for its centrality is the connection the transform makes between temporal data and frequency information. A time-varying sinusoidal signal may have complicated patterns of repetition, and yet be made up of only a few frequencies. A signal with ten thousand data points may be describable in terms of two or three frequency values. We know how the frequencies combine, so we can reconstruct the signal.

Where do physical oscillations come from? The harmonic potential shows up far more than its springy roots would imply. Any local minimum in a complicated potential has an effective spring constant, and that effective spring constant determines a time scale of oscillation, i.e. a period (whose inverse defines the frequency). If we know the frequencies associated with a signal, then, we know something about the local minimum defined by some complicated potential.

There is also a human element to the Fourier transform. Much of the information we process can be described in terms of frequencies – both light and sound come to mind. So our ability to isolate and manipulate the individual frequencies of physiological input can lead to interesting changes in perception.

We'll start by looking at the Fourier transform and Fourier series as decompositions – any function can be (uniquely) decomposed into a basis made up of sine and cosine. The Fourier series is used when the function of interest is defined on a finite domain, or is repetitive. The Fourier transform is defined for all functions, and allows us to "decompose" functions on $-\infty \longrightarrow \infty$. The Fourier series is an infinite sum, while the Fourier transform is an integral (a very special type of integral, and that is why this chapter is placed as it is, directly after our study of integral approximation).

After a review of both the discrete and continuous projection of functions onto the complex exponential basis, and the utility of these, we'll look at how the Fourier transform can be computed numerically. Since a numerical Fourier transform is defined on a grid of values, the distinction between the Fourier series and the (discrete) Fourier transform will be blurred. Finally, we will look at how the numerical Fourier transform can be computed efficiently, culminating in a discussion of the aptly named "Fast Fourier Transform," or FFT. All of the explicit mathematical material here can be found in [3, 40, 41] (for example).

## 7.1 Fourier transform

The Fourier transform of a function $p(t)$ is defined to be:

$$P(f) = \int_{-\infty}^{\infty} p(t)e^{2\pi i f t} dt, \tag{7.1}$$

and the "inverse" Fourier transform is:

$$p(t) = \int_{-\infty}^{\infty} P(f)e^{-2\pi i f t} df. \tag{7.2}$$

These are the conventions we'll use, but there are a wide variety of normalizations and other variables that can be introduced in the definition – the particulars depend on context.

The first thing to notice about the above is the implicit equality of the "inverse" Fourier transform of $P(f)$ and $p(t)$ itself. This has implications – let's take a moment and see how the equality comes about. Explicitly, inserting $P(f)$ from (7.1) into (7.2) gives:

$$\int_{-\infty}^{\infty} \left[ \int_{-\infty}^{\infty} p(t')e^{2\pi i f t'} dt' \right] e^{-2\pi i f t} df = \int_{-\infty}^{\infty} p(t') \left[ \int_{-\infty}^{\infty} e^{2\pi i f(t'-t)} df \right] dt'. \tag{7.3}$$

The integral inside the square brackets is interesting – it has truncated form:

$$\int_{-F}^{F} e^{2\pi i f(t-t')} df = \frac{\sin(2\pi F(t-t'))}{\pi(t-t')}, \tag{7.4}$$

and it is the limit as $F \longrightarrow \infty$ that appears in (7.3). What does the function on the right in (7.4) look like in that case? Well, take $t = t'$, then

$$\frac{\sin(2\pi F(t-t'))}{\pi(t-t')} \bigg|_{t=t'} = 2F \tag{7.5}$$

and the value at $t = t'$ will go to infinity as $F$ goes to infinity. There is also a compelling property of the integral of (7.4) – let $T = t - t'$, then for any value of

$F$, we have

$$\int_{-\infty}^{\infty} \frac{\sin(2\pi FT)}{\pi T} dT = 1. \tag{7.6}$$

---

### The Dirac delta function

In fact it is the case that the function (7.4), for $F \longrightarrow \infty$ defines the delta distribution: $\delta(t - t')$, a clever function that is useful – the key elements of its definition are:

$$\delta(t - t') = \begin{cases} \infty & t = t' \\ 0 & t \neq t' \end{cases} \tag{7.7}$$

with the property that:

$$\int_{-\infty}^{\infty} \delta(t - t')dt = 1 = \int_{-\infty}^{\infty} \delta(t - t')dt', \tag{7.8}$$

and given the highly localized non-zero value of the $\delta$, we need only integrate over a small region:

$$\int_{t'-\epsilon}^{t'+\epsilon} \delta(t - t')dt = 1$$

$$\int_{t'}^{t'+\epsilon} \delta(t - t')dt = \frac{1}{2}, \tag{7.9}$$

where the second line is an expression of the symmetry of the delta function, $\delta(t' - t) = \delta(t - t')$. The most important property of the delta function, for us, is the integral of an arbitrary function $q(t)$ times the delta:

$$\int_{-\infty}^{\infty} \delta(t - t')q(t')dt' = q(t) \tag{7.10}$$

so that the $\delta$ picks out the value $q(t)$ from the integral.

The result of limit-taking and comparison with these fundamental properties of the $\delta$ function allows us to identify the limit, as $F \longrightarrow \infty$, of (7.4) with the delta function:

$$\int_{-\infty}^{\infty} e^{2\pi i f(t-t')}df = \delta(t - t'). \tag{7.11}$$

---

We can return to (7.3) with (7.11) in place to finish the job:

$$\int_{-\infty}^{\infty} p(t')\delta(t - t')dt' = p(t). \tag{7.12}$$

So the "inverse Fourier transform" of $P(f)$ is indeed $p(t)$.

**Example**

Take the function:

$$p(t) = A\cos(2\pi \bar{f} t) = \frac{A}{2}\left(e^{2\pi i \bar{f} t} + e^{-2\pi i \bar{f} t}\right) \quad (7.13)$$

for some definite value of $\bar{f}$ and arbitrary amplitude $A$. The Fourier transform is:

$$P(f) = \frac{1}{2} A \int_{-\infty}^{\infty} \left(e^{2\pi i (f+\bar{f})t} + e^{2\pi i (f-\bar{f})t}\right) dt \quad (7.14)$$

and the two terms in the integrand are of precisely the form (7.11), so

$$P(f) = \frac{1}{2} A \left(\delta(f + \bar{f}) + \delta(f - \bar{f})\right). \quad (7.15)$$

Taking the inverse Fourier transform returns the original cosine function with the frequency $\bar{f}$. The Fourier transform is telling us the frequency of a function of time – a cosine becomes a pair of "points" on the $P(f)$ side, symmetrically located at $f = -\bar{f}$ and $f = \bar{f}$.

There are a variety of properties of the Fourier transform that are mimicked in its discrete form. For example, real functions (which will be our primary concern) have Fourier transform with the property that $P(-f) = P(f)^*$. To see this, take:

$$P(f) = \int_{-\infty}^{\infty} p(t) e^{2\pi i f t} dt. \quad (7.16)$$

The complex conjugate of this function is:

$$P(f)^* = \int_{-\infty}^{\infty} p(t) e^{-2\pi i f t} dt = \int_{-\infty}^{\infty} p(t) e^{2\pi i (-f)t} dt = P(-f). \quad (7.17)$$

In the cosine example above, we have a real function that is also even (meaning $\cos(-x) = \cos(x)$), and that implies that $P(-f) = P(f)$ (can you see why from (7.17)?), a relation borne out by (7.15).

### Fourier transform of differential equations

The Fourier transform can be used to turn partial differential equations into ordinary ones (and ordinary differential equations into algebraic ones). A good example from [3] is to start with the one-dimensional wave equation in conservative form:

$$\frac{\partial u(x, t)}{\partial t} + \frac{\partial}{\partial x}(vu(x, t)) = 0 \quad (7.18)$$

with $u(x, 0) = g(x)$ given, and constant $v$. If we Fourier transform the temporal variable,

$$U(x, f) = \int_{-\infty}^{\infty} u(x, t) e^{2\pi i f t} dt \qquad (7.19)$$

then the inverse transform is:

$$u(x, t) = \int_{-\infty}^{\infty} U(x, f) e^{-2\pi i f t} df, \qquad (7.20)$$

and we can write the temporal derivative of $u(x, t)$ easily:

$$\frac{\partial u}{\partial t} = \int_{-\infty}^{\infty} U(x, f)(-2\pi i f) e^{-2\pi i f t} df. \qquad (7.21)$$

Using these in the wave equation, we have

$$\int_{-\infty}^{\infty} \left[ -2\pi i f U(x, f) + v \frac{\partial U(x, f)}{\partial x} \right] e^{-2\pi i f t} df = 0. \qquad (7.22)$$

The term in square brackets must vanish (by the independence of the exponential factors),

$$\frac{\partial U}{\partial x} = \frac{2\pi i f}{v} U \longrightarrow U(x, f) = F(f) e^{2\pi i f \frac{x}{v}}, \qquad (7.23)$$

where $F(f)$ is an arbitrary function of $f$. To understand the role of $F(f)$, take (7.20) with our solution inserted, and evaluate at $t = 0$:

$$u(x, 0) = \int_{-\infty}^{\infty} F(f) e^{2\pi i \frac{f}{v} x} df = \int_{-\infty}^{\infty} (F(-vz)v) e^{-2\pi i z x} dz = g(x), \qquad (7.24)$$

using $z = -f/v$ – but now we see that $F(-vz)v = G(z)$, the Fourier transform of $g(x)$ (with respect to its spatial argument, now). Then

$$u(x, t) = \int_{-\infty}^{\infty} F(f) e^{2\pi i f(\frac{x}{v} - t)} df = \int_{-\infty}^{\infty} (F(-vz)v) e^{-2\pi i z(x - vt)} dz \qquad (7.25)$$

$$= g(x - vt),$$

the usual result.

## 7.2 Power spectrum

The Fourier transform returns a continuous set of complex values that can be thought of as the coefficients associated with the decomposition of a function into the basis $e^{2\pi i f t}$, where $f$ is the frequency of the "mode." The size of these coefficients (the magnitude of the value $P(f)$ for some $f$) tells us about the "amount" of that frequency present in a signal.

One important application of the Fourier transform is in finding the power spectrum of a signal. The "total power" associated with a signal $p(t)$ is defined to be:

$$P \equiv \int_{-\infty}^{\infty} p(t)^* p(t) dt \qquad (7.26)$$

or, roughly, the area under the curve of $p(t)$'s magnitude (squared). If we take the Fourier transform of $p(t)$:

$$P(f) = \int_{-\infty}^{\infty} p(t) e^{2\pi i f t} dt, \qquad (7.27)$$

then we can compute the total area under $P(f)^* P(f)$ – that is

$$\begin{aligned}
\tilde{P} &= \int_{-\infty}^{\infty} \left[ \int_{-\infty}^{\infty} p(\bar{t})^* e^{-2\pi i f, \bar{t}} d\bar{t} \right] \left[ \int_{-\infty}^{\infty} p(t) e^{2\pi f t} dt \right] df \\
&= \int_{-\infty}^{\infty} \int_{-\infty}^{\infty} p(\bar{t})^* p(t) \underbrace{\left[ \int_{-\infty}^{\infty} e^{2\pi i f (t - \bar{t})} df \right]}_{= \delta(t - \bar{t})} dt \, d\bar{t} \\
&= \int_{-\infty}^{\infty} p(t)^* p(t) dt \\
&= P
\end{aligned} \qquad (7.28)$$

so that $P$, the total power associated with the signal, can be found either from $p(t)$ or $P(f)$ – this is known as Parseval's theorem.

We will be working with the Fourier transform, so that we can define the "power in an interval $df$ around $f$" as $dP = P(f)^* P(f) df$. That will tell us, roughly, how much of the signal exists in a range of frequencies, and we can use this to filter out certain frequencies, for example. So the "power spectrum" of a signal will be a plot of $|P(f)|^2$ versus $f$.

## 7.3 Fourier series

The fundamental result for Fourier series is that any periodic function $q(t)$ can be decomposed into an infinite set of exponentials. The requirement is $q(t + T) = q(t)$ for period $T$ – then we only need $q(t)$ defined for $t \in [0, T]$. The Fourier series expression in this case is:

$$q(t) = \sum_{n=-\infty}^{\infty} \alpha_n e^{2\pi i n \frac{t}{T}} \qquad (7.29)$$

for complex coefficients $\{\alpha_n\}_{n=-\infty}^{\infty}$. This equation represents a decomposition of the function $q(t)$ into an orthogonal basis set[1] – for integer $m$ and $n$, we have:

$$\frac{1}{T} \int_0^T e^{2\pi i(n-m)\frac{t}{T}} dt = \delta_{nm} \tag{7.31}$$

and the "Kronecker delta" is defined to be $\delta_{nm} = 1$ if $n = m$, else 0. Note the similarity between the Kronecker delta (both its definition and its integral representation) and the Dirac delta *function* from (7.11) – a provocative coincidence?

The orthogonality relation (7.31) allows us to identify the coefficients $\{\alpha_n\}_{n=-\infty}^{\infty}$ by integration – hit both sides of (7.29) with $\frac{1}{T} e^{-2\pi i m \frac{t}{T}}$ and integrate from $0 \longrightarrow T$:

$$\frac{1}{T} \int_0^T q(t) e^{-2\pi i m \frac{t}{T}} dt = \alpha_m. \tag{7.32}$$

So associated with each function $q(t)$ is an infinite set of coefficients $\{\alpha_n\}_{n=-\infty}^{\infty}$ (complex, in general) that decompose the function into the (complex) exponential basis. As we shall see in a moment, there is a relation between the set of coefficients $\{\alpha_n\}_{n=-\infty}^{\infty}$ and the values obtained as approximations to $P(f)$ from the discrete Fourier transform.

## 7.4 Discrete Fourier transform

Now let's return to a hypothetical continuous function $p(t)$ defined on a domain $t \in [0, T]$ where either the function $p(t)$ is periodic, or it is zero outside of this domain. If we sample the function with a time step $\Delta t$, so that we have a set of $N$ values $\{p_j = p(t_j)\}_{j=0}^{N-1}$ with $t_j = j\Delta t$ for $j = 0 \longrightarrow N - 1$ and $N\Delta t = T$, then the sum that approximates the Fourier transform integral is (using the simple rectangle approximation from Section 6.2):

$$P(f) = \int_{-\infty}^{\infty} p(t) e^{2\pi i f t} dt \approx \sum_{j=0}^{N-1} p_j e^{2\pi i f t_j} \Delta t. \tag{7.33}$$

We'll write $t_j = j\Delta t$ and then make a grid of frequency values: $f_k = k\Delta f$, where $k \in [-N/2, N/2]$ so that we don't lose any information. How should we choose $\Delta f$? Suppose we had one half cycle of a sine wave in an interval $\Delta t$ – then $p(t_1) = p(t_2) = 0$ (as does every other value on the grid). In fact, any number of half cycles in an interval $\Delta t$ evaluates to the same grid function (namely zero at all

---

[1] The orthogonality here is with respect to the inner product defined by:

$$a(t) \cdot b(t) \equiv \int_0^T a(t)^* b(t) dt, \tag{7.30}$$

for two complex functions $a(t)$ and $b(t)$.

grid points). So the maximum frequency we can resolve (the minimum resolvable period) corresponds to putting a half cycle of the wave in an interval:

$$2\pi f_{max} \Delta t = \pi \longrightarrow \boxed{f_{max} = \frac{1}{2\Delta t}}, \tag{7.34}$$

and this maximum frequency is called the "Nyquist frequency." Our maximum frequency for sampling is: $f_{N/2} = \frac{N}{2}\Delta f = \frac{1}{2\Delta t}$ and using this in (7.34) gives an expression for the frequency spacing:

$$\boxed{\Delta f = \frac{1}{N\Delta t} = \frac{1}{T}}. \tag{7.35}$$

Now we can rewrite (7.33) in terms of the discrete points $P_k$ approximating the Fourier transform values $P(f_k)$ (we remove the factor of $\Delta t$ from the integrand to make a dimensionless Fourier transform – you can reintroduce it to set the time scale),

$$P_k = \sum_{j=0}^{N-1} p_j e^{2\pi i f_k j \Delta t} = \sum_{j=0}^{N-1} p_j e^{2\pi i \frac{kj}{N}}, \tag{7.36}$$

using the fact that $f_k = k\Delta f = \frac{k}{N\Delta t}$. Notice, from the periodicity of the exponential,

$$e^{2\pi i \frac{(k+N)j}{N}} = e^{2\pi i \frac{kj}{N}} e^{2\pi i j} = e^{2\pi i \frac{kj}{N}}, \tag{7.37}$$

so that $P_{k+N} = P_k$. This allows us to embed the frequencies associated with $k = -N/2 \longrightarrow -1$ in the frequencies $k = N/2 \longrightarrow N-1$, and suggests the following ordering for the discrete Fourier transform:

$$P_k = \sum_{j=0}^{N-1} p_j e^{2\pi i \frac{kj}{N}} \quad \text{for } k = 0 \longrightarrow N-1 \tag{7.38}$$

with the understanding that:

$$\mathbf{P} \doteq \begin{pmatrix} P_0 = P(0) = P_N \\ P_1 = P(f_1) \\ \vdots \\ P_{N/2} = P(f_{N/2}) = P(f_{-N/2}) \\ P_{N/2+1} = P(f_{-N/2+1}) \\ \vdots \\ P_{N-1} = P(f_{-1}) \end{pmatrix}. \tag{7.39}$$

We obtain the negative frequency values together with the positive ones.

With this ordering, we can approximate the inverse discrete Fourier transform as:

$$p_j = \frac{1}{N} \sum_{k=0}^{N-1} P_k e^{-2\pi i \frac{jk}{N}} \tag{7.40}$$

and recover the approximation to $p(t_j)$ for $j = 0 \longrightarrow N - 1$.[2] In this form, we can make the connection between $P_k$ and the coefficients $\{\alpha_k\}_{k=-\infty}^{\infty}$ appearing in the Fourier series (7.29) – if we write $j = t_j/\Delta t$ with $N\Delta t = T$, then the above is:

$$p_j = \frac{1}{N} \sum_{k=0}^{N-1} P_k e^{-2\pi i k \frac{t_j}{N\Delta t}} = \frac{1}{N} \sum_{n=0}^{N-1} P_n e^{-2\pi i n \frac{t_j}{T}}. \tag{7.41}$$

The exponential appearing in the sum on the right is precisely the discretized form of $e^{-2\pi i n t/T}$ appearing in (7.29) (never mind the minus sign in the current expression, we sum over both positive and negative values for $n$, disguised by our ordering). The discrete Fourier transform approximations $P_k = P(f_k)$ then play the role of the decomposition coefficients $\{\alpha_k\}_{k=-N/2}^{N/2}$ in (7.29) as can be seen by comparing (7.36) with the discretized form of (7.32).

## 7.5 Recursion

The discrete Fourier transform:

$$P_k = \sum_{j=0}^{N-1} p_j e^{2\pi i \frac{kj}{N}} \tag{7.42}$$

has, owing to its periodicity, a natural recursive structure. We can exploit this, following [12, 34, 40], to generate a method to calculate the transform quickly.

Take $N = 2^n$, we can split $P_k$ into an even sum and an odd sum:

$$
\begin{aligned}
P_k &= \sum_{j=0,2,4,\ldots}^{N-2} p_j e^{2\pi i \frac{kj}{N}} + \sum_{j=1,3,5,\ldots}^{N-1} p_j e^{2\pi i \frac{kj}{N}} \\
&= \sum_{J=0}^{\frac{1}{2}N-1} p_{2J} e^{2\pi i \frac{kJ}{N/2}} + \sum_{J=0}^{\frac{1}{2}N-1} p_{2J+1} e^{2\pi i \frac{kJ}{N/2}} e^{2\pi i \frac{k}{N}} \\
&= \sum_{J=0}^{\frac{1}{2}N-1} p_{2J} e^{2\pi i \frac{kJ}{N/2}} + e^{2\pi i \frac{k}{N}} \sum_{J=0}^{\frac{1}{2}N-1} p_{2J+1} e^{2\pi i \frac{kJ}{N/2}}
\end{aligned}
\tag{7.43}
$$

---

[2] If you're wondering where the $1/N$ in (7.40) comes from, work Problem 7.11.

where the second line comes from defining $j = 2J$ in the first sum, and $j = 2J + 1$ in the second. Now each of the sums is itself a Fourier transform of half of the original data $p_j$, the first sum is the Fourier transform of the even data, the second is the Fourier transform of the odd data. We can define:

$$
\begin{aligned}
P_k^e &\equiv \sum_{j=0}^{\frac{1}{2}N-1} p_{2j} e^{2\pi i \frac{kj}{N/2}} \\
P_k^o &\equiv \sum_{j=0}^{\frac{1}{2}N-1} p_{2j+1} e^{2\pi i \frac{kj}{N/2}},
\end{aligned}
\tag{7.44}
$$

and then defining $\omega \equiv e^{2\pi i/N}$, we can write the recursion as

$$
P_k = P_k^e + \omega^k P_k^o.
\tag{7.45}
$$

Notice that each of the sums $P_k^e$ and $P_k^o$ is of length $2^{n-1}$, and thus can be split into even and odd portions themselves. We can continue dividing and using the recursion until we eventually end up with a single data point. We then have the "one-point" Fourier transform – if we have one data point ($N = 1$), the Fourier transform of it is itself:

$$
P_0 = \sum_{j=0}^{0} e^{2\pi i 0} p_0 = p_0.
\tag{7.46}
$$

---

**Example with $n = 2$**

We have a set of data: $\{p_0, p_1, p_2, p_3\}$, and our Fourier transform is:

$$
P_k = \sum_{j=0}^{3} p_j \omega^{kj} \quad \text{for } k = 0 \longrightarrow 3.
\tag{7.47}
$$

Then:

$$
P_k = \underbrace{\left(p_0 + p_2\omega^{2k}\right)}_{\equiv P_k^e} + \omega^k \underbrace{\left(p_1 + p_3\omega^{2k}\right)}_{\equiv P_k^o}
$$

$$
= \underbrace{p_0}_{\equiv P_k^{ee}} + \omega^{2k} \underbrace{p_2}_{\equiv P_k^{oe}} + \omega^k \left[ \underbrace{p_1}_{\equiv P_k^{eo}} + \omega^{2k} \underbrace{p_3}_{\equiv P_k^{oo}} \right].
\tag{7.48}
$$

The exponential prefactors are determined only by $k$, and are written as powers of $\omega$, so our final form is:

$$
\begin{aligned}
P_k &= p_0 + \omega^{2k} p_2 + \omega^k \left(p_1 + \omega^{2k} p_3\right) \\
&= \left(P_k^{ee} + \omega^{2k} P_k^{oe}\right) + \omega^k \left(P_k^{eo} + \omega^{2k} P_k^{oo}\right).
\end{aligned}
\tag{7.49}
$$

At the end of the recursion, we have individual elements, so it is clear when a recursive function should return (when it gets an input of length 1). There is, suggested in the above, a clever ordering of the original input vector that allows for easy application of the recursion – this is discussed in, for example, [40]. If we know to group, in this case, $p_0$ with $p_2$, and $p_1$ with $p_3$, then we can just multiply and add. The process amounts to knowing what the final single-value of a particular sequence of even/odd splittings will be, and using that as the "one-point" transform. We will ignore this ordering as it provides a refinement of the recursive structure, but is not fundamental to understanding the FFT.

## 7.6 FFT algorithm

We will take the recursion implied above at face value, forming successive splittings of the data and recombining them as we go. We need, then, to loop through the values of $k$ from $k = 0 \longrightarrow N - 1$ to set the value of $P_k$ given the even and odd sequences $P_k^e$ and $P_k^o$. That process can be simplified slightly by noting that:

$$P_{k+N/2} = P_{k+N/2}^e + \omega^{k+N/2} P_{k+N/2}^o = P_k^e + \omega^k e^{\pi i} P_k^o = P_k^e - \omega^k P_k^o, \quad (7.50)$$

so that we actually only need to loop through half the values of $k$ (lucky thing, since the $P_k^e$ and $P_k^o$ have only half the number of entries of the original data). In (7.50), we have used the periodicity of the $P_k^e$ and $P_k^o$, namely: $P_{k+N/2}^e = P_k^e$ and $P_{k+N/2}^o = P_k^o$ – these are themselves discrete transforms of length $N/2$, so the argument in (7.37) applies with $N \longrightarrow N/2$.

The code, modified from [12], to carry out the FFT is shown in Implementation 7.1 – all we have to do is provide a list (of length $N = 2^n$) called `invec`. The recursive function either splits the data in two and calls itself on each half, or, if the data consists of only one point, it returns the value it is given (the one point transform (7.46)). As the successive recursive calls return, `FFT` must put together the lists, reassembling according to (7.45) and (7.50) (this is done inside the `For` loop in Implementation 7.1).

In terms of the timing of the method, think of a single pass through the recursion itself – if we have $N$ elements in the vector, then the timing is given recursively by:

$$T(N) = 2T(N/2) + O(N) \quad (7.51)$$

where $T(N)$ is the "time" it takes to execute the function `FFT` for a vector of length $N$. The term $2T(N/2)$ denotes the fact that we call `FFT` twice, once for

**Implementation 7.1** Fast Fourier Transform

```
FFT[invec_] := Module[{k, elist, olist, netlist, eft, oft, n, omega},
  If[Length[invec] == 1,
    Return[invec];
  ];
  n = Length[invec];
  omega = 1;
  elist = Table[invec[[j]], {j, 1, n, 2}];
  olist = Table[invec[[j]], {j, 2, n, 2}];
  eft = FFT[elist];
  oft = FFT[olist];
  netlist = Table[0, {j, 0, n - 1}];
  For[k = 0, k ≤ n / 2 - 1, k = k + 1,
    netlist[[k + 1]] = eft[[k + 1]] + omega oft[[k + 1]];
    netlist[[k + n / 2 + 1]] = eft[[k + 1]] - omega oft[[k + 1]];
    omega = omega Exp[2 Pi I / n];
  ];
  Return[netlist];
]
```

even and odd lists, each of length $N/2$. The $O(N)$ appearing in (7.51) is there to account for the $N$ operations we must perform in addition to calling FFT, namely the reconstruction found in the For loop of Implementation 7.1. In order to solve the recursion, let $N = 2^n$, and then (7.51) can be written as:

$$
\begin{aligned}
T(n) &= 2T(n-1) + O(2^n) \\
&= 2\big(T(n-2) + O(2^{n-1})\big) + O(2^n) \\
&= 4T(n-2) + 2O(2^n) \\
&= 8T(n-3) + 3O(2^n) \\
&= 2^k T(n-k) + kO(2^n)
\end{aligned}
\tag{7.52}
$$

or, working our way all the way to $T(0)$ by setting $k = n$,

$$
T(n) = 2^n T(0) + nO(2^n).
\tag{7.53}
$$

The dominant term is $nO(2^n)$, so we say $T(n) = O(n2^n)$ or, in terms of the original length $N = 2^n$,

$$
T(N) = O(N \log_2(N)).
\tag{7.54}
$$

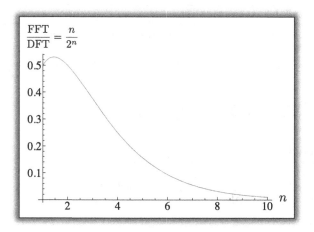

Figure 7.1 Timing ratio of the FFT to the DFT for input vectors of length $2^n$.

How many operations would be required if we simply started with the definition of the discrete Fourier transform (7.42)? We have $N$ values of $k$ to compute, and each one involves a sum of $N$ elements, so that to carry out the summation for all values of $k$ would require $N^2$ operations. If we think of $N = 2^n$, then the brute force approach would have timing $2^{2n}$ while the fast Fourier transform requires $n2^n$. The ratio of speed increase for the FFT is:

$$\frac{\text{Time for FFT}}{\text{Time for DFT}} = \frac{n2^n}{2^{2n}} = \frac{n}{2^n}, \tag{7.55}$$

and a plot of this timing ratio is shown in Figure 7.1. Even at modest values of $n$, the FFT takes a tiny fraction of the time required by the direct, discrete Fourier transform.

### *The inverse fast Fourier transform*

We can use virtually identical code to generate the elements of the inverse Fourier transform from (7.40). The only differences are the sign of the exponand in $\omega$, and the final division by $N$. To homogenize the treatment, we can write a single recursive function that takes either sign as input, and then calls FFT with whatever sign is appropriate (we'll denote the function that computes the inverse iFFT just to distinguish it).

**Example**

We can take signals with known frequency decompositions and check the performance of our FFT. To make contact with the continuous Fourier transform, let's take the function:

$$p(t) = \cos(2\pi(5)t) \tag{7.56}$$

on $t \in [0, 1]$. We'll discretize with $N = 2^{10}$ points, so that $\Delta t = 1/N$ and $\Delta f = 1$. Since $p(t)$ is periodic, $p(0) = p(1)$ so we need only include one of those points – for that reason, our discrete vector of data goes from $0 \longrightarrow 1 - \Delta t$. Plots of the function and its power spectrum (calculated using FFT) are shown in Figure 7.2.

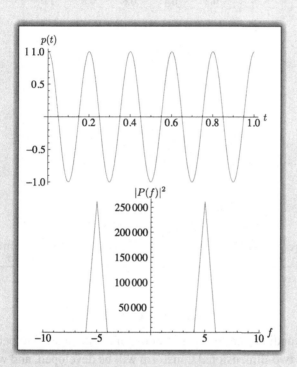

Figure 7.2 The function $p(t)$ from (7.56), along with its power spectrum calculated using Implementation 7.1. We expect delta spikes for $|P(f)|^2$ at $f = -5$ and $5$ – we get triangular approximations.

Note that to plot the Fourier transform requires some manipulation – from our ordering (7.39), we must rotate the entries of our vector to get the negative frequency values. The Mathematica command RotateRight shifts the upper half of the FFT output into the lower half easily. Then we attach, to each entry, the correct $f_j = j\Delta f$ for $j = -N/2 \longrightarrow N/2$ to generate plots like Figure 7.2.

We can add some damping to the original function – let:

$$p(t) = \cos(2\pi(5)t)\, e^{-4t}, \tag{7.57}$$

then the peaks in the Fourier transform broaden as shown in Figure 7.3.

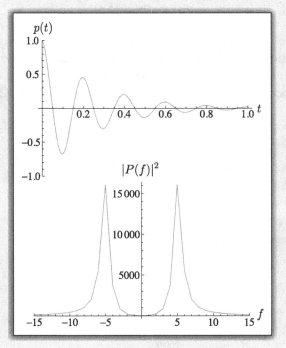

Figure 7.3  The function $p(t)$ from (7.57) and its power spectrum.

## 7.7 Applications

We'll start with audio filtering as an example of using the FFT to perform concrete tasks. Then, the two-dimensional transform will be developed, and written in terms of successive one-dimensional FFTs. Finally, we can use a "spectrogram" to look at the change in frequency dependence of a signal as a function of time.

### 7.7.1 Sound input

Mathematica's Import command gives direct access to the pressure data associated with sound input. We can import a "wave" file (the name refers to a specific format), and access the waveform directly. Using our FFT, we can Fourier transform this time series of data, and look at the relevant frequencies that make up the data.

From the discretized form of the power spectrum, we know that the coefficients that make up the discrete Fourier transform have the property that the "total power" of the signal is approximately

$$P \approx \sum_{j=0}^{N-1} |P_j|^2 \tag{7.58}$$

and so each coefficient in the Fourier transform (more properly, the ratio $|P_j|^2/P$) tells us the amount of total power that is associated with each frequency. Remember that we have the usual frequency ordering, so we have to be a little careful with the individual terms in the sum above.

As a final point – the FFT is optimized to work with input vectors that are of length $2^n$, and there are a variety of ways to generate, from data that is not of this length, appropriate $2^n$-sized vectors. In the current example, we will simply truncate our sample at length $2^n$, but one can also pad. For the sound input shown in the top plot of Figure 7.4, we have taken $N = 2^{16}$ points of sound data sampled at 41 kHz, so that the total time taken up by our sample is $T = N/41\,000\,\text{Hz} \approx 1.6\,\text{s}$. We have $\Delta t = T/N = 1/41\,000\,\text{Hz}$, and then $\Delta f = \frac{1}{Ndt} \approx 0.63\,\text{Hz}$. In the lower plot of Figure 7.4, we see the power spectrum for this sound data.

### 7.7.2 Low pass filter

We can perform filtering in frequency using the FFT – for our sound example, we could, for example, specify a frequency cutoff, and zero the components of the FFT above this cutoff, then perform an inverse FFT to recover a temporal signal. Since we have cut the frequencies higher than some provided cutoff, while allowing the lower frequencies to "pass" through, this filter is called a "low pass" filter. The code to implement the low pass filter is shown in Implementation 7.2 – you provide the input data `invec`, the frequency above which to cut, `cutabove`, and the total time associated with the data, `T`. This last input is important – remember that as we have it set up, the FFT is basically time-independent, so that in order to associate a frequency with a particular index in the FFT vector, we must know the total time of the sample.

There are a few indexing tricks that are performed in the low pass filter shown in Implementation 7.2, just to deal with Mathematica's "start-with-one" indexing. More important is the assumption that is made within the loop – sound data is real data, so that the Fourier components should satisfy $P(-f) = P(f)^*$. If you refer to the ordering in (7.39), you'll see that we are enforcing this relation exactly before transforming back (the second line in the `For` loop performs the conjugation and assignment). That ensures that our output waveform will be real, and playable

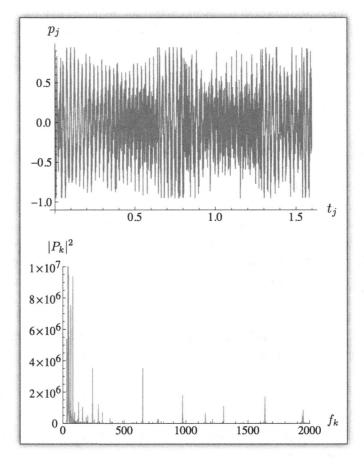

Figure 7.4  The input waveform (top) and the power spectrum components (the absolute values of the coefficients in the Fourier transform, squared).

**Implementation 7.2** Low pass filter

```
LoPass[invec_, cutabove_, T_] := Module[{outvec, NN, dt, df, cutindex, FTdata, index},
  NN = Length[invec];
  dt = N[T / NN];
  df = N[1 / (NN dt)];
  FTdata = FFT[N[invec]];
  cutindex = Round[cutabove / df];
  outvec = Table[0.0, {j, 1, Length[FTdata]}];
  outvec[[1]] = FTdata[[1]];
  For[index = 2, index ≤ cutindex, index = index + 1,
   outvec[[index]] = FTdata[[index]];
   outvec[[Length[FTdata] - (index - 2)]] = Conjugate[FTdata[[index]]];
  ];
  outvec = iFFT[outvec];
  Return[outvec];
 ]
```

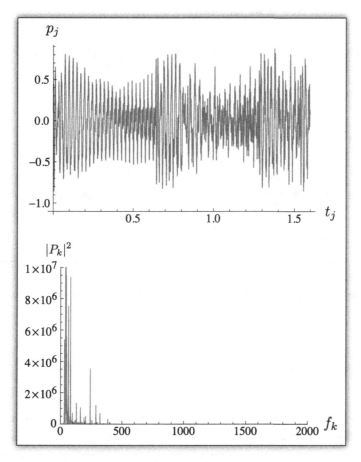

Figure 7.5 The power spectrum is filtered so that all frequencies above 440 Hz are cut, then the inverse Fourier transform gives us the top sound waveform.

using Mathematica's ListPlay command. Listen to the samples in the chapter notebook to hear a very bassey version of the Gravitea-Time song.

We can also "see" (in a synesthetic twist) the difference in the Fourier transform and post-low-pass sound sample. Using a frequency cut that allows only $f < 440$ Hz to pass, the plots analagous to Figure 7.4 are shown in Figure 7.5.

### 7.7.3 Two-dimensional Fourier transform

How should we define the Fourier transform of a function of more than one variable? Let's start by taking a function $u(x, y) = p(x)q(y)$,[3] a product of two

---

[3] We are using spatial variable names now, and that makes sense for the two-dimensional transform, since we don't generally have two time-like directions in which to decompose.

separate functions. Now if we let:

$$P(v_x) = \int_{-\infty}^{\infty} p(x) e^{2\pi i v_x x} dx \quad Q(v_y) = \int_{-\infty}^{\infty} q(y) e^{2\pi i v_y y} dy, \quad (7.59)$$

then we have a sensible form for $U(v_x, v_y)$, just

$$U(v_x, v_y) = P(v_x) P(v_y) = \left( \int_{-\infty}^{\infty} p(x) e^{2\pi i v_x x} dx \right) \left( \int_{-\infty}^{\infty} q(y) e^{2\pi i v_y y} dy \right).$$
$$(7.60)$$

We take the Fourier transform of the two independent functions separately – the variables $v_x$ and $v_y$ replace $f$ as the transformed variable(s), and these are referred to as "wave numbers" in this spatial context.

---

**Example**
The simplest decomposable function, from a Fourier transform point of view, would be

$$u(x, y) = u_0 e^{-2\pi i f_x x} e^{-2\pi i f_y y}, \quad (7.61)$$

and the Fourier transform is:

$$U(v_x, v_y) = u_0 \delta(v_x - f_x) \delta(v_y - f_y). \quad (7.62)$$

This expression ties together the values of $f_x$ and $f_y$ – the Fourier transform is zero unless *both* $v_x = f_x$ and $v_y = f_y$.

---

From this multiplicatively separable motivation, we can define the general two-dimensional (and higher) Fourier transform via:

$$U(v_x, v_y) = \int_{-\infty}^{\infty} \left( \int_{-\infty}^{\infty} u(x, y) e^{2\pi i v_y y} dy \right) e^{2\pi i v_x x} dx. \quad (7.63)$$

The inverse Fourier transform is then

$$u(x, y) = \int_{-\infty}^{\infty} \int_{-\infty}^{\infty} U(v_x, v_y) e^{-2\pi i v_x x} e^{-2\pi i v_y y} dv_x dv_y. \quad (7.64)$$

As for the computation of these functions, think of the two-dimensional integrals we computed in Section 6.4 – we just integrate in each direction separately. In (7.63), the integral in parentheses is itself a function of $x$ only, and then we integrate in $x$. If we take an array of values, $u_{jk} \equiv u(x_j, y_k)$, then we can accomplish the double integration by Fourier transforming first in $y$ (so performing an FFT on the rows of $u_{jk}$, viewed as a matrix), then in $x$ (FFTs of the resulting columns). The code to carry this out for two-dimensional data is shown in Implementation 7.3. You can see there that we act on each row with the FFT, and then transpose the output matrix, so that now the rows are oriented in the "$x$"-direction, we perform the FFT on each of these rows, transpose the resulting table and return it as the two-dimensional Fourier transform.

---

**Implementation 7.3** Two-dimensional Fourier transform

```
FFT2D[uxy_] := Module[{FFTy, FFTret},
  FFTy = Table[FFT[uxy[[k]]], {k, 1, Length[uxy]}];
  FFTy = Transpose[FFTy];
  FFTret = Table[FFT[FFTy[[k]]], {k, 1, Length[FFTy]}];
  FFTret = Transpose[FFTret];
  Return[FFTret];
]
```

---

**Example**

Take the function

$$u(x, y) = \sin(2\pi(4)x)\sin(2\pi(8)y) \tag{7.65}$$

as an example. In Figure 7.6, we plot the function itself, and the power spectrum associated with its two-dimensional Fourier transform (computed numerically). The repetition in the power spectrum comes from the real-valued data, just as in the one-dimensional case, but now we have reflections about both the $\nu_x$ and $\nu_y$ axes.

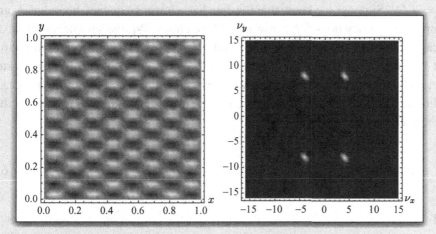

Figure 7.6 The test function (7.65) (left), along with the power spectrum of its Fourier transform (right).

### 7.7.4 Spectrogram

Many signals exhibit time-dependent frequencies – think of a function like:

$$p(t) = A \cos(2\pi(f_0 t)t) \tag{7.66}$$

where the frequency is itself a function of time, $f(t) = f_0 t$. How can we capture signals of this type using our Fourier transform? Given discretized data, we know how to compute the FFT, but probing the time-dependence of the frequencies means we have to generate multiple FFTs – we can do this by chopping up the signal into small pieces, calculating the Fourier transform of each piece, and then arraying these transforms from left to right as functions of time. An easy way to accomplish this ordering is to make contour strips of the power spectrum, and place them next to one another. This provides a visualization of the shift from one frequency to another as time goes on and is called a "spectrogram." These appear in a variety of contexts, notably the analysis of speech data (humans cause different resonant frequencies to be excited at different times – that's called "talking").

One has to be a little careful in setting up the spectrogram – the plan is to sample segments of the input data, itself discrete, and so we have to make sure that the local Fourier transform captures all the relevant frequencies. Suppose we have $N$ data points for the function $p(t)$, taking us from $t = 0 \longrightarrow T - \Delta t$ (again, we lose nothing in truncating the right-hand point, because of $T$ periodicity) so that $\Delta t = T/N$. Suppose we know that the data has a lower frequency bound of $f_\ell$ and an upper frequency bound $f_u$. How should we choose $N_s$, the sample length, so as to capture all frequencies between $f_\ell$ and $f_u$? The easiest approach is to require that $\Delta f_s = f_\ell$ and $\frac{1}{2}N_s \Delta f_s = f_u$, i.e. we take $f_\ell$ as our (sample) frequency step size, and demand that we include the upper frequency at the top of the range. In addition, we want $N = \alpha N_s$ with $\alpha \in \mathbb{Z} > 1$ so we get $\alpha$ samples to put together in the spectrogram.

$$\frac{1}{2}N_s f_\ell = f_u \longrightarrow N_s = \frac{2f_u}{f_\ell}, \tag{7.67}$$

and

$$\frac{1}{N_s \Delta t} = f_\ell \longrightarrow T = \frac{\alpha}{f_\ell}. \tag{7.68}$$

**Example**

Take the continuous function

$$p(t) = \cos(2\pi(4)t)(T - t) + \cos(2\pi(8)t)t, \tag{7.69}$$

where the signal changes linearly from sinusoidal oscillation with frequency 4 Hz, to 8 Hz. We'll set $\alpha = 16$, and, in order to capture all of the data, $f_\ell = 1$ Hz, $f_u = 16$ Hz, then:

$$N_s = \frac{2 \times 16}{1} = 32 \quad T = 16 \quad N = \alpha N_s = 512. \tag{7.70}$$

Now we discretize the function $p(t)$, Fourier transform blocks of size 32, take the power spectrum (and reduce to only the positive frequencies, since the data is real) and lay them down in strips on a contour plot – the result is shown in Figure 7.7.

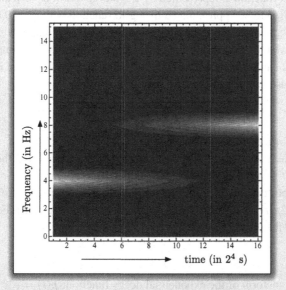

Figure 7.7 An example of a spectrogram, calculated using time-samples of length 32. We have tuned the sampling so as to clearly include the relevant frequencies, 4 and 8 Hz here.

Our choice to make individual, non-overlapping strips is motivated by simplicity and clarity. In fact, a spectrogram is generally defined using a sliding window (Fourier transform the first $M$ elements of the data, from $0 \longrightarrow M$, then move over one and Fourier transform the data from $1 \longrightarrow M + 1$, etc.).

## Further reading

1. Arfken, George B. & Hans J. Weber. *Mathematical Methods for Physicists*. Academic Press, 2001.
2. Cormen, Thomas H., Charles E. Leiserson, & Ronald L. Rivest. *Introduction to Algorithms*. The MIT Press, 1990.
3. Knuth, Donald. *Art of Computer Programming Vol 2. Semi-Numerical Algorithms*. Addison-Wesley, 1997.
4. Press, William H., Saul A. Teukolsky, William T. Vetterling, & Brian P. Flannery. *Numerical Recipes in C*. Cambridge University Press, 1996.
5. Riley, K. F., M. P. Hobson, & S. J. Bence. *Mathematical Methods for Physics and Engineering*. Cambridge University Press, 2002.

## Problems

### Problem 7.1

On a grid, the delta function is represented by a spike, so that if $q_0 = 0$ and $q_{\pm 1} = \pm \Delta q$, we would expect $\delta(q)$ to be approximated by a grid function with $\delta_j = 0$ for all $j \neq 0$ and $\delta_0 = \alpha$ where $\alpha$ is a number representing the infinite value that the delta should take on at zero. Given (7.8) what do you expect $\alpha$ to be in terms of $\Delta q$ (use the simplest integral approximation you can)? Check your prediction using the example from Figure 7.2 (the raw data for this case can be found in the chapter notebook).

### Problem 7.2

The Fourier transform of cosine was given in (7.15).

(a) Find the Fourier transform of $p(t) = A \sin(2\pi \bar{f} t)$. Note that it has the property that $P(-f) = P(f)^*$ (to be expected, since the signal $p(t)$ is real).

(b) What is the Fourier transform of a sinusoidal signal with a phase shift $\phi$:
$p(t) = A \sin(2\pi \bar{f} t + \phi)$? Verify that you recover the correct, sine, Fourier transform for $\phi = 0$ and the cosine case for $\phi = \pi/2$.

### Problem 7.3

Find the Fourier transform of the exponential decay function: $p(t) = e^{-at}$ for $t \geq 0$ (that is, assume $p(t) = 0$ for $t < 0$, and $\alpha \in \mathbb{R}$). Sketch the function $P(f)^* P(f)$ (the power spectrum).

### Problem 7.4

What is the Fourier transform of a Gaussian, $p(t) = \dfrac{1}{\sqrt{2\pi \sigma^2}} e^{-\frac{t^2}{2\sigma^2}}$?

### Problem 7.5

What is the Fourier transform of $p(t) = \delta(t - t')$? Sketch the power spectrum, $P(f)^* P(f)$. Notice that on the temporal side, the function is highly localized, while on the frequency side, it is the opposite of highly localized.

## Problem 7.6

What are the Fourier series coefficients, $\{\alpha_m\}_{m=-\infty}^{\infty}$ for:

(a) A pure cosine function $q(t) = A\cos(2\pi nt/T)$? What is the relation between $\alpha_m$ and $\alpha_{-m}$?

(b) A pure sine function: $q(t) = A\sin(2\pi nt/T)$? Again find the relation between the $+m$ and $-m$ coefficients.

## Problem 7.7

Start with the Fourier series written in terms of sine and cosine:

$$q(t) = \sum_{m=-\infty}^{\infty} \alpha_m(\cos(2\pi mt/T) + i\sin(2\pi mt/T)). \tag{7.71}$$

Write the sum entirely in terms of the positive (and zero) values of $m$ (so your sum will include $\alpha_m$ and $\alpha_{-m}$ as coefficients), and then show that:

(a) If $q(t) = -q(-t)$ (an antisymmetric function), then the full Fourier series collapses to the sine series:

$$q(t) = \sum_{m=1}^{\infty} \beta_m \sin(2\pi mt/T). \tag{7.72}$$

Express the $\beta_m$ in terms of the $\alpha_m$.

(b) If $q(t) = q(-t)$ (a symmetric function), only the cosine terms contribute:

$$q(t) = \beta_0 + \sum_{m=1}^{\infty} \beta_m \cos(2\pi mt/T). \tag{7.73}$$

Why does the cosine series get the $m = 0$ case? Express $\beta_0$ and all the $\beta_m$ in terms of $\alpha_0$ and $\alpha_m$.

## Problem 7.8

Find the coefficients $\{\alpha_m\}_{m=-\infty}^{\infty}$ in the Fourier series representation of a periodic square wave, with

$$q(t) = \begin{cases} A & 0 \le t \le T/2 \\ -A & T/2 \le t \le T \end{cases} \tag{7.74}$$

using (7.32). This is an antisymmetric function in $t$; write it in terms of a sine series as in (7.72). Using your coefficients in the sine series expression, generate the truncated function:

$$q_N(t) = \sum_{m=1}^{N} \beta_m \sin(2\pi mt/T) \tag{7.75}$$

with $A = 1$, $T = 1$, and plot this function for $t \in [0, T]$ and $N = 10, 50,$ and $100$.

**Problem 7.9**

Find the coefficients $\{\alpha_m\}_{m=-\infty}^{\infty}$ in the Fourier series representation of a periodic triangle wave, with

$$q(t) = \begin{cases} \frac{4A}{T}(T-t) - 3A & 0 \le t \le T/2 \\ \frac{4A}{T}(T+t) - 7A & T/2 \le t \le T \end{cases} \tag{7.76}$$

using (7.32). Verify that $\alpha_{-m} = \alpha_m$ and write the resulting cosine series (7.73).

**Problem 7.10**

If you look at the definition of $P(f)$, it should be clear that $P(0)$ is just the integral of $p(t)$, and similarly, $p(0) = \int_{-\infty}^{\infty} P(f)df$. For the discrete Fourier transform, the analogous statements can be generated using a rectangular approximation for the integrals. What, then, is the discrete Fourier transform of a constant (discrete) function $q(t_j) = A$ for $t_j = j\Delta t \in [0, T]$ (use (7.36), i.e., don't worry about $\Delta t$)? What is the (discrete) function, $q(t_j)$, whose discrete Fourier transform is $P(f_j) = \alpha$, a constant (use (7.40))?

**Problem 7.11**

By discretizing $p(t) = \int_{-\infty}^{\infty} P(f)e^{-2\pi i f t}df$, show that the inverse Fourier transform is indeed given by (7.40) (use a grid in $t$ space with the usual $\Delta t$, and your grid for frequency with $\Delta f = \frac{1}{N\Delta t}$ as usual) – note that you must re-introduce the factor of $\Delta t$ that we got rid of in defining $P_k$.

## Lab problems

**Problem 7.12**

Implement the discrete Fourier transform using the rectangular approximation (6.38) for the integral, as in (7.33). Take a signal $p(t) = \sin(2\pi(5)t)$ and make a vector of data with $N = 2^8$, $\Delta t = 2/N$ (giving a frequency spacing of $\Delta f = 1/2$). Calculate the Fourier transform of this data using your (slow) method – check against the FFT output. Make a power spectrum that shows peaks in the correct locations. Now replace the second half of your data vector with zeroes, so this time the entries of the data vector are

$$p_j = \begin{cases} \sin(2\pi(5)j\Delta t) & j \le 2^7 \\ 0 & j > 2^7 \end{cases}, \tag{7.77}$$

and again go out to $N = 2^8$ with $\Delta t = 2/N$. So you've just taken the original signal and replaced half of it with zero. What does the power spectrum look like now? This problem is meant to reassure you that "padding" data with zeroes (which is done to achieve data vectors of size $2^n$, for example) doesn't change much.

**Problem 7.13**

Download the data file "ch7.dat" from the book's website. If you Import it, you will find a list of elements – the first entry, for each element, is the "time," $t_j$ (in seconds, equally spaced with some $\Delta t$), and the second is a piece of data constructed from a linear combination of sines and cosines. Find the frequencies (in Hz) of each of these oscillatory functions and determine, for each frequency, whether it comes from a sine or cosine term. With what weight does each term occur (that is, what is the coefficient out front)? You have now completely reconstructed the function that generated the data; check that your function matches the data on the grid.

## Problem 7.14

Use the FFT to find the Fourier transform of the triangle wave described by (7.76). Take $A = 2, n = 2^9$ and $T = 1$ – plot the difference between your FFT values and the series coefficients you obtained in Problem 7.9.

## Problem 7.15

Rewrite the recursive FFT so that it can be used to generate the inverse Fourier transform. This amounts to introducing a minus sign in the appropriate location, and dividing by $N$ (the size of the input vector). Be careful *where* you do that division (remember, the recursive structure means that FFT gets called with vectors of all lengths from $N$ down to 1, dividing by two each time). Test your inverse Fourier transform on the transform from the previous problem and ensure that you recover the triangle wave with correct magnitude (you may need to use the Chop command to get rid of some small imaginary detritus); try plotting the result of the inverse FFT.

## Problem 7.16

Modify Implementation 7.2 to make a "band pass" filter, allowing only the frequencies between $f_\ell \longrightarrow f_h$ to pass (all other components of $P(f_k)$ are zero). Download the "GraviteaTime.wav" data from the website and load it as in the chapter notebook. Try out your band pass filter using $f_\ell = 440$ Hz and $f_h = 880$ Hz. Since the data is not of length $2^n$, you should find $n$ for which $2^n$ is closest to the actual data size, and then take only the first $2^n$ elements of the data (the FFT requires data to be a power of two).

## Problem 7.17

Use the FFT to find the Fourier transform of the square wave (7.74) with amplitude $A = 2$ – take $N = 2^9$ and set $\Delta t$ so that $\Delta f = 1/4$. Compare this with the coefficients you got for the Fourier series in Problem 7.8; be careful, in that series, the coefficients are indexed by $m$, an integer, corresponding to the integer indices in your FFT (i.e. the raw $-N/2 \longrightarrow N/2$, rather than the frequencies $-N\Delta f/2 \longrightarrow N\Delta f/2$). You should notice that the negative and positive frequencies are flipped; this is a result of the minus sign that occurs in the FFT exponand, absent from the Fourier series definition (7.29) (a formality of the definition, just multiply your series coefficients by $-1$).

(a) Plot the exact Fourier series coefficients together with the (imaginary portion of) your FFT values.

(b) Use your band pass filter to allow only $0 \longrightarrow 10$ Hz to pass; this should recover the oscillatory step that you generated by direct truncation in Problem 7.8.

**Problem 7.18**

Find the two-dimensional Fourier transform of:

$$u(x, y) = \sin\left(2\pi(16)\left((x - 1/2)^2 + (y - 1/2)^2\right)\right),\qquad(7.78)$$

using $N_x = 64 = N_y$ points in the $x$ and $y$ directions, with $\Delta x = \Delta y = \frac{1}{N_x}$ (so that the region is square with $x, y \in [0, 1]$). Plot the power spectrum associated with the Fourier transform with the correct wave numbers along the axes.

**Problem 7.19**

Download the image "ch7image.mat" from the book website. Read it in using the `Import` command, and view it as a matrix. Assuming the width and height are both unity, take the two-dimensional Fourier transform of this image matrix and plot its power spectrum.

**Problem 7.20**

We can make a time-varying signal with specific frequencies (think of a piece of music, where notes are played for finite intervals, then changed) and explore that signal using the spectrogram – let the signal be:

$$p(t) = \begin{cases} 2\cos(2\pi\,4t) + \cos(2\pi\,8t) & t < 4 \\ 4\cos(2\pi\,10t) & 4 \leq t < 8 \\ 2\cos(2\pi\,6t) & 8 \leq t < 12 \\ \cos(2\pi\,3t) + \cos(2\pi\,7t) & t \geq 12 \end{cases}.\qquad(7.79)$$

Take $T = 16$ (s) as the total time, and set $\Delta t = 1/32$. Make the discretized signal (for $t = 0 \longrightarrow T - \Delta t$), and send it through the `SpectroGram` function from the chapter notebook (or write your own) with a low frequency of 1 Hz and a high of 16 Hz.

# 8

# Harmonic oscillators

This chapter is an intermission. So far, we have studied the numerical solution of differential equations (both ordinary and partial) and approximation of integrals. Those two broad topics cover a lot of physical applications. The following four chapters deal with linear algebra, in particular, the numerical process of matrix inversion (which we have used already to solve the discrete form of Poisson's problem in Chapter 4) and finding the eigenvalues and eigenvectors of a matrix. This latter subject has not yet appeared directly, and the current chapter is meant, in part, to introduce the need for a numerical method that calculates the eigenvalues and eigenvectors of a matrix.

Specifically, we will study the solution to the linear equation of motion:

$$\ddot{\mathbf{X}} = \mathbb{Q}\mathbf{X} + \mathbf{A} \tag{8.1}$$

(with appropriate initial conditions) for a vector $\mathbf{X}(t) \in \mathbb{R}^n$, a symmetric matrix $\mathbb{Q} \in \mathbb{R}^{n \times n}$ and a constant vector $\mathbf{A} \in \mathbb{R}^n$. This equation of motion comes from a set of $n$ balls connected by springs in one dimension, or as the linearization of the ball and spring problem in higher dimensions. In fact, this same equation of motion represents the linearization of almost any potential about an equilibrium, and so its solution describes the local dynamics for an array of physical problems.

## 8.1 Physical motivation

We can think specifically of ball and spring models; those lead pretty directly to quadratic potentials and linear forces, so are a natural starting point in developing an example of an equation of motion of the form (8.1). But there are other places in which linear(ized) forces naturally arise.

### *8.1.1  Linear chains*

A set of balls and springs can be used to approximate continuous distributions of mass that have local interactions that look "springy." We can take a continuum limit of the balls and springs to obtain the wave equation describing the longitudinal motion of disturbances in mass density for a continuous spring, like a slinky. So by studying the one-dimensional oscillation of balls and springs, we can get a discrete (in the sense of non-continuum) idea of the motion of a continuous system.

In higher dimensions, the equations of motion for balls connected by springs has additional, nonlinear contributions, as we shall see. But it is possible to linearize those equations of motion to obtain (8.1), and in that setting, we can study the transverse oscillation of chains made up of balls and springs. In addition, we can make "sheets" of balls connected by springs and use those to study discrete approximations to stretched membranes.

### *8.1.2  Molecular dynamics*

Molecular dynamics (MD) refers to solving Newton's second law for a particular set of forces. In this approximation to microscopic motion, we take atoms (or collections of atoms) and represent them as point particles with appropriate mass and charge (for ions). The motion of the atoms is governed by electrostatic interaction, and a collection of effective forces. If there is a surrounding environment, there can be additional contributions to approximate its influence (sometimes random forces are included, for example).

The full MD potential has some familiar elements – bonds between atoms have an effect that can be mimicked by quadratic Hooke's law potentials. Chemical bonding is intrinsically quantum mechanical, but from a macroscopic point of view, bonds are well-approximated by (pretty stiff) springs. There are also "angle" oscillations, Van der Waals interaction, "dihedral" angle oscillations, and on and on. Most of the components in the potential lead to complicated and expensive (to calculate) nonlinear forces, but we know how to handle those in a Verlet setting.[1] The real problem is the broad range of time scale – bond oscillations occur at the femto-second level, while the large-scale structural motion of proteins occurs on the order of seconds. So using a time step that is small enough to capture the bond oscillations means we cannot run out far enough to see the large-scale motion.

Since the crystallographic states of some proteins are known, there are well-defined equilibrium configurations, and so linearizing the full MD potential and

---

[1] Verlet and its variants are preferred for this type of problem since they conserve energy well, and only require one evaluation of the forces per step.

probing the motion near those equilibria is an option (that leads to (8.1)). Another approach is the "elastic network model," in which the MD potential is thrown out entirely and replaced by springs connecting all atoms (within a certain radius) together (also resulting in (8.1)). In either case, the hope is to gain relevant information by replacing the natural, nonlinear (effective) forces in the original problem with simpler ones.

### 8.1.3 Harmonic approximation

The harmonic oscillator potential provides a tractable model for more complicated physics since it leads to linear forces in the vicinity of minima (of the potential). Almost any physical force can be written in terms of a quadratic potential if we are close to the equilibrium configuration (defined as that configuration for which the total force is zero).

Think of the Taylor expansion of a potential $U(x)$ about a minimum, $x_0$:

$$U(x) = U(x_0) + \left. \frac{\partial U}{\partial x} \right|_{x=x_0} (x - x_0) + \frac{1}{2} \left. \frac{\partial^2 U}{\partial x^2} \right|_{x=x_0} (x - x_0)^2 + \cdots \qquad (8.2)$$

The first term is a constant – that sets the scale of the potential energy, but will not change the equations of motion. The second term, involving the first derivative, is zero by the assumption that $x_0$ is a minimum. That gives the second derivative term as the first dynamical contribution. The second derivative of $U$, evaluated at $x_0$ tells us the effective spring constant to use – we can calculate the period of oscillation associated with motion near this minimum in the usual way: for a spring with constant $k$, the period of motion is $T = 2\pi \sqrt{m/k}$. In the context of (8.2), the "spring constant" is $U''(x_0)$, so $T = 2\pi \sqrt{m/U''(x_0)}$.

As a pictorial example in one dimension, consider the quartic potential $U$ shown in Figure 8.1. At the two local minima, we compute the second derivative of $U$ and take that to be the $k$ in a local quadratic approximation: $\bar{U}(x) = \frac{1}{2}kx^2$. To the extent that this is a good approximation, we can probe the local dynamics of a complicated potential using this exactly solvable, simplified, quadratic potential.

## 8.2 Three balls and two springs

Let's move on to a specific setting, where we can see the details of (8.1) emerge as a relevant equation of motion. Consider the following physical system in one dimension: three balls of equal mass $m$ connected by two identical springs with spring constant $k$ and equilibrium spacing $a$. The system, shown in Figure 8.2, can be described by three coordinates, the $x$ locations of the three balls as a function of time: $x_{-1}(t)$, $x_0(t)$ and $x_1(t)$. Our goal is to find the trajectories of these balls

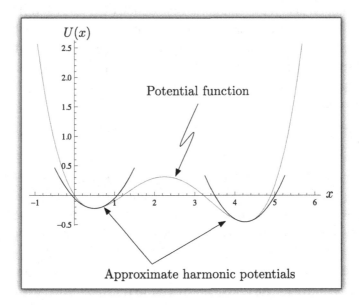

Figure 8.1 A quartic potential with two local minima. The parabolas that approximate the potential near each of these minima are drawn on top. In the vicinity of
the minima of the actual potential, then, we can use the quadratic approximation.

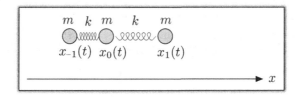

Figure 8.2 Three balls are connected by two springs.

given some initial configuration and velocity (or, alternatively, initial and final
configurations).

### 8.2.1 Equations of motion

Referring to Figure 8.2, we can construct the equation of motion for each of the
balls using a linear restoring force (Hooke's law). For the ball on the left, with
coordinate labelled by $x_{-1}(t)$, we have one spring. The separation distance of
interest here is the difference $x_0(t) - x_{-1}(t)$. We compare this separation distance
to the equilibrium length of the spring – if $x_0(t) - x_{-1}(t)$ is *less than* $a$, then the
force on the left-most ball will be to the left, so should be negative. We can get the

sign right by setting:

$$m\ddot{x}_{-1}(t) = k(x_0(t) - x_{-1}(t) - a), \tag{8.3}$$

and it is clear that if $x_0(t) - x_{-1}(t)$ is *greater* than $a$, the force (right-hand side of the above) is positive, indicating that it points to the right.

For the ball on the right, the same qualitative analysis holds – we have one spring, the force will depend on $x_1(t) - x_0(t) - a$, and when this quantity is less than zero (indicating that $x_1(t) - x_0(t) < a$), we want a force on the ball that points to the right, so

$$m\ddot{x}_1(t) = -k(x_1(t) - x_0(t) - a). \tag{8.4}$$

Finally, for the middle ball, we can either think about each of the springs connected to it individually, or appeal to Newton's third law – the force due to the spring on the left must be equal and opposite to the force of the spring on the left on the left-most ball, and similarly for the spring on the right. This observation immediately gives:

$$m\ddot{x}_0(t) = -k(x_0(t) - x_{-1}(t) - a) + k(x_1(t) - x_0(t) - a). \tag{8.5}$$

---

**Spring Lagrangian**

In the interest of checking our equations of motion (8.3), (8.4), (8.5), we will generate the same equations using the Euler–Lagrange approach. The Lagrangian for this problem, $L = T - U$, is:

$$L = \frac{1}{2}m\left(\dot{x}_{-1}^2 + \dot{x}_0^2 + \dot{x}_1^2\right) - \left[\frac{1}{2}k(x_1 - x_0 - a)^2 + \frac{1}{2}k(x_0 - x_{-1} - a)^2\right]. \tag{8.6}$$

The equations of motion are given, as usual, by

$$\left(\frac{d}{dt}\frac{\partial L}{\partial \dot{x}_0} - \frac{\partial L}{\partial x_0}\right) = 0, \tag{8.7}$$

for the $x_0$ coordinate, similarly for the other two. All together, they look like:

$$m\ddot{x}_{-1} = k(x_0 - x_{-1} - a)$$
$$m\ddot{x}_0 = -k(x_0 - x_{-1} - a) + k(x_1 - x_0 - a) \tag{8.8}$$
$$m\ddot{x}_1 = -k(x_1 - x_0 - a).$$

---

Define the vector **X** by

$$\mathbf{X} = \begin{pmatrix} x_{-1}(t) \\ x_0(t) \\ x_1(t) \end{pmatrix}, \tag{8.9}$$

then we can write (8.3), (8.4), and (8.5), in terms of vectors and a matrix $\mathbb{Q}$ defined by:

$$\underbrace{\begin{pmatrix} \ddot{x}_{-1}(t) \\ \ddot{x}_0(t) \\ \ddot{x}_1(t) \end{pmatrix}}_{\ddot{\mathbf{X}}} = \underbrace{\begin{pmatrix} -\frac{k}{m} & \frac{k}{m} & 0 \\ \frac{k}{m} & -\frac{2k}{m} & \frac{k}{m} \\ 0 & \frac{k}{m} & -\frac{k}{m} \end{pmatrix}}_{\mathbb{Q}} \underbrace{\begin{pmatrix} x_{-1}(t) \\ x_0(t) \\ x_1(t) \end{pmatrix}}_{\mathbf{X}} + \underbrace{\begin{pmatrix} -\frac{ka}{m} \\ 0 \\ \frac{ka}{m} \end{pmatrix}}_{\mathbf{A}}. \qquad (8.10)$$

In terms of vectors and matrices, then, we have

$$\ddot{\mathbf{X}} = \mathbb{Q}\mathbf{X} + \mathbf{A}. \qquad (8.11)$$

We now have the desired equation coming from a concrete physical system. The approach we will take in solving (8.11) applies to any equation of this form, but we will continue to refer to the simple toy problem as we go.

### 8.2.2 Solution

How should we go about solving (8.11)? It's great that we can write it in such compact form, but is it useful? Yes, there are a variety of routes that we can take, the first is to define the notion of matrix exponentiation: $e^{\mathbb{Q}t}$ is a natural candidate for expressing $\mathbf{X}(t)$. That approach is tied directly to the second route (see Problem 8.8): Turn this set of three coupled linear ODEs into three *un*-coupled linear ODEs by a change of variables. The question of how to define the "new" coordinates is answered by the eigenvectors and eigenvalues of the matrix $\mathbb{Q}$. In this case, with only three balls, finding the eigenvalues and eigenvectors of $\mathbb{Q}$ is relatively easy (if tedious). But what if we had a hundred balls? The pattern of the corresponding matrix $\mathbb{Q}$ is relatively clear, so we can imagine constructing $\mathbb{Q} \in \mathbb{R}^{100 \times 100}$, but finding its eigenvalues and eigenvectors?

Remember, if we want to make a physical prediction about this system, we must be able to find these values, and a method for finding the eigenvalues and eigenvectors of a matrix is given in Chapter 10. We'll work "by hand" for now, to show the logic, and to demonstrate that knowing the eigenvalues and eigenvectors solves the problem (and provides physically interesting information), but it is important to remember that just because you can theoretically solve a problem if you *had* the relevant numbers doesn't mean they are easy to obtain.

The key to the de-coupling is our ability to solve the "eigenvalue problem" for the matrix $\mathbb{Q}$ – we need to find the set of vectors $\{\mathbf{v}_i\}_{i=1}^3$ (in this case, but in general there will be $n$ eigenvectors for $\mathbb{Q} \in \mathbb{R}^{n \times n}$) and numbers $\{\lambda_i\}_{i=1}^3$ such that:

$$\mathbb{Q}\mathbf{v}_i = \lambda_i \mathbf{v}_i \quad \text{for } i = 1, 2, 3. \qquad (8.12)$$

The vector $\mathbf{v}_i$ is evidently very special, and indeed is specialized to a particular matrix $\mathbb{Q}$ – it is a vector that, when multiplied by $\mathbb{Q}$, is scaled by a number $\lambda_i$. Because $\mathbb{Q}$ is symmetric, we expect to find three orthogonal eigenvectors $\{\mathbf{v}_i\}_{i=1}^3$ with three real eigenvalues. Suppose we have them, $\mathbf{v}_1$, $\mathbf{v}_2$, and $\mathbf{v}_3$ with $\lambda_1$, $\lambda_2$ and $\lambda_3$ satisfying (8.12). We can summarize the defining statement of eigenvectors/values in matrix form by introducing

$$\mathbb{V} = \left[\, \mathbf{v}_1 \,\middle|\, \mathbf{v}_2 \,\middle|\, \mathbf{v}_3 \,\right], \tag{8.13}$$

where the columns of $\mathbb{V}$ are made up of the vectors $\mathbf{v}_1$, $\mathbf{v}_2$, and $\mathbf{v}_3$. Then from the definition of matrix–matrix multiplication, we have:

$$\mathbb{Q}\mathbb{V} = \left[\, \lambda_1\mathbf{v}_1 \,\middle|\, \lambda_2\mathbf{v}_2 \,\middle|\, \lambda_3\mathbf{v}_3 \,\right] = \mathbb{V}\mathbb{L}$$

$$\mathbb{L} \doteq \begin{pmatrix} \lambda_1 & 0 & 0 \\ 0 & \lambda_2 & 0 \\ 0 & 0 & \lambda_3 \end{pmatrix}. \tag{8.14}$$

Since the eigenvectors of a symmetric matrix are orthogonal and normalizable: $\mathbf{v}_1 \cdot \mathbf{v}_1 = 1$, $\mathbf{v}_1 \cdot \mathbf{v}_2 = 0$, $\mathbf{v}_1 \cdot \mathbf{v}_3 = 0$, $\mathbf{v}_2 \cdot \mathbf{v}_2 = 1$, $\mathbf{v}_2 \cdot \mathbf{v}_3 = 0$, and $\mathbf{v}_3 \cdot \mathbf{v}_3 = 1$, we have the statement, again in matrix form, that $\mathbb{V}^T\mathbb{V} = \mathbb{I}$ (so that the inverse of $\mathbb{V}$ is $\mathbb{V}^T$, and $\mathbb{V}$ is an "orthogonal" matrix).

The orthogonality of $\mathbb{V}$ makes it easy to invert the first equation in (8.14):

$$\mathbb{Q}\mathbb{V} = \mathbb{V}\mathbb{L} \longrightarrow \mathbb{Q} = \mathbb{V}\mathbb{L}\mathbb{V}^T, \tag{8.15}$$

which provides a "decomposition" of $\mathbb{Q}$ in terms of an orthogonal matrix $\mathbb{V}$ and a diagonal matrix $\mathbb{L}$. Think of the utility of such a decomposition in our current problem:

$$\ddot{\mathbf{X}} = \mathbb{Q}\mathbf{X} + \mathbf{A} \longrightarrow \ddot{\mathbf{X}} = \mathbb{V}\mathbb{L}\mathbb{V}^T\mathbf{X} + \mathbf{A}. \tag{8.16}$$

Keep in mind that $\mathbb{V}$ is independent of time, and has inverse $\mathbb{V}^T$, so we can multiply through on the left by $\mathbb{V}^T$ and write

$$\frac{d^2}{dt^2}(\mathbb{V}^T\mathbf{X}) = \mathbb{L}(\mathbb{V}^T\mathbf{X}) + \mathbb{V}^T\mathbf{A}. \tag{8.17}$$

If we define $\mathbf{\Psi} \equiv \mathbb{V}^T\mathbf{X}$ and $\mathbf{B} \equiv \mathbb{V}^T\mathbf{A}$, then the above can be written as:

$$\ddot{\mathbf{\Psi}} = \mathbb{L}\mathbf{\Psi} + \mathbf{B}. \tag{8.18}$$

The advantage? Well, we have decoupled the system – $\mathbb{L}$ is diagonal, so these three equations read:

$$\ddot{\Psi}_1 = \lambda_1 \Psi_1 + B_1$$
$$\ddot{\Psi}_2 = \lambda_2 \Psi_2 + B_2 \qquad (8.19)$$
$$\ddot{\Psi}_3 = \lambda_3 \Psi_3 + B_3.$$

Each of these three, now independent, equations of motion, has solution ($i = 1, 2, 3$):

$$\Psi_i = \begin{cases} \alpha_i e^{\sqrt{\lambda_i}t} + \beta_i e^{-\sqrt{\lambda_i}t} - \frac{B_i}{\lambda_i} & \lambda_i \neq 0 \\ \alpha_i t + \beta_i + \frac{1}{2}B_i t^2 & \lambda_i = 0 \end{cases}. \qquad (8.20)$$

Finally, we need to set the coefficients $\{\alpha_i, \beta_i\}$ in (8.20) – these are fixed by the initial conditions (or boundary conditions). Suppose we are given the initial position, $\mathbf{X}(0)$ and velocity $\dot{\mathbf{X}}(0)$ (the initial position and velocity of each of the three particles is provided). When we move to $\boldsymbol{\Psi}$, we get:

$$\boldsymbol{\Psi}(0) = \mathbb{V}^T \mathbf{X}(0) \qquad \dot{\boldsymbol{\Psi}}(0) = \mathbb{V}^T \dot{\mathbf{X}}(0) \qquad (8.21)$$

and we can use these constant values to set the undetermined coefficients from (8.20). Once $\boldsymbol{\Psi}(t)$ is fixed, we can recover the "real" space in which motion occurs, $\mathbf{X}(t)$, since $\boldsymbol{\Psi} \equiv \mathbb{V}^T \mathbf{X} \longrightarrow \mathbf{X} = \mathbb{V}\boldsymbol{\Psi}$.

---

**Energy motivation**

The quadratic potential energy for this pair of springs also focuses our attention on the decomposition of $\mathbf{X}$ in terms of the eigenvectors of $\mathbb{Q}$. The potential energy is

$$U = \frac{1}{2}\left(k(x_0 - x_{-1} - a)^2 + k(x_1 - x_0 - a)^2\right), \qquad (8.22)$$

and we can write this expression in terms of the matrix $\mathbb{Q}$ and vector $\mathbf{A}$ from (8.10):

$$U = -\frac{1}{2}m\left(\mathbf{X}^T \mathbb{Q}\mathbf{X} + \frac{m^2}{k^2}\mathbf{A}^T \mathbb{Q}\mathbf{A}\right) + \frac{m^2}{k}\mathbf{X}^T \mathbb{Q}\mathbf{A} \qquad (8.23)$$

where the minus sign and factors of $m$ and $k$ just come from the original definitions of $\mathbb{Q}$ and $\mathbf{A}$ in (8.10). Now if we set $\mathbf{X} = \mathbb{V}\boldsymbol{\Psi}$, then we can write the energy as

$$U = -\frac{1}{2}m\left(\boldsymbol{\Psi}^T \mathbb{V}^T \mathbb{Q}\mathbb{V}\boldsymbol{\Psi} + \frac{m^2}{k^2}\mathbf{A}^T \mathbb{Q}\mathbf{A}\right) + \frac{m^2}{k}\boldsymbol{\Psi}^T \mathbb{V}^T \mathbb{Q}\mathbf{A}$$

$$= -\frac{1}{2}m\boldsymbol{\Psi}^T \mathbb{L}\boldsymbol{\Psi} + \frac{m^2}{k}\boldsymbol{\Psi}^T \mathbb{L}\mathbb{V}^T \mathbf{A} - \frac{1}{2}\frac{m^3}{k^2}\mathbf{A}\mathbb{Q}\mathbf{A}. \qquad (8.24)$$

In this form, we see that the equations of motion will automatically be decoupled in the $\Psi_1$, $\Psi_2$ and $\Psi_3$ "generalized coordinates." The first term represents the potential energy of three uncoupled harmonic oscillators, the second term, linear in $\boldsymbol{\Psi}$ will give back constant offsets for each of the equations of motion (representing constant forces, note that $\mathbb{V}^T \mathbf{A}$ is just the projection of $\mathbf{A}$ into $\boldsymbol{\Psi}$ space, what we've called $\mathbf{B}$) and the final term is a constant that will not change the equations of motion (we're re-setting the zero point energy of the potential).

So we could have started with the potential energy itself, and asked for what set of coordinates the potential energy could be viewed as three uncoupled oscillators – that would have provided the motivation for our interest in the eigenvectors/values of $\mathbb{Q}$.

## 8.3 Solution for a particular case

The first step to a solution is to find the eigenvectors and eigenvalues of $\mathbb{Q}$ from (8.10).[2] These are:

$$\lambda_1 = 0 \quad \mathbf{v}_1 = \frac{1}{\sqrt{3}} \begin{pmatrix} 1 \\ 1 \\ 1 \end{pmatrix}$$

$$\lambda_2 = -\frac{3k}{m} \quad \mathbf{v}_2 = \frac{1}{\sqrt{6}} \begin{pmatrix} 1 \\ -2 \\ 1 \end{pmatrix} \tag{8.25}$$

$$\lambda_3 = -\frac{k}{m} \quad \mathbf{v}_3 = \frac{1}{\sqrt{2}} \begin{pmatrix} -1 \\ 0 \\ 1 \end{pmatrix}.$$

Turning to a specific setting – let $k/m = 1$ m/s$^2$ and $a = 1$ m$^3$. In this case:[4]

$$\mathbb{Q} \doteq \begin{pmatrix} -1 & 1 & 0 \\ 1 & -2 & 1 \\ 0 & 1 & -1 \end{pmatrix}, \tag{8.26}$$

and this has, from our formulae (8.14) and (8.25):

$$\mathbb{V} \doteq \begin{pmatrix} \frac{1}{\sqrt{3}} & \frac{1}{\sqrt{6}} & -\frac{1}{\sqrt{2}} \\ \frac{1}{\sqrt{3}} & -\sqrt{\frac{2}{3}} & 0 \\ \frac{1}{\sqrt{3}} & \frac{1}{\sqrt{6}} & \frac{1}{\sqrt{2}} \end{pmatrix} \tag{8.27}$$

---

[2] To find these manually, we take the relation $\mathbb{Q}\mathbf{v} = \lambda\mathbf{v}$, and note that $(\mathbb{Q} - \lambda\mathbb{I})\mathbf{v} = 0$, which means that $\mathbf{v}$ must be in the null space of $\mathbb{Q} - \lambda\mathbb{I}$ (or $\mathbf{v} = 0$). For a null space to exist, we must have zero determinant, so that $\mathrm{Det}(\mathbb{Q} - \lambda\mathbb{I}) = 0$, giving us a polynomial in $\lambda$ for which we can find roots. Once we have $\lambda$, simple elimination can be used to solve the linear equation $\mathbb{Q}\mathbf{v} = \lambda\mathbf{v}$ for the unknown elements of $\mathbf{v}$. The numerical approach we will see in Chapter 10 treats the problem very differently.

[3] Or imagine a dimensionless scheme in which $t = \sqrt{m/kq}$ and $x = ay$, so that $y''(q) = -(y(q) - 1)$.

[4] The matrix in (8.26) has an interesting structure when compared with, for example, (4.54).

with the first column corresponding to the (normalized) eigenvector with eigenvalue $\lambda_1 = 0$, the second column is the eigenvector with eigenvalue $\lambda_2 = -3$, and the third column is the eigenvector with eigenvalue $\lambda_3 = -1$. The vectors $\mathbf{A}$ and $\mathbf{B}$ are:

$$\mathbf{A} \doteq \begin{pmatrix} -1 \\ 0 \\ 1 \end{pmatrix} \qquad \mathbf{B} \equiv \mathbb{V}^T \mathbf{A} \doteq \begin{pmatrix} 0 \\ 0 \\ \sqrt{2} \end{pmatrix}. \tag{8.28}$$

In this case, our $\Psi_1(t)$, $\Psi_2(t)$ and $\Psi_3(t)$ read:

$$\begin{aligned} \Psi_1(t) &= \alpha_1 t + \beta_1 \\ \Psi_2(t) &= \alpha_2 \cos(\sqrt{3}t) + \beta_2 \sin(\sqrt{3}t) \\ \Psi_3(t) &= \alpha_3 \cos(t) + \beta_3 \sin(t) + \sqrt{2}, \end{aligned} \tag{8.29}$$

and then the solution for $x_{-1}(t)$, $x_0(t)$, and $x_1(t)$ in terms of the six constants of integration $\{\alpha_i, \beta_i\}_{i=1}^3$ from the definition of $\Psi$ is:

$$\begin{aligned} x_{-1}(t) &= \frac{\alpha_1 t + \beta_1}{\sqrt{3}} - \frac{1}{\sqrt{2}}\left(\sqrt{2} + \alpha_3 \cos t + \beta_3 \sin t\right) \\ &\quad + \frac{1}{\sqrt{6}}\left(\alpha_2 \cos(\sqrt{3}t) + \beta_2 \sin(\sqrt{3}t)\right) \\ x_0(t) &= \frac{\alpha_1 t + \beta_1}{\sqrt{3}} - \sqrt{\frac{2}{3}}\left(\alpha_2 \cos(\sqrt{3}t) + \beta_2 \sin(\sqrt{3}t)\right) \\ x_1(t) &= \frac{\alpha_1 t + \beta_1}{\sqrt{3}} + \frac{1}{\sqrt{2}}\left(\sqrt{2} + \alpha_3 \cos t + \beta_3 \sin t\right) \\ &\quad + \frac{1}{\sqrt{6}}\left(\alpha_2 \cos(\sqrt{3}t) + \beta_2 \sin(\sqrt{3}t)\right). \end{aligned} \tag{8.30}$$

Now we just have to set the initial conditions. Suppose we start with all three masses at rest, and the initial configuration (keeping in mind that the equilibrium position is now at $a = 1$ m): $x_{-1}(0) = -3/2$, $x_0(0) = 1/4$, $x_1(0) = 3/4$. This set of initial data can be solved by considering the set of six equations: $\dot{\mathbf{X}}(0) = 0$ which tells us that $\alpha_1 = \beta_2 = \beta_3 = 0$ and

$$\mathbf{X}(0) \doteq \begin{pmatrix} -\frac{3}{2} \\ \frac{1}{4} \\ \frac{3}{4} \end{pmatrix} \longrightarrow \beta_1 = -\frac{1}{2\sqrt{3}} \alpha_2 = -\frac{5}{4\sqrt{6}} \alpha_3 = \frac{1}{4\sqrt{2}}. \tag{8.31}$$

## The normal modes

The independent collective motion of the particles represented by $\{\Psi_i(t)\}_{i=1}^N$ are called the "normal modes" of the oscillatory system. We can characterize these special solutions by finding the real-space (meaning positions) movement of the "pure" $\Psi_1(t)$, $\Psi_2(t)$, and $\Psi_3(t)$ solutions separately for our three ball problem. That is, we will find $\mathbf{X}(t)$ that comes from $\Psi_1(t) = \alpha t + \beta$ with $\Psi_2(t) = \Psi_3(t) = 0$ and then do the same for the other two. Recall the general solution:

$$\Psi_1(t) = \alpha_1 t + \beta_1$$

$$\Psi_2(t) = \alpha_2 \cos(\sqrt{|\lambda_2|}t) + \beta_2 \sin(\sqrt{|\lambda_2|}t) \qquad (8.32)$$

$$\Psi_3(t) = \alpha_3 \cos(\sqrt{|\lambda_3|}t) + \beta_3 \sin(\sqrt{|\lambda_3|}t)$$

(all of our eigenvalues are less than or equal to zero, hence the use of oscillatory solutions and absolute values). Note that the vector $\mathbf{A}$ has been discarded, since it is not needed for the characterization of the modes.

Setting $\Psi_2(t) = \Psi_3(t) = 0$ corresponds to setting the constants $\alpha_2 = \beta_2 = \alpha_3 = \beta_3 = 0$. Then $\mathbf{X}(t) = \mathbb{V}\Psi$ is:

$$\mathbf{X} \doteq \frac{\alpha_1 t + \beta_1}{\sqrt{3}} \begin{pmatrix} 1 \\ 1 \\ 1 \end{pmatrix} \qquad (8.33)$$

and this is clearly just the rigid motion of the entire set of balls down the $x$ axis – what we would normally call center of mass motion (associated, naturally, with the non-oscillatory, null eigenvalue).

Setting $\Psi_1(t) = \Psi_3(t) = 0$ corresponds to the choice $\alpha_1 = \beta_1 = \alpha_3 = \beta_3 = 0$, and we are left with $\Psi_2(t) = \alpha_2 \cos(\sqrt{|\lambda_2|}t) + \beta_2 \sin(\sqrt{|\lambda_2|}t)$. When we project back: $\mathbf{X} = \mathbb{V}\Psi$, gives

$$\mathbf{X} \doteq \frac{1}{\sqrt{6}} \left( \alpha_2 \cos(\sqrt{3}t) + \beta_2 \sin(\sqrt{3}t) \right) \begin{pmatrix} 1 \\ -2 \\ 1 \end{pmatrix}. \qquad (8.34)$$

This corresponds to the two outer masses moving to the right, while the center mass moves to the left (look at just the vector portion to see this – obviously, there are magnitude issues associated with the time-dependent function out front).

Finally, we set $\Psi_1(t) = \Psi_2(t) = 0$ and leave $\Psi_3(t) = \alpha_3 \cos(\sqrt{|\lambda_3|}t) + \beta_3 \sin(\sqrt{|\lambda_3|}t)$; now $\mathbf{X} = \mathbb{V}\Psi$ is:

$$\mathbf{X} \doteq \frac{1}{\sqrt{2}} (\alpha_3 \cos(t) + \beta_3 \sin(t)) \begin{pmatrix} -1 \\ 0 \\ 1 \end{pmatrix}. \qquad (8.35)$$

Our final vector leaves the center mass fixed and moves the left mass to the left, the right mass to the right. The motion of each of these three modes is displayed in Figure 8.3.

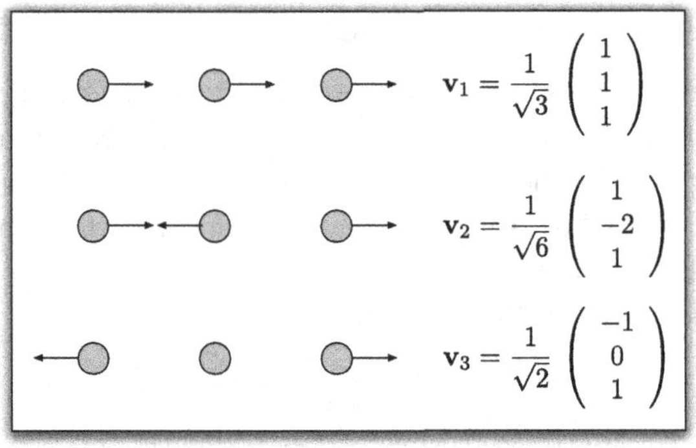

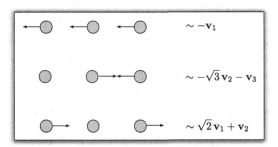

Figure 8.3 The motion of the three balls under the influence of pure $\Psi_1(t)$ (top), pure $\Psi_2(t)$ (middle), and pure $\Psi_3(t)$ (bottom).

Notice that the vectors in (8.33), (8.34), and (8.35) are parallel to the eigenvectors of $\mathbb{Q}$, the columns in (8.27). We could make a picture like Figure 8.3 just by being given $\mathbb{Q}$, or in more general settings, the connectivity of the springs in the system. Normal modes are intrinsic to the structure of the quadratic energy governing harmonic oscillators. In addition, the temporal evolution of each mode, and its energy, is associated with the eigenvalues of $\mathbb{Q}$ – so that without solving the equations of motion, or even writing them down, we can say a lot about the system.

While the motion of these masses is time-dependent, we can think of the vector portion of (8.33), (8.34), and (8.35) as determining the relative motion while the time-dependent functions sitting out front give us "decomposition" coefficients at any time $t$. The idea is that any valid motion of the system can be made up out of these three fundamental motions, and in Figure 8.4 a couple of potential motions and how we can use these "basis motions" to describe them are shown.

Figure 8.4 Some examples of the decomposition of generic relative motion of three balls into the normal modes.

## 8.4 General solution

We have been looking at a specific case with three balls in one dimension. But any time we have an equation of the form (8.1), the same approach we have used here will work. Start with $\mathbb{Q} \in \mathbb{R}^{n \times n}$, and the equation of motion with initial conditions:

$$\ddot{\mathbf{X}} = \mathbb{Q}\mathbf{X} + \mathbf{A} \quad \mathbf{X}(0) = \mathbf{X}_0 \quad \dot{\mathbf{X}}(0) = \mathbf{V}_0. \tag{8.36}$$

Decompose $\mathbb{Q} = \mathbb{V}\mathbb{L}\mathbb{V}^T$, diagonalize using $\mathbf{\Psi} = \mathbb{V}^T\mathbf{X}$:

$$\ddot{\mathbf{\Psi}} = \mathbb{L}\mathbf{\Psi} + \mathbf{B} \quad \mathbf{\Psi}(0) = \mathbb{V}^T\mathbf{X}_0 \equiv \mathbf{\Psi}^0 \quad \dot{\mathbf{\Psi}}(0) = \mathbb{V}^T\dot{\mathbf{X}}_0 \equiv \dot{\mathbf{\Psi}}^0. \tag{8.37}$$

For all eigenvectors having non-zero eigenvalue, the solution reads:

$$\Psi_i(t) = \frac{1}{2\lambda_i}e^{-t\sqrt{\lambda_i}}\left[ B_i\left(-1 + e^{t\sqrt{\lambda_i}}\right)^2 + \Psi_i^0\left(1 + e^{2t\sqrt{\lambda_i}}\right)\right.$$
$$\left. + \dot{\Psi}_i^0\sqrt{\lambda_i}\left(-1 + e^{2t\sqrt{\lambda_i}}\right)\right], \tag{8.38}$$

and for all eigenvectors having zero eigenvalue, the solution is:

$$\Psi_i(t) = \frac{1}{2}B_i t^2 + \Psi_i^0 + t\dot{\Psi}_i^0. \tag{8.39}$$

With the elements of $\mathbf{\Psi}$ in hand, we recover the positions via $\mathbf{X} = \mathbb{V}\mathbf{\Psi}$.

In the chapter notebook, you will find the function SolveIt which takes the matrix $\mathbb{Q}$, vector $\mathbf{A}$, and initial positions and velocities, calculates the decomposition of $\mathbb{Q}$, and solves for each entry of $\mathbf{\Psi}$ using either (8.38) or (8.39) as appropriate. Once again, we have a solution written in terms of simple functions, but in order to evaluate these, we must have the eigenvalues and eigenvectors of $\mathbb{Q}$.

## 8.5 Balls and springs in D = 3

It is interesting that the one-dimensional ball-and-spring system that led us to the generic form (8.36) (which then takes on a life of its own) does not have linear equations of motion in three dimensions. To see this, think of three masses $m$ with positions $\mathbf{x}_1(t)$, $\mathbf{x}_2(t)$, and $\mathbf{x}_3(t)$. If the first mass is connected to each of the other ones by a spring with spring constant $k$ and equilibrium length $a$, then the force on the first mass is, referring to Figure 8.5:

$$\mathbf{F}_1 = -k(r_{12} - a)\hat{\mathbf{r}}_{12} - k(r_{13} - a)\hat{\mathbf{r}}_{13} \tag{8.40}$$

where $\mathbf{r}_{12} \equiv \mathbf{x}_1 - \mathbf{x}_2$ (the vector pointing from $\mathbf{x}_2$ to $\mathbf{x}_1$), $r_{12}$ is its magnitude, and $\hat{\mathbf{r}}_{12} \equiv \mathbf{r}_{12}/r_{12}$ is the associated unit vector, and similarly for $\mathbf{r}_{13} \equiv \mathbf{x}_1 - \mathbf{x}_3$.

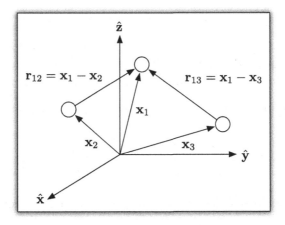

Figure 8.5 The center mass, at $\mathbf{x}_1$, is connected by springs to the masses at $\mathbf{x}_2$ and $\mathbf{x}_3$.

The force here is nonlinear – the appearance of $ka\hat{\mathbf{r}}_{12}$ (for example) establishes that. So we cannot use the linear formulation of the problem directly. What we could do is solve the equations of motion numerically using an ordinary differential equation solver of the sort described in Chapter 2. If we are viewing these masses as atoms or molecules, in a molecular dynamics setting, then the springs represent bonds, and there will be other forces acting between the atoms. These will not typically be linear forces, and so we *must* use a numerical solution in order to find the location of all particles as functions of time.

### *Linearized force*

Another approach is to linearize the forces about an equilibrium configuration and study the resulting oscillatory motion in the vicinity of that equilibrium. Take a single spring (with constant $k$) connecting two masses, located at $\mathbf{x}_1$ and $\mathbf{x}_2$ – call the equilibrium position of the two masses, where there is no spring force, $\mathbf{q}_1$ and $\mathbf{q}_2$ respectively, as shown in Figure 8.6. Define the separation vector $\mathbf{r}_{12} \equiv \mathbf{x}_1 - \mathbf{x}_2$, and the equilibrium separation vector $\mathbf{q}_{12} \equiv \mathbf{q}_1 - \mathbf{q}_2$. Then we can write the force on the first particle (as in (8.40))

$$\mathbf{F}_1 = -k(r_{12} - q_{12})\hat{\mathbf{r}}_{12} = -k\left(1 - \frac{q_{12}}{r_{12}}\right)\mathbf{r}_{12}, \qquad (8.41)$$

where the second equality displays the nonlinearity explicitly.

Suppose that each particle is close to its equilibrium location, let $\mathbf{x}_1 = \mathbf{q}_1 + \boldsymbol{\epsilon}_1$ and $\mathbf{x}_2 = \mathbf{q}_2 + \boldsymbol{\epsilon}_2$ with $\boldsymbol{\epsilon}_1$ and $\boldsymbol{\epsilon}_2$ small. Then $\mathbf{r}_{12} = \mathbf{q}_{12} + \boldsymbol{\epsilon}_{12}$ for infinitesimal separation vector $\boldsymbol{\epsilon}_{12} \equiv \boldsymbol{\epsilon}_1 - \boldsymbol{\epsilon}_2$. In order to linearize $\mathbf{F}_1$ in (8.41), we insert $\mathbf{r}_{12}$ in

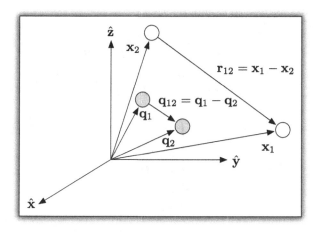

Figure 8.6 Two masses (located at $\mathbf{x}_1$ and $\mathbf{x}_2$) are connected by a spring – the equilibrium positions of the masses are $\mathbf{q}_1$ and $\mathbf{q}_2$.

terms of $\mathbf{q}_{12}$ and $\boldsymbol{\epsilon}_{12}$ and keep only terms linear in $\boldsymbol{\epsilon}_{12}$. Expanding the term $1/r_{12}$ that shows up,

$$
\begin{aligned}
\frac{1}{r_{12}} = \left[r_{12}^2\right]^{-\frac{1}{2}} &= \left[q_{12}^2 + 2\boldsymbol{\epsilon}_{12} \cdot \mathbf{q}_{12} + \epsilon_{12}^2\right]^{-\frac{1}{2}} \\
&= \left[q_{12}^2\left(1 + \frac{2\boldsymbol{\epsilon}_{12} \cdot \mathbf{q}_{12} + \epsilon_{12}^2}{q_{12}^2}\right)\right]^{-\frac{1}{2}} \\
&\approx \frac{1}{q_{12}}\left(1 - \frac{\boldsymbol{\epsilon}_{12} \cdot \mathbf{q}_{12}}{q_{12}^2}\right),
\end{aligned}
\tag{8.42}
$$

and inserting this in (8.41)

$$
\mathbf{F}_1 \approx -k\left[\frac{\boldsymbol{\epsilon}_{12} \cdot \mathbf{q}_{12}}{q_{12}^2}\right](\mathbf{q}_{12} + \boldsymbol{\epsilon}_{12}) \approx -k\left[\frac{\boldsymbol{\epsilon}_{12} \cdot \mathbf{q}_{12}}{q_{12}^2}\right]\mathbf{q}_{12},
\tag{8.43}
$$

again, keeping only the term linear in $\boldsymbol{\epsilon}_{12}$.

Finally, we'd like to write this force, and the equal and opposite form attached to the second particle, in terms of the position vector $\mathbf{X} \in \mathbb{R}^6$ here (two particles in three dimensions) – that will allow us to pull out the matrix $\mathbb{Q}$ relevant to this linearization. We can write the linearized $\mathbf{F}_1$ in (8.43) as (referring to the components of $\mathbf{q}_{12}$ as $q_{12x}$, $q_{12y}$ and $q_{12z}$):

$$\mathcal{Q}_1 \equiv -\frac{k}{q_{12}^2} \begin{pmatrix} q_{12x}^2 & q_{12x}q_{12y} & q_{12x}q_{12z} \\ q_{12x}q_{12y} & q_{12y}^2 & q_{12y}q_{12z} \\ q_{12x}q_{12z} & q_{12y}q_{12z} & q_{12z}^2 \end{pmatrix}$$

$$\mathbf{F}_1 = \begin{pmatrix} \mathcal{Q}_1 & -\mathcal{Q}_1 \end{pmatrix} \begin{pmatrix} \epsilon_{1x} \\ \epsilon_{1y} \\ \epsilon_{1z} \\ \epsilon_{2x} \\ \epsilon_{2y} \\ \epsilon_{2z} \end{pmatrix} \tag{8.44}$$

where the matrix appearing in $\mathbf{F}_1$ has two three-by-three $\mathcal{Q}_1$ blocks sitting next to each other (giving three rows and six columns). If we want to embed the force acting on the second particle, $\mathbf{F}_2$, together with $\mathbf{F}_1$ in the natural way, we can expand the matrix:

$$\mathbf{F} \equiv \begin{pmatrix} \mathbf{F}_1 \\ \mathbf{F}_2 \end{pmatrix} = \underbrace{\begin{pmatrix} \mathcal{Q}_1 & -\mathcal{Q}_1 \\ -\mathcal{Q}_1 & \mathcal{Q}_1 \end{pmatrix}}_{\equiv \mathcal{Q}} \begin{pmatrix} \epsilon_{1x} \\ \epsilon_{1y} \\ \epsilon_{1z} \\ \epsilon_{2x} \\ \epsilon_{2y} \\ \epsilon_{2z} \end{pmatrix}. \tag{8.45}$$

The infinitesimal vector can be written in terms of the actual locations $\mathbf{x}_1$ and $\mathbf{x}_2$ and the equilibrium ones to give:

$$\mathbf{X} \equiv \begin{pmatrix} \mathbf{x}_1 \\ \mathbf{x}_2 \end{pmatrix}$$

$$\mathbf{Q} \equiv \begin{pmatrix} \mathbf{q}_1 \\ \mathbf{q}_2 \end{pmatrix} \tag{8.46}$$

$$\mathbf{F} = \mathcal{Q}(\mathbf{X} - \mathbf{Q}) \equiv \mathcal{Q}\mathbf{X} + \mathbf{A},$$

for $\mathbf{A} \equiv \mathcal{Q}\mathbf{Q}$. Using this force, for unit mass, in Newton's second law gives (8.1).

If we have more masses and springs, we just build the $\mathcal{Q}_1$ blocks for each connected pair and construct the full $\mathcal{Q}$ by appropriate placement of $\mathcal{Q}_1$ and its negative. The function MakeQ in the chapter notebook provides an example of building $\mathcal{Q}$ given spring connectivity information and equilibrium positions.

## Further reading

1. Allen, M. P. & D. J. Tildesley. *Computer Simulations of Liquids.* Oxford University Press, 1987.
2. Arfken, George B. & Hans J. Weber. *Mathematical Methods for Physicists.* Academic Press, 2001.
3. Goldstein, Herbert, Charles Poole, & John Safko. *Classical Mechanics.* Addison-Wesley, 2002.
4. Rapaport, D. C. *The Art of Molecular Dynamics Simulation.* Cambridge University Press, 1995.

## Problems

### Problem 8.1

Given the potential $U(x) = U_0 \cos(x^2)$ in one dimension, find the minima for $x \in [0, \pi]$ and give the period of oscillation for each (assume the particle undergoing the oscillatory motion has mass $m$, and note that for this potential, $x$ is unitless).

### Problem 8.2

Given a cubic force in one dimension: $F = \alpha(x - a)(x - b)(x - c)$, find the coefficient $k$ in the linearized approximation about $b$ (Hint: set $x = b + \epsilon$ and find the term linear in $\epsilon$). How could you ensure that the motion under the influence of this linearized force is oscillatory?

### Problem 8.3

We have been focused on the harmonic oscillator, which is relevant for a potential *well* – for any starting conditions, the particle is trapped in the vicinity of the equilibrium position (taken to be zero). If we instead take our potential to be a maximum at zero, motion is away from the zero for almost all values of initial conditions, and we call such equilibria "unstable." Solve an example of the "instanton" equation:

$$m\ddot{x} = kx \tag{8.47}$$

for $x(t)$ given the initial conditions $x(0) = 0$, $\dot{x}(0) = \epsilon > 0$ but small. Show that a particle under the influence of this force can be found arbitrarily far from the equilibrium position, unless $\epsilon = 0$ exactly.

### Problem 8.4

Develop the appropriate matrix form of the equations of motion for $n$ balls connected by $n - 1$ springs in one dimension, all with constant $k$ and equilibrium $a$ (assume all balls have unit mass). Write a function that provides both the matrix $\mathbb{Q}$ and the vector $\mathbf{A}$ appearing in $\ddot{\mathbf{X}} = \mathbb{Q}\mathbf{X} + \mathbf{A}$. Do you recognize the structure of the matrix $\mathbb{Q}$?

## Problem 8.5

In (8.14), the association between the column vector form of the matrix product $\mathbb{VL}$ and the product itself is made – check this connection by explicitly multiplying $\mathbb{V}$ by $\mathbb{L}$ on the right and showing that you can then partition the matrix into three columns as shown.

## Problem 8.6

For the discretized form of Newton's second law with a linear restoring force, we have an eigenvalue problem for zeroed boundaries:

$$\left(x_{j+1} - 2x_j + x_{j-1}\right) = -\Delta t^2 \lambda x_j, \tag{8.48}$$

where $\lambda$ is the unknown eigenvalue. The matrix form of this equation, for $x_j \approx x(t_j)$, can be written as $\ddot{\mathbf{X}} = -\mathbb{Q}\mathbf{X}$ where $-\mathbb{Q}$ is identical (modulo the factor of $\Delta t^2$, and the first and last diagonal entries) to the multiple-particle matrix you found in Problem 8.4.

(a) Set $x_j \equiv A e^{\pi i k (j \Delta t)}$ (for integer $k$, and using $t_j \equiv j\Delta t$) and insert in the above. Find $\lambda$ for this ansatz. How many distinct values for $\lambda$ are there?

(b) Set $N = 100$ with $\Delta t = 1/(N+1)$ and find the eigenvalues of your matrix using the built-in function `Eigenvalues` (technically, to define the eigenvalue problem properly, you'll want the eigenvalues of the *negative* of your matrix). Plot the eigenvalues predicted in part (a) on top of these. Eigenvalue ordering does not matter here, so just sort both the eigenvalues determined by `Eigenvalues` and your analytical ones.

## Problem 8.7

For a mass on a spring with damping, we had (4.13). If we work on a discrete grid with $s_j = j\Delta s$, then a discretized form of this equation is:

$$q_{j+1} - 2q_j + q_{j-1} = -\tilde{\omega}^2 \Delta s^2 q_j - \frac{1}{2}\Delta s\left(q_{j+1} - q_{j-1}\right). \tag{8.49}$$

We can think of this equation as an eigenvalue problem for determining $\tilde{\omega}^2$ (just write everything except the $\tilde{\omega}^2 \Delta s^2 q_j$ on the left). The matrix we would make here is not symmetric, and it could have eigenvalues with imaginary components. Show that the continuous version of this problem, $-[q''(s) + q'(s)] = \omega^2 q(s)$ with $q(0) = q(1) = 0$ has eigenvalues $\omega^2 \in \mathbb{R}$.

## Problem 8.8

If we take the series definition of the exponential:

$$e^q = \sum_{j=0}^{\infty} \frac{q^j}{j!} \tag{8.50}$$

for constant $q$, then we can extend the definition to matrices by replacing $q$ with $\mathbb{Q}$ and understanding $\mathbb{Q}^j$ to mean matrix multiplication. Show that for an invertible, symmetric matrix $\mathbb{Q}$ with decomposition $\mathbb{Q} = \mathbb{V}\mathbb{L}\mathbb{V}^T$ (for orthogonal matrix $\mathbb{V}$ and $\mathbb{L}$ diagonal with the eigenvalues of $\mathbb{Q}$ along the diagonal) we have:

$$e^{\mathbb{Q}} = \mathbb{V}\mathbb{M}\mathbb{V}^T \tag{8.51}$$

where $\mathbb{M}$ is a diagonal matrix with entries: $M_{ii} = e^{L_{ii}}$, the exponential of the eigenvalues.

## Problem 8.9

Write the solution to

$$\ddot{\Psi} = \sigma\Psi + a \quad \text{and} \quad \ddot{\Psi} = a \tag{8.52}$$

for constants $\sigma$ and $a$ in terms of the initial and final values of $\Psi$ ($\Psi_0$ at time $t = 0$, $\Psi_T$ at some end time $t = T$).

## Problem 8.10

The solver from the chapter notebook takes an equilibrium configuration of $N$ points in three dimensions, and a "connectivity" list of the form $\{\{i_1, j_1\}, \{i_2, j_2\}\ldots\}$ describing the connections between balls (numbered in some manner).

(a) We'll take the equilibrium configuration to be a square grid of integer spaced balls, 10 on a side (as shown in Figure 8.7). Write a function that generates this grid, returning a vector of positions (in three dimensions) – use the ordering shown in Figure 8.7 for your vector, so that the first element in the vector is $\{1, 1, 0\}$, the second is $\{1, 2, 0\}$, the ninety first is $\{10, 1, 0\}$, etc. (note that the first ball, then, is at $x = y = 1$, not zero).

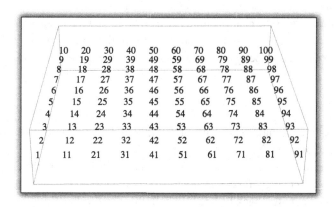

Figure 8.7 Grid of balls, $10 \times 10$, placed at integer locations in the $x - y$ plane, all have height $z = 0$.

(b) Write a function (call it MakeNNQ) that connects nearest neighbors, so that for example, ball 1 is connected to 2 and 11. Write a similar function that connects next-nearest neighbors (called MakeNNNQ). In each case, express the output of your function as a list of pairs: $\{\{1, 2\}, \{1, 11\}, \ldots\}$. Hint: Forget the integer numerology you might use in this $10 \times 10$ case; just check distances.

## Lab problems

### Problem 8.11

Using your RK4 routine, solve Newton's equations of motion for three balls (8.8) with $k = 1$ N/m, $m = 1$ kg, $a = 1$ m. Use the following initial conditions:

$$x_{-1}(0) = -1 \text{ m} \quad x_0(0) = 0 \quad x_1(0) = 1 \text{ m}$$
$$\dot{x}_{-1}(0) = 0 \quad \dot{x}_0(0) = -0.5 \text{ m/s} \quad \dot{x}_1(0) = 0.5 \text{ m/s}.$$

(8.53)

Make a movie of the output using $\Delta t = 0.01$ for 500 steps.

### Problem 8.12

Write a function that generates the matrix $\mathbb{Q}$ and vector $\mathbf{A}$ from Problem 8.4 given the parameter $n$ (the number of balls) – assume all spring constants are the same, $k = 1$ N/m, and take $m = 1$ kg, $a = 1$ m. Using the Solveit function from the chapter notebook, find the solution for $n = 20$ balls that start from rest. For the initial spacing, put the first ball at $-0.5$ m, and the rest at 1 m, 2 m, 3 m, etc. What is the velocity of the third ball at time $t = 0.5$ s? Make a movie of the motion for $t = 0 \longrightarrow 50$ s in steps of $\Delta t = 0.1$ s. Save the initial positions and the positions at time $t = 10$ s for use in Problem 8.17.

### Problem 8.13

Modify your matrix generator from the previous problem so that the springs connecting the balls get stiffer along the chain: Take $k_n = n$ N/m, so that the first spring has $k_1 = 1$ N/m, etc. Using the same $n = 20$ balls, and the same initial conditions (as in the previous problem), again find the velocity of the third ball at $t = 0.5$ s and make the associated movie.

### Problem 8.14

Set up the $10 \times 10$ configuration of balls on a grid as in Problem 8.10. This defines the equilibrium configuration, $\bar{\mathbf{X}}$.
(a) Make the nearest neighbor connectivity list, and use MakeQ from the chapter notebook to generate $\mathbb{Q}$. Take $\bar{\mathbf{X}}$ as the initial configuration, but move ball "one" to $\{0.5, 0.5, 0\}$ (i.e. pull it back by a half in both the $x$ and $y$ directions – remember from Problem 8.10 that the equilibrium position of the first ball is $x = y = 1$). Take all balls at rest for the initial velocities. Now solve using Solveit, and make a movie of the motion for $t = 0 \longrightarrow 20$ s in steps of $\Delta t = 0.4$ s.
(b) Do the same thing using the next-nearest neighbor connectivity. Note the difference between the motion when compared with that coming from nearest neighbor connectivity. What is the position of the first ball at time $t = 2.0$ s?

**Problem 8.15**

Write a Verlet solver that does not make any linear approximation in the spring setting, so that the force between two particles connected by a spring appears in the form (8.40) (with $k = 1$ N/m). All you need to do is use Implementation 2.1 with an appropriate function to calculate the accelerations. Use the next-nearest neighbor connectivity list, and generate the forces exactly for each particle. In order to compare with part (b) of the previous problem, set the equilibrium lengths for the nearest neighbor springs to 1, and the equilibrium lengths for the next-nearest neighbor springs to $\sqrt{2}$ (an easy way to establish the equilibrium lengths is to save them in the connectivity list, so that your next-nearest neighbor connectivity generator from Problem 8.10 could be augmented to include the equilibrium length of the spring: $\{i_1, j_1, a_1\}$). For your time step in Verlet, use $\Delta t = 0.1$ s – assemble the output into a movie that goes from $t = 0 \longrightarrow 20$ s in steps of 0.4 s as before. Find the position of the first ball at time $t = 2.0$ s and compare with the previous problem.

**Problem 8.16**

Use your Verlet solver with a nearest neighbor connectivity (equilibrium length 1) to reproduce part (a) of Problem 8.14 – using the same initial conditions found there, make the equivalent movie ($t = 0 \longrightarrow 20$ s in steps of $\Delta t = 0.4$ s) – notice the fundamentally different behavior that comes from the linearized approximation used in Problem 8.14.

**Problem 8.17**

Modify the solver in the chapter notebook so that it is a "boundary-value" solver – i.e. rather than providing the initial position and velocity, we give the initial and final positions. To do this, rewrite the solution for $\Psi_i(t)$ in terms of $\Psi_0$ (the vector Psi0 as in the SolveIt function) and $\Psi_T$ (which will replace DPsi0 from the chapter notebook) where $T$ is some specified total time – you should have the appropriate expressions from Problem 8.9. Test your solver using Problem 8.12 as follows – use the $\mathbb{Q}$ and $\mathbf{A}$ from that problem and take the positions at time $t = 0$ and 10 s (that should have been saved). If you run your new solver on this boundary data, you should find that all initial velocities are zero – how close do you come?

**Problem 8.18**

Using your new boundary condition solver, take the next-nearest neighbor configuration from Problem 8.14 (with the same $\bar{\mathbf{X}}$). Let $\mathbf{X}(0) = \bar{\mathbf{X}}$ so that the configuration is in equilibrium at $t = 0$. For the final configuration, move ball 45 to $\{5.5, 5.5, 0\}$. Give the system a total time of $T = 20$ s and make a movie of the solution in steps of 0.4 s. What is the initial velocity of ball 1?

# 9

# Matrix inversion

There are deep reasons for why linear (with the emphasis on *linear*) algebra plays such an important role in physics. For one thing, it is a subject that stresses, but does not limit us to, linear transformations. To the extent that linearity (and its hallmark: superposition) is a property that we allow, even if only in approximation, the language used to express the associated physics, whether finite or not, will be linear algebra. Of course, aside from the "harmonic approximation" to potentials from the last chapter, we have already seen the utility of matrices in describing discretized differential operators. These are *linear* differential operators (like the Laplacian, Helmholtz, or D'Alembertian operators) whose finite-dimensional approximations are precisely finite-dimensional matrices. So from the point of view of solving physical problems (whether they are ODE, PDE, or "other"), we have an interest in linear algebra.

There are not so many unrelated topics in the subject. The two major goals of applied linear algebra are: 1. Invert, or come as close to inverting as is mathematically possible, a linear system: given $\mathbb{A} \in \mathbb{R}^{n \times m}$, $\mathbf{b} \in \mathbb{R}^n$, find $\mathbf{x} \in \mathbb{R}^m$ such that $\|\mathbb{A}\mathbf{x} - \mathbf{b}\|$ is minimized,[1] and 2. Find the eigenvalues and eigenvectors associated with a matrix $\mathbb{A} \in \mathbb{R}^{n \times n}$ – i.e. find $\{\lambda_i\}_{i=1}^n$ with $\lambda_i \in \mathbb{C}$ and $\{\mathbf{v}_i\}_{i=1}^n$ with $\mathbf{v}_i \in \mathbb{C}^n$ such that $\mathbb{A}\mathbf{v}_i = \lambda_i \mathbf{v}_i$ (known as the "eigenvalue problem"). We will work with real matrices, but all of the techniques we discuss have natural extensions for complex matrices. Our main topic in this chapter will be matrix inversion: given $\mathbb{A} \in \mathbb{R}^{n \times n}$ and $\mathbf{b} \in \mathbb{R}^n$, find $\mathbf{x} \in \mathbb{R}^n$ such that $\mathbb{A}\mathbf{x} = \mathbf{b}$. We'll look at the extension to the non-square case in Chapter 10, where we also study the eigenvalue problem.

---

[1] This problem, when $\mathbb{A}$ is square, is just matrix inversion. When $\mathbb{A}$ is not square, it is the "least squares" problem if we use the usual Pythagorean norm, so that, for a vector $\mathbf{w}$, $\|\mathbf{w}\| \equiv \sqrt{\mathbf{w} \cdot \mathbf{w}}$, but other norms lead to different solutions to the minimization problem.

## 9.1 Definitions and points of view

We'll start with a review of matrix-vector multiplication, and the partitioning of that multiplication into individual elements and columns. We know how to describe a matrix $\mathbb{A} \in \mathbb{R}^{n \times m}$ by specifying its entries:

$$\mathbb{A} \doteq \begin{pmatrix} A_{11} & A_{12} & \cdots & A_{1m} \\ A_{21} & A_{22} & \cdots & A_{2m} \\ \vdots & \vdots & \ddots & \vdots \\ A_{n1} & A_{n2} & \cdots & A_{nm} \end{pmatrix}. \tag{9.1}$$

The addition and subtraction of matrices, and multiplication-by-a-scalar operations all act element-by-element.

If we have a vector $\mathbf{y} \in \mathbb{R}^m$, then matrix-vector multiplication is defined as usual:

$$\mathbb{A}\mathbf{y} \doteq \begin{pmatrix} \sum_{j=1}^{m} A_{1j} y_j \\ \sum_{j=1}^{m} A_{2j} y_j \\ \vdots \\ \sum_{j=1}^{m} A_{nj} y_j \end{pmatrix}. \tag{9.2}$$

Another, sometimes useful view of the above product is to think of the matrix $\mathbb{A}$ partitioned into column vectors,

$$\mathbb{A} \doteq [\, \mathbf{a}_1 \quad | \quad \mathbf{a}_2 \quad | \quad \mathbf{a}_3 \quad | \quad \cdots \quad | \quad \mathbf{a}_m \,], \tag{9.3}$$

where each of the column vectors $\{\mathbf{a}_i\}_{i=1}^{m} \in \mathbb{R}^n$. For example, $\mathbf{a}_2$ is:

$$\mathbf{a}_2 \doteq \begin{pmatrix} A_{12} \\ A_{22} \\ \vdots \\ A_{n2} \end{pmatrix}. \tag{9.4}$$

From this column-oriented point of view, we can think of a matrix-vector product as a weighted sum of the columns of $\mathbb{A}$, that is:

$$\mathbb{A}\mathbf{y} = \sum_{j=1}^{m} y_j \mathbf{a}_j \tag{9.5}$$

highlighting the role of matrix-vector multiplication as a linear combination of vectors (the columns of $\mathbb{A}$) multiplied by scalars (the entries of $\mathbf{y}$). Then a natural question, especially from an inversion point of view, is: Are the columns of $\mathbb{A}$ linearly independent? And if so, what linear combinations can we actually produce from the columns of $\mathbb{A}$?

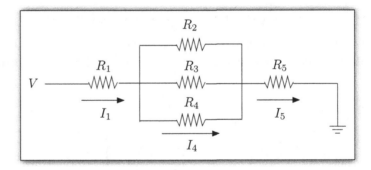

Figure 9.1 A circuit with five arbitrary (but specified) resistors connected as shown. We apply a known voltage $V$ on the left. The problem is to find the current flowing through each resistor.

## 9.2 Physical motivation

We have encountered matrix inversion in a variety of contexts already, notably in the finite difference solution of ODEs and PDEs as in Chapter 4. But we can set up any linear set of equations as a matrix inverse problem. To start off, then, we'll take a look at some of the more common physical problems that benefit directly from the ability to find $\mathbf{x}$ solving $\mathbb{A}\mathbf{x} = \mathbf{b}$ given $\mathbb{A}$ and $\mathbf{b}$.

### 9.2.1 Circuits

Electrical circuits often provide sets of algebraic relations that can be formulated in terms of matrix-vector multiplication. Consider the circuit in Figure 9.1 – we have five resistors, with, we will assume, numerical values that are provided. On the left, we apply a voltage $V$, and the right goes to ground. Current flows out of the voltage source, through the resistor network, and out to ground. Our goal will be to find the current through each resistor (call the current flowing through the first resistor $I_1$, through the second, $I_2$, and so on); our tool will be Kirchhoff's laws.

The "wires" in the circuit diagram are perfect conductors, so are equipotentials. Each resistor provides a "voltage drop" of $IR$, from Ohm's law – these two observations can be combined to follow a particular path of current through the circuit and obtain a linear equation (this is basically the statement of Kirchhoff's loop rule). Take the current path that goes from the voltage source, through resistors one, two and five. We have:

$$V - I_1 R_1 - I_2 R_2 - I_5 R_5 = 0, \tag{9.6}$$

we start at potential $V$, and drop across three resistors down to zero (ground).

Taking the other two obvious paths, we can add two additional equations:

$$V - I_1 R_1 - I_3 R_3 - I_5 R_5 = 0$$
$$V - I_1 R_1 - I_4 R_4 - I_5 R_5 = 0. \tag{9.7}$$

We have three equations, but five unknowns. The remaining two equations come from the conservation of current (no current is created or destroyed in the process of going through the circuit – this leads to Kirchhoff's junction rule). The first immediate consequence of current conservation is that $I_5 = I_1$ – all the current that left the voltage source must eventually flow into ground. The second observation we can make is that all of the current leaving the voltage source must take one of the three paths through the parallel network of resistors, so

$$I_1 = I_2 + I_3 + I_4. \tag{9.8}$$

Now there are five equations and five unknowns. Since $I_5 = I_1$, we can simplify, leaving us with four equations and four unknowns – rewriting these, preparatory to matrix formulation, we have

$$I_1(R_1 + R_5) + I_2 R_2 = V$$
$$I_1(R_1 + R_5) + I_3 R_3 = V$$
$$I_1(R_1 + R_5) + I_4 R_4 = V \tag{9.9}$$
$$I_1 - I_2 - I_3 - I_4 = 0.$$

To set this up as an $\mathbb{A}\mathbf{x} = \mathbf{b}$ problem, let

$$\mathbf{x} \doteq \begin{pmatrix} I_1 \\ I_2 \\ I_3 \\ I_4 \end{pmatrix} \qquad \mathbf{b} \doteq \begin{pmatrix} V \\ V \\ V \\ 0 \end{pmatrix}. \tag{9.10}$$

Constructing $\mathbb{A}$ takes a little thought, and you should check the result:

$$\mathbb{A} \doteq \begin{pmatrix} R_1 + R_5 & R_2 & 0 & 0 \\ R_1 + R_5 & 0 & R_3 & 0 \\ R_1 + R_5 & 0 & 0 & R_4 \\ 1 & -1 & -1 & -1 \end{pmatrix}. \tag{9.11}$$

Now, if we knew how to construct $\mathbb{A}^{-1}\mathbf{b}$, we would have the current through each resistor. Mathematica has a built-in matrix inversion routine called `Inverse` that we use just to demonstrate the idea (remember, we're going to see how to invert matrices in this chapter, so the use of `Inverse` is not technically cheating). You should practice with this built-in routine by constructing (9.11) for $R_1 = 100\,\Omega$,

$R_2 = 200\,\Omega$, $R_3 = 1000\,\Omega$, $R_4 = 500\,\Omega$, and $R_5 = 100\,\Omega$, setting $V = 9$ V and inverting to find the current (which will then be in amperes).

### 9.2.2 Fitting data

Fitting functions to data provides a good example of the need for matrix inversion. The data could be experimental, or the output of some other numerical routine. The idea is to describe the discrete data in terms of continuous functions of one sort or another.

Suppose we have a set of data $\{d_j\}_{j=1}^n$, the voltage output from a circuit sampled at discrete temporal intervals, for example. We'll assume that the data is indexed by some fixed grid spacing, $\Delta t$, so that $d_j$ is the value of the data at "time" $t_j = j\Delta t$. Now take a set of functions $\{f_k(t)\}_{k=1}^n$ – these will be the functions that we use to fit the data, so they should be chosen based on physical insight into the type of data we are fitting. If, for example, the output voltage is clearly oscillatory, we might take $f_k(t) = \sin(2\pi k t)$. Our starting point is the assumption that:

$$d_j = \sum_{k=1}^n \alpha_k f_k(t_j) \quad \text{for } j = 1 \longrightarrow n, \tag{9.12}$$

with constants $\{\alpha_k\}_{k=1}^n$, i.e. we assume that the data can be captured by a linear combination of our functions $\{f_k(t)\}_{k=1}^n$. We want to find the particular linear combination, the set of coefficients appearing in (9.12), $\{\alpha_k\}_{k=1}^n$. Rewrite (9.12) as a matrix equation as follows (referring to the interpretation of matrix-vector multiplication from (9.5): we are taking a linear combination of the columns of a matrix)

$$\underbrace{\begin{pmatrix} d_1 \\ d_2 \\ \vdots \\ d_n \end{pmatrix}}_{\equiv \mathbf{b}} = \underbrace{\begin{pmatrix} f_1(t_1) & f_2(t_1) & \cdots & f_n(t_1) \\ f_1(t_2) & f_2(t_2) & \cdots & f_n(t_2) \\ \vdots & \vdots & \cdots & \vdots \\ f_1(t_n) & f_2(t_n) & \cdots & f_n(t_n) \end{pmatrix}}_{\equiv \mathbb{A}} \underbrace{\begin{pmatrix} \alpha_1 \\ \alpha_2 \\ \vdots \\ \alpha_n \end{pmatrix}}_{\equiv \mathbf{x}}, \tag{9.13}$$

where the matrix on the right is a square matrix. If the matrix is invertible, then we can find the vector of coefficients by inversion; we want $\mathbf{x} = \mathbb{A}^{-1}\mathbf{b}$ given $\mathbb{A}$ and $\mathbf{b}$.

### Polynomial interpolation

As an example, let's take some fake data, generate the matrix and vectors found in (9.13), and verify that we recover the correct linear combination. We'll make our data out of a known linear combination of polynomials, say

$$d(t) = 1 - 2t + 3t^2 - 4t^3 + 5t^4 - 6t^5 + 7t^6, \tag{9.14}$$

and we'll take a fixed grid, $t_j = j \Delta t$, so that our data is really $\{d_j = d(t_j)\}_{j=1}^n$. Take the functions $\{f_k(t)\}_{k=1}^n$ to be $f_k(t) = t^{k-1}$. It is clear that to correctly match $d(t)$, we must have $n = 7$, and so we must have only seven points in our temporal grid. Now we can make the matrix associated with (9.13). In this case, where we have a square matrix of increasing polynomials evaluated on a grid of the same size as the maximum polynomial power, $\mathbb{A}$ is called a "Vandermonde" matrix and takes the form

$$\mathbb{A} \doteq \begin{pmatrix} 1 & t_1 & t_1^2 & t_1^3 & t_1^4 & t_1^5 & t_1^6 \\ 1 & t_2 & t_2^2 & t_2^3 & t_2^4 & t_2^5 & t_2^6 \\ 1 & t_3 & t_3^2 & t_3^3 & t_3^4 & t_3^5 & t_3^6 \\ 1 & t_4 & t_4^2 & t_4^3 & t_4^4 & t_4^5 & t_4^6 \\ 1 & t_5 & t_5^2 & t_5^3 & t_5^4 & t_5^5 & t_5^6 \\ 1 & t_6 & t_6^2 & t_6^3 & t_6^4 & t_6^5 & t_6^6 \\ 1 & t_7 & t_7^2 & t_7^3 & t_7^4 & t_7^5 & t_7^6 \end{pmatrix}. \tag{9.15}$$

We construct $\mathbf{b}$ out of the function (9.14) evaluated at our seven grid points, and solve $\mathbb{A}\mathbf{x} = \mathbf{b}$ for $\mathbf{x}$, recovering:

$$\alpha_1 = 1 \, \alpha_2 = -2 \, \alpha_3 = 3 \, \alpha_4 = -4 \, \alpha_5 = 5 \, \alpha_6 = -6 \, \alpha_7 = 7. \tag{9.16}$$

This example works out well, since we tailored our function to the polynomial basis used in the Vandermonde matrix – a plot of the function $d(t)$ (or, equivalently in this case, $\sum_{k=1}^7 \alpha_k t^{k-1}$ after we have performed the matrix inversion to find the coefficients), together with the grid points used to generate the solution, is shown in Figure 9.2.

This procedure of polynomial fitting by matrix inversion is identical in content (although different in spirit) to the explicit construction of interpolating polynomials from Section 6.3. All we are doing is matching a set of data with a polynomial of some degree, and either approach will yield the same polynomial.

There are two problems with the type of data fitting described by (9.13). The first problem is that, assuming the matrix $\mathbb{A}$ is invertible, we *will* get a solution. But that solution, remember, is guaranteed only to match the data at the points $t_j$; what happens in between those points may or may not be relevant. As an example, suppose our data was made from a discretization of $d(t) = \sin(16\pi t)$. We'll keep our polynomial basis, in fact, we'll just use (9.15) on seven sampled points of $d(t)$. The result is shown in Figure 9.3. While our polynomial matches the data at the seven sampled points, it does not capture the overall behavior of $d(t)$ away from the sampled points. This problem can be fixed by astute choice of basis functions; we need not use polynomials as our set $\{f_k(t)\}_{k=1}^n$.

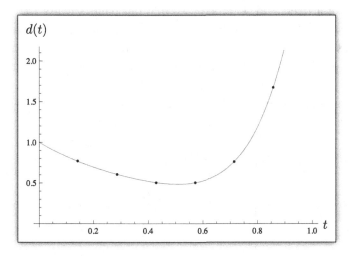

Figure 9.2 A plot of $d(t)$ from (9.14) (solid) and the points used to construct **b** (the data points used in the fit, the one at $t = 1.0$ is not shown).

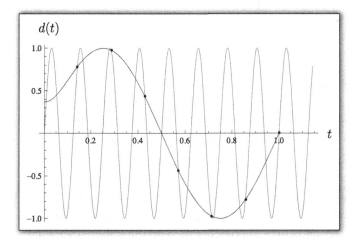

Figure 9.3 In gray, the function $d(t) = \sin(16\pi t)$, with sampled points marked. In black, the polynomial obtained by inverting (9.15) using $\{d_j = d(t_j)\}_{j=1}^n$ to construct the vector **b** – this polynomial goes through all seven points, but cannot capture the oscillatory structure away from the grid points.

The second problem is that the matrix $\mathbb{A}$ appearing in (9.13) must be square. If our sampled data has $n = 10\,000$, then we have to cook up ten thousand functions $f_k(t)$, and generate a large matrix for inversion, even though the sampled data may just consist of a single, simple function. As an example, suppose $d(t) = \sin(16\pi t)$, and you have cleverly decided to take the functions $f_k(t) = \sin(2\pi k t)$. Then it is clear that only $\alpha_8 = 1$ is non-zero, and yet you must still go through the procedure of making a $10\,000 \times 10\,000$ matrix out of irrelevant functions $f_k$ where $k \neq 8$. That is a huge waste, and we will see how to avoid this situation for linear problems in Chapter 10, and for the more general case in Chapter 12.

## 9.3 How do you invert a matrix?

Beyond the obvious paper and pencil-pushing routine, how does one invert a gigantic matrix? How does `Mathematica` do it? More important, is there a particularly efficient way to invert matrices? In some cases, the answer is simple – we know how to invert diagonal matrices, just take the inverse of the diagonal entries, and there are similarly easy ways to invert matrices with a lot of structure.

Most matrix inversion methods involve the construction of very special matrix decompositions that exist for invertible matrices (and have analogues for non-invertible matrices) – these decompositions reduce the original arbitrary matrix to a product of matrices with simple inverses. We'll take a look at some canonical matrix decompositions and how they facilitate inversion. This is just to give a flavor, so when someone comes up to you on the street and asks you what's going on under a particular hood, you have a notion – the actual machinations involved in constructing stable, efficient matrix decompositions are vast (see [19], the standard for exhaustive listings).

### 9.3.1 Easy inversions

Our target will be factorizations of matrices into other matrices; the goal is to factor the hard-to-invert $\mathbb{A}$ into easy-to-invert products. What matrices do we know that are easy to invert? Diagonal matrices are the easiest: the inverse of a diagonal matrix is just the diagonal matrix with the inverse entries (i.e. if a matrix $\mathbb{D}$ has the single entry $d_{ii}$ in its $i$th row, then the inverse matrix $\mathbb{D}^{-1}$ has the single entry $1/d_{ii}$ in its $i$th row). Orthogonal matrices, ones with the property that $\mathbb{Q}^T \mathbb{Q} = \mathbb{I}$, are easy to invert – for these, taking the transpose produces the inverse.

Less obvious, perhaps, are the lower and upper triangular matrices. An upper triangular matrix has entries above, and on, its diagonal. If we have an upper triangular matrix $\mathbb{R} \in \mathbb{R}^{n \times n}$, and we're given a vector $\mathbf{b} \in \mathbb{R}^n$, then we can find $\mathbf{x} \in \mathbb{R}^n$ recursively. Think of the form of the equation $\mathbb{R}\mathbf{x} = \mathbf{b}$:

$$\begin{pmatrix} R_{11} & R_{12} & R_{13} & \cdots & & R_{1n} \\ 0 & R_{22} & R_{23} & \cdots & & R_{2n} \\ 0 & 0 & \ddots & \vdots & & \vdots \\ 0 & 0 & \cdots & R_{(n-1)(n-1)} & & R_{(n-1)n} \\ 0 & 0 & \cdots & 0 & & R_{nn} \end{pmatrix} \begin{pmatrix} x_1 \\ x_2 \\ \vdots \\ x_{n-1} \\ x_n \end{pmatrix} = \begin{pmatrix} b_1 \\ b_2 \\ \vdots \\ b_{n-1} \\ b_n \end{pmatrix}. \tag{9.17}$$

Start from the bottom, we have the equation:

$$R_{nn}x_n = b_n \longrightarrow x_n = \frac{b_n}{R_{nn}}, \tag{9.18}$$

so we know $x_n$. One up from the bottom, our matrix-vector equation reads:

$$R_{(n-1)(n-1)}x_{n-1} + R_{(n-1)n}x_n = b_{n-1}, \tag{9.19}$$

but since we know $x_n$, we can solve for $x_{n-1}$:

$$x_{n-1} = \frac{1}{R_{(n-1)(n-1)}}\left(b_{n-1} - R_{(n-1)n}x_n\right). \tag{9.20}$$

We continue up the tower: for each row $j$, we know all the values for $x_k$ with $k > j$, and we can use that knowledge to solve for $x_j$. It may be algebraically tedious (providing just the sort of repetitive task at which computers excel), but it is straightforward (meaning that we can easily generate instructions for a computer to follow). Our first approach to general matrix inversion will be to develop a factorization for $\mathbb{A} \in \mathbb{R}^{n \times n}$ that gives us both an orthogonal matrix, and an upper triangular matrix, each of which is easy to invert.

### 9.3.2 The QR decomposition

The basic idea is that for any nonsingular $\mathbb{A} \in \mathbb{R}^{n \times n}$, we can factor it into an orthogonal matrix,[2] $\mathbb{Q}$, and an upper triangular matrix $\mathbb{R} \in \mathbb{R}^{n \times n}$: $\mathbb{A} = \mathbb{Q}\mathbb{R}$ with both $\mathbb{Q}$ and $\mathbb{R}$ nonsingular. Here, solving the inverse problem amounts to inverting $\mathbb{Q}$ and $\mathbb{R}$. For $\mathbb{A}\mathbf{x} = \mathbf{b}$, we take:

$$\mathbb{Q}\mathbb{R}\mathbf{x} = \mathbf{b} \longrightarrow \mathbb{R}\mathbf{x} = \mathbb{Q}^T\mathbf{b} \equiv \mathbf{c}, \tag{9.21}$$

and then we just do an upper triangular inversion of $\mathbb{R}\mathbf{x} = \mathbf{c}$ as described above. The nice property of $\mathbb{Q}$ is its orthogonality – we are generating a "natural" orthonormal basis for the space spanned by the columns of $\mathbb{A}$.

### 9.3.3 QR exists (for square $\mathbb{A}$)

Take $\mathbb{A} \in \mathbb{R}^{n \times n}$ (invertible) – the idea for the QR decomposition is to progressively generate a set of orthonormal vectors that spans $\mathbb{R}^n$ and write the column vectors of $\mathbb{A}$ in terms of these – the coefficients appearing in those linear combinations form precisely the columns of the matrix $\mathbb{R}$ as suggested by (9.5) (each column of $\mathbb{R}$, when multiplied by $\mathbb{Q}$, defines a different linear combination of the basis vectors that are the columns of $\mathbb{Q}$).

---

[2] Orthogonality means that $\mathbb{Q}^T\mathbb{Q} = \mathbb{I}$, i.e. that the columns of $\mathbb{Q}$ form an orthonormal basis for $\mathbb{R}^n$ for square $\mathbb{Q}$.

To prove the existence of the factorization, we basically construct the two matrices. We will take a set of vectors, and successively construct maximally independent vectors that span the same space while keeping track of the original vectors' decomposition in the resulting basis. As an example that provides a sketch of the proof, if we had three vectors $\mathbf{a}_1$, $\mathbf{a}_2$, and $\mathbf{a}_3$ that are linearly independent, and we want a basis for the three-dimensional space spanned by these (and the coefficients needed to reproduce $\mathbf{a}_1$, $\mathbf{a}_2$, and $\mathbf{a}_3$ in that basis), we start with the first vector and normalize it:

$$\mathbf{Q}_1 = \frac{\mathbf{a}_1}{R_{11}}, \qquad R_{11} = a_1 \tag{9.22}$$

where $a_1 \equiv \sqrt{\mathbf{a}_1 \cdot \mathbf{a}_1}$. $\mathbf{Q}_1$ is a unit vector pointing in the direction of $\mathbf{a}_1$, and we have defined the linear combination of $\mathbf{Q}_1$ that gives $\mathbf{a}_1$, namely $\mathbf{a}_1 = R_{11}\mathbf{Q}_1$. Now take $\mathbf{a}_2$ – we'll project out all of the $\mathbf{Q}_1$ direction:

$$\mathbf{q}_2 = \mathbf{a}_2 - \underbrace{(\mathbf{Q}_1 \cdot \mathbf{a}_2)}_{\equiv R_{12}} \mathbf{Q}_1, \qquad R_{22} = q_2. \tag{9.23}$$

This second vector is clearly orthogonal to $\mathbf{Q}_1$, and we can normalize it $\mathbf{Q}_2 = \mathbf{q}_2/R_{22}$. Once again, we have the correct linear combination in the definition of the coefficients $R_{12}$ and $R_{22}$: $\mathbf{a}_2 = R_{12}\mathbf{Q}_1 + R_{22}\mathbf{Q}_2$ from (9.23). Finally, the third vector is just $\mathbf{a}_3$ with the two directions $\mathbf{Q}_1$ and $\mathbf{Q}_2$ projected out

$$\mathbf{q}_3 = \mathbf{a}_3 - \underbrace{(\mathbf{Q}_1 \cdot \mathbf{a}_3)}_{\equiv R_{13}} \mathbf{Q}_1 - \underbrace{(\mathbf{Q}_2 \cdot \mathbf{a}_3)}_{\equiv R_{23}} \mathbf{Q}_2, \qquad R_{33} = q_3, \tag{9.24}$$

and we can perform the final normalization: $\mathbf{Q}_3 = \mathbf{q}_3/R_{33}$ to obtain the linearly independent, normalized, set. The vector $\mathbf{a}_3$ can be written in terms of the three basis vectors: $\mathbf{a}_3 = R_{13}\mathbf{Q}_1 + R_{23}\mathbf{Q}_2 + R_{33}\mathbf{Q}_3$, from (9.24).

If we view the vectors $\mathbf{Q}_j$ as the columns of a matrix $\mathbb{Q}$, then it is clear by construction that $\mathbb{Q}^T\mathbb{Q} = \mathbb{I}$, so $\mathbb{Q}$ is orthogonal. In addition, the three equations:

$$\mathbf{a}_1 = R_{11}\mathbf{Q}_1$$
$$\mathbf{a}_2 = R_{12}\mathbf{Q}_1 + R_{22}\mathbf{Q}_2 \tag{9.25}$$
$$\mathbf{a}_3 = R_{13}\mathbf{Q}_1 + R_{23}\mathbf{Q}_2 + R_{33}\mathbf{Q}_3$$

can be written in matrix form by taking $\mathbb{A}$ to have $\mathbf{a}_1$, $\mathbf{a}_2$, and $\mathbf{a}_3$ as its columns, and defining the matrix:

$$\mathbb{R} \doteq \begin{pmatrix} R_{11} & R_{12} & R_{13} \\ 0 & R_{22} & R_{23} \\ 0 & 0 & R_{33} \end{pmatrix} \tag{9.26}$$

then (referring again to (9.5)):

$$A = \mathbb{Q}\mathbb{R}. \tag{9.27}$$

Now, generally, we will be given a matrix $A \in \mathbb{R}^{n \times n}$ (invertible), and we view its column vectors as the collection $\{\mathbf{a}_i\}_{i=1}^n$. Instead of stopping at $\mathbf{Q}_3$, we continue to orthogonalize and normalize until we have $\mathbf{Q}_n$. The procedure of adding basis vectors one at a time, and recovering the decomposition of the vector $\mathbf{a}_j$ out of the $j$ basis vectors, is what gives the matrix $\mathbb{R}$ its upper triangular structure. The algorithm that carries out this procedure, called "Gram–Schmidt," is shown in Algorithm 9.1 (see [14, 19, 45] for discussions of this algorithm, and its modified form).[3]

---

**Algorithm 9.1** Gram–Schmidt

---

  **for** $j = 1 \to n$ **do**
    $\mathbf{q}_j \leftarrow \mathbf{a}_j$
    **for** $i = 1 \to j - 1$ **do**
      $R_{ij} \leftarrow \mathbf{Q}_i \cdot \mathbf{a}_j$
      $\mathbf{q}_j \leftarrow \mathbf{q}_j - R_{ij}\mathbf{Q}_i$
    **end for**
    $R_{jj} \leftarrow \sqrt{\mathbf{q}_j \cdot \mathbf{q}_j}$
    $\mathbf{Q}_j = \frac{\mathbf{q}_j}{R_{jj}}$
  **end for**

---

**Example**
In the interests of transparency, let's take the matrix:

$$A = \begin{pmatrix} 1 & 2 \\ 3 & 4 \end{pmatrix} \tag{9.28}$$

and compute its QR decomposition.

Our first step is to take the first column of $A$ and normalize it. For

$$\mathbf{a}_1 \doteq \begin{pmatrix} 1 \\ 3 \end{pmatrix} \tag{9.29}$$

we have $\sqrt{\mathbf{a}_1 \cdot \mathbf{a}_1} = \sqrt{10}$ – this provides the $\mathbb{R}$ entry: $R_{11} = \sqrt{10}$. The normalized vector $\mathbf{a}_1$ becomes the first column in $\mathbb{Q}$:

$$\mathbf{Q}_1 = \frac{1}{\sqrt{10}} \begin{pmatrix} 1 \\ 3 \end{pmatrix}. \tag{9.30}$$

---

[3] This is the "classical" Gram–Schmidt iteration, and proceeds exactly as described. It is not numerically stable, meaning here that it will fail to produce, for example, orthogonal $\mathbb{Q}$ for some matrices – see Problems 9.13 and 9.14 for an exploration of the "modified" Gram–Schmidt procedure.

Next, we project out all $\mathbf{Q}_1$ components in $\mathbf{a}_2$:

$$\mathbf{q}_2 = \mathbf{a}_2 - \underbrace{(\mathbf{Q}_1 \cdot \mathbf{a}_2)}_{=R_{12}} \mathbf{Q}_1$$

$$= \begin{pmatrix} 2 \\ 4 \end{pmatrix} - \frac{7}{5} \begin{pmatrix} 1 \\ 3 \end{pmatrix} \tag{9.31}$$

$$= \begin{pmatrix} \frac{3}{5} \\ -\frac{1}{5} \end{pmatrix}$$

where the value of $R_{12}$ is defined in the above equation as the dot product of $\mathbf{Q}_1$ with $\mathbf{a}_2$: $R_{12} = \frac{14}{\sqrt{10}}$. Finally, we normalize $\mathbf{q}_2$ to find $\mathbf{Q}_2$, the second column of $Q$ – the magnitude of $\mathbf{q}_2$ also provides the $R_{22}$ entry: $R_{22} = \sqrt{\mathbf{q}_2 \cdot \mathbf{q}_2} = \sqrt{\frac{2}{5}}$, and the normalized vector is:

$$\mathbf{Q}_2 = \sqrt{\frac{5}{2}} \begin{pmatrix} \frac{3}{5} \\ -\frac{1}{5} \end{pmatrix}. \tag{9.32}$$

Putting it all together, we have:

$$\begin{pmatrix} 1 & 2 \\ 3 & 4 \end{pmatrix} = \begin{pmatrix} \frac{1}{\sqrt{10}} & \frac{3}{\sqrt{10}} \\ \frac{3}{\sqrt{10}} & -\frac{1}{\sqrt{10}} \end{pmatrix} \begin{pmatrix} \sqrt{10} & \frac{14}{\sqrt{10}} \\ 0 & \sqrt{\frac{2}{5}} \end{pmatrix}, \tag{9.33}$$

where the first matrix is $Q$, the second is $R$. You can check the product by hand, but again, `Mathematica` provides a routine to perform this special decomposition, called, predictably: `QRDecomposition` (note that the output orthogonal matrix provided by that routine is the transpose of our $Q$).

For $A \in \mathbb{R}^{n \times m}$ (usually taken arbitrarily with $n > m$, so long skinny matrices), we can still define $Q \in \mathbb{R}^{n \times m}$ and $R \in \mathbb{R}^{m \times m}$ such that $A = QR$. The upper triangular matrix is still invertible, provided $A$ has "full column rank," i.e. its columns are linearly independent. It is possible to find $Q$ even for noninvertible and/or non-spanning $A$. In this case, $R$ is more difficult to define and use, but the utility is still clear: $Q$ forms an orthonormal basis for the space spanned by the columns of $A$.

### 9.3.4 The LU decomposition

Any invertible matrix can be decomposed into the product of an upper triangular (all non-zero entries above the diagonal, with diagonal included) and a lower triangular (non-zero entries below, and on, the diagonal) matrix:

$$A = LU. \tag{9.34}$$

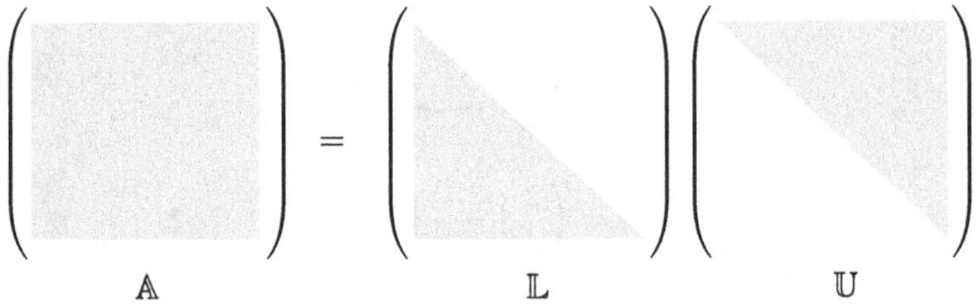

Figure 9.4 The decomposition of $\mathbb{A}$ into the product of a lower triangular ($\mathbb{L}$) and upper triangular ($\mathbb{U}$) matrix.

The mental picture to keep in mind is shown in Figure 9.4.

Let's start with the advertisement. For the matrix inverse problem

$$\mathbb{A}\mathbf{x} = \mathbf{b}, \tag{9.35}$$

if the LU decomposition exists, so that we have $\mathbb{L}$ and $\mathbb{U}$ with $\mathbb{A} = \mathbb{L}\mathbb{U}$, then we can split the problem into two "easy" inversions:

$$\mathbb{L}\mathbf{y} = \mathbf{b} \quad \mathbb{U}\mathbf{x} = \mathbf{y}. \tag{9.36}$$

Operationally, we now have two matrix inverse problems, first we solve $\mathbb{L}\mathbf{y} = \mathbf{b}$ for unknown $\mathbf{y}$, then we have to solve $\mathbb{U}\mathbf{x} = \mathbf{y}$ for unknown $\mathbf{x}$ (at that stage, we will know $\mathbf{y}$). Think about the first part of the problem, solving for $\mathbf{y}$ – written out, we have something like:

$$\begin{pmatrix} L_{11} & 0 & 0 & 0 & 0 \\ L_{21} & L_{22} & 0 & 0 & 0 \\ L_{31} & L_{32} & L_{33} & 0 & 0 \\ L_{41} & L_{42} & L_{43} & L_{44} & 0 \\ & & & & \ddots \end{pmatrix} \begin{pmatrix} y_1 \\ y_2 \\ y_3 \\ y_4 \\ \vdots \end{pmatrix} = \begin{pmatrix} b_1 \\ b_2 \\ b_3 \\ b_4 \\ \vdots \end{pmatrix}. \tag{9.37}$$

Then the solution, for $\mathbf{y}$, is:

$$y_1 = \frac{b_1}{L_{11}}$$

$$y_2 = \frac{b_2 - L_{21}y_1}{L_{22}}$$

$$y_3 = \frac{b_3 - L_{31}y_1 - L_{32}y_2}{L_{33}} \tag{9.38}$$

$$y_j = \frac{1}{L_{jj}}\left(b_j - \sum_{i=1}^{j-1} L_{ji}y_i\right),$$

where the last line represents the solution for the $j$th entry – notice that by the time we are calculating this entry, all previous ones are known, so the right-hand side is a straightforward computation. Indeed, all we do is loop through the indices, storing the answer for each $y_j$ until we get to the end. The procedure is just the top-down version of the upper triangular inversion scheme described in Section 9.3.1.

So if we know $\mathbb{L}$ and $\mathbf{b}$, it is relatively easy to find $\mathbf{y}$, and then we just do the same thing for the upper triangular problem: $\mathbb{U}\mathbf{x} = \mathbf{y}$. The combination of factoring the matrix $\mathbb{A}$ into $\mathbb{L}\mathbb{U}$ and then solving in this manner is referred to as "Gaussian elimination" (see [31]). There are a variety of matrices for which this type of approach will not work (not because of linear algebra, but because of the numerical structure of the matrix, and the associated rounding errors that a computer will inevitably make) – one hallmark of a "difficult" matrix is one in which certain entries are orders of magnitude larger than others. We can see the crux of the problem even for diagonal matrices: Suppose you have a diagonal matrix with entries of order 1, except for a very small entry of order $10^{-13}$ – this matrix is numerically uninvertible because it has a "zero" entry, and therefore negligible determinant. The same goes for diagonal matrices with entries of order 1 except for a single $10^{13}$ entry – now, by comparison, the rest of the diagonal elements are zero and we have a non-trivial one-dimensional subspace of the full $\mathbb{R}^n$, not invertible.

Caveats aside, the procedure can be corrected for these special cases, and we will focus on the bulk of matrices, for which LU will work well. All we need to do is establish the existence of the decomposition $\mathbb{A} = \mathbb{L}\mathbb{U}$, and then we'll move on to the construction of $\mathbb{L}$ and $\mathbb{U}$. Existence is most easily proven via induction.

---

**Induction proofs**

We review the idea behind "proof-by-induction" (see, for example, [33]). There are three basic ingredients to an inductive proof of a proposition indexed by $N$:

- Assume the proposition is true for $N \in \mathbb{Z}$.
- Using this assumption, prove that it is true for $N + 1$.
- Show that the proposition is true for "some" $N$.

Inductive proofs are most useful when there is a natural integer counting going on, and when you already know the answer, so to speak. That is, generally induction is used to verify propositions rather than generate (or nullify) them.

As an example, let's prove, for integer $N$, that

$$S(N) \equiv \sum_{j=1}^{N} j^2 = \frac{1}{6}N(N + 1)(2N + 1), \qquad (9.39)$$

and right off the bat, notice the signposts for induction: 1. Someone has told us the answer, and 2. $N$ is an integer.

Start by assuming (9.39) is true for an integer $N$, and we'll show that it is true for $N + 1$. Consider:

$$S(N + 1) = \sum_{j=1}^{N+1} j^2 = \sum_{j=1}^{N} j^2 + (N + 1)^2 = S(N) + (N + 1)^2, \tag{9.40}$$

where we used the inductive assumption in the final equality. Now performing the sum and writing the result in terms of $P \equiv N + 1$,

$$S(N + 1) = \frac{1}{6} N(N + 1)(2N + 1) + (N + 1)^2$$

$$= \frac{1}{6}(1 + N)(2 + N)(3 + 2N) \tag{9.41}$$

$$= \frac{1}{6} P(P + 1)(2P + 1)$$

the desired form. We have established that if (9.39) is true for $N$, it is also true for $N + 1$. Now we just need an initial $N$ to get the process started. Take $N = 1$,

$$\sum_{j=1}^{1} 1^2 = 1 = \frac{1}{6}(1)(2)(3) = 1, \tag{9.42}$$

and our induction is complete.

### 9.3.5 LU exists

Proving existence for matrix factorizations is fun, specially when the proof follows an inductive pattern. We will prove the existence of the LU factorization explicitly (following [14]).

**To show:** For $\mathbb{A} \in \mathbb{R}^{n \times n}$ invertible,[4] there exists upper triangular $\mathbb{L}$ and lower triangular $\mathbb{U} \in \mathbb{R}^{n \times n}$ such that $\mathbb{A} = \mathbb{L}\mathbb{U}$ (as a corollary, it is clear that $\mathbb{L}$ and $\mathbb{U}$ are each separately invertible).

**Proof by induction:** Take $n = 1$, then $\mathbb{A} = 1 \times A_{11}$ is a valid LU decomposition. The lower triangular $\mathbb{L}$ is one-by-one with an entry 1 on its diagonal, while the upper triangular $\mathbb{U}$ is one-by-one with $A_{11}$ on its diagonal. So the inductive assumption holds for $n = 1$.

Now assume true for $\mathbb{B} \in \mathbb{R}^{n \times n}$, i.e. $\mathbb{B} = \mathbb{L}\mathbb{U}$, and let us write a generic invertible matrix $\mathbb{A} \in \mathbb{R}^{(n+1) \times (n+1)}$ in block form:

$$\mathbb{A} = \begin{pmatrix} \mathbb{B} & \mathbf{b} \\ \mathbf{c}^T & \delta \end{pmatrix} \tag{9.43}$$

---

[4] With all leading principal submatrices invertible, see Section 9.3.7.

with $\mathbf{b}, \mathbf{c} \in \mathbb{R}^n$, and $\delta \in \mathbb{R}$. Let $\mathbf{u} = \mathbb{L}^{-1}\mathbf{b}$ ($\mathbb{L}^{-1}$ exists since $\mathbb{B}$ is nonsingular by assumption), $\mathbf{v}^T = \mathbf{c}^T \mathbb{U}^{-1}$ ($\mathbb{U}^{-1}$ exists since $\mathbb{B}$ is nonsingular), and $\eta = -\mathbf{v}^T\mathbf{u} + \delta$, then we can factor the right-hand side of (9.43)

$$\mathbb{A} = \begin{pmatrix} \mathbb{L} & 0 \\ \mathbf{v}^T & 1 \end{pmatrix} \begin{pmatrix} \mathbb{U} & \mathbf{u} \\ 0 & \eta \end{pmatrix}. \tag{9.44}$$

By construction, then, $\mathbb{A}$ has a decomposition in terms of a lower triangular and upper triangular product – the only potential snag would be $\eta = 0$, which it cannot be, since $\mathbb{A}$ is invertible by assumption (and $\det(\mathbb{A}) = \det(\mathbb{L})\det(\mathbb{U})$ – the determinant of a lower/upper triangular matrix is just the product of the diagonal elements, so if $\eta = 0$, then $\det(\mathbb{A}) = 0$) so we are done.

Almost any numerical matrix package will include an LU factorization routine. Using Mathematica's `LUDecomposition` routine, we can demonstrate the factorization with the example from (9.28):

$$\mathbb{A} = \begin{pmatrix} 1 & 2 \\ 3 & 4 \end{pmatrix} = \begin{pmatrix} 1 & 0 \\ 3 & 1 \end{pmatrix} \begin{pmatrix} 1 & 2 \\ 0 & -2 \end{pmatrix}. \tag{9.45}$$

---

**Example**

Before moving on to the actual computation of the LU factorization, let's do an example of the first step in the solution process for $\mathbb{A}\mathbf{x} = \mathbf{b}$, now that we have the LU decomposition of a manageable matrix. Remember that we take the pair:

$$\mathbb{L}\mathbf{y} = \mathbf{b} \quad \mathbb{U}\mathbf{x} = \mathbf{y} \tag{9.46}$$

and solve the left equation first, then use the (now known) $\mathbf{y}$ to solve the equation on the right. Take $\mathbb{A}$ as in (9.45), so we know $\mathbb{L}$. Suppose we were given $\mathbf{b} = \begin{pmatrix} 1 \\ 2 \end{pmatrix}$. Then we need to find $y_1$ and $y_2$ in:

$$\begin{pmatrix} 1 & 0 \\ 3 & 1 \end{pmatrix} \begin{pmatrix} y_1 \\ y_2 \end{pmatrix} = \begin{pmatrix} 1 \\ 2 \end{pmatrix}. \tag{9.47}$$

The top equation tells us that $y_1 = 1$, and the second equation, which reads: $3y_1 + y_2 = 2$ can be solved by inserting the known value of $y_1$: $3 + y_2 = 2 \longrightarrow y_2 = -1$. So we conclude:

$$\mathbf{y} = \begin{pmatrix} 1 \\ -1 \end{pmatrix}. \tag{9.48}$$

Now we're ready to carry out the upper-triangular inverse to solve for $\mathbf{x}$ in $\mathbb{U}\mathbf{x} = \mathbf{y}$.

### 9.3.6 Computing the LU decomposition

The inductive proof that establishes the existence of an LU factorization does not directly inform its construction. The simplest way to construct the matrices $\mathbb{L}$ and $\mathbb{U}$, from [45] (for example), is not the most efficient, but makes the procedure clear.

Define a matrix $\mathbb{L}_k$ to have ones along the diagonal, and non-zero entries below the diagonal in its $k$th column (all other entries are zero). For example,

$$\mathbb{L}_2 \doteq \begin{pmatrix} 1 & 0 & 0 & 0 \\ 0 & 1 & 0 & 0 \\ 0 & a & 1 & 0 \\ 0 & b & 0 & 1 \end{pmatrix}, \tag{9.49}$$

with arbitrary $a$ and $b$, is a four-by-four matrix with this structure.

These matrices have the property that multiplication respects the lower triangular form, so that $\mathbb{L}_j \mathbb{L}_k$ ($j < k$) results in a matrix that is lower triangular, with ones along the diagonal, and that has the $j$th column of $\mathbb{L}_j$ and $k$th column of $\mathbb{L}_k$ (if $j = k$, we get back an $\mathbb{L}_j$ matrix with the sum of the non-zero, below-diagonal, terms). Again, working in four dimensions to demonstrate the pattern:

$$\begin{pmatrix} 1 & 0 & 0 & 0 \\ 0 & 1 & 0 & 0 \\ 0 & a & 1 & 0 \\ 0 & b & 0 & 1 \end{pmatrix} \begin{pmatrix} 1 & 0 & 0 & 0 \\ 0 & 1 & 0 & 0 \\ 0 & 0 & 1 & 0 \\ 0 & 0 & c & 1 \end{pmatrix} = \begin{pmatrix} 1 & 0 & 0 & 0 \\ 0 & 1 & 0 & 0 \\ 0 & a & 1 & 0 \\ 0 & b & c & 1 \end{pmatrix}. \tag{9.50}$$

The inverse of an $\mathbb{L}_k$ matrix is also an $\mathbb{L}_k$ matrix, with entries negated below the diagonal; for the example matrix in (9.49), we have the inverse, denoted here as $\hat{\mathbb{L}}_k$:

$$\hat{\mathbb{L}}_2 \equiv \begin{pmatrix} 1 & 0 & 0 & 0 \\ 0 & 1 & 0 & 0 \\ 0 & -a & 1 & 0 \\ 0 & -b & 0 & 1 \end{pmatrix}. \tag{9.51}$$

With these properties in place, our project is to successively upper-triangularize a matrix $\mathbb{A}$ by multiplying $\mathbb{A}$ by these $\mathbb{L}_k$ matrices so as to zero the portion below the diagonal – the idea is to construct a sequence of multiplications:

$$\mathbb{L}_{n-1} \mathbb{L}_{n-2} \dots \mathbb{L}_2 \mathbb{L}_1 \mathbb{A} = \mathbb{U}, \tag{9.52}$$

such that $\mathbb{L}_1 \mathbb{A}$ gives a matrix with zeroes below the diagonal in the first column, $\mathbb{L}_2 \mathbb{L}_1 \mathbb{A}$ gives a matrix with zeroes below the diagonal for the first two columns, etc. Then the product, $\left( \prod_{i=n-1}^{1} \mathbb{L}_i \right) \mathbb{A}$ will be upper triangular, indeed *the* upper triangular matrix $\mathbb{U}$ as indicated in (9.52). The lower triangular matrix, $\mathbb{L}$, is just the inverse of the $\mathbb{L}_k$ products on the left of (9.52) – that product is itself lower triangular, and the inverse of a lower triangular matrix is lower triangular, so we will get $\mathbb{A} = \mathbb{L}\mathbb{U}$ in the end. You will show in Problem 9.5 how to move $\mathbb{L}_{n-1}\mathbb{L}_{n-2}\dots\mathbb{L}_1$ to the right-hand side of (9.52) given the individual inverses of the product constituents, allowing us to generate both $\mathbb{U}$ and $\mathbb{L}$ in one pass.

How can we construct the $\mathbb{L}_1$ matrix to get the process started? Well, remember this matrix, when multiplying $\mathbb{A}$, is meant to zero the first column, and it need do no more than that. Denote the entries of $\mathbb{A}$ by $a_{ij}$, and think of the first column vector of $\mathbb{A}$ – this is a vector in $\mathbb{R}^n$, and we want to construct $\mathbb{L}_1$ that will zero everything below the first entry (when multiplying $\mathbb{A}$). It is relatively clear that the following will work:

$$\mathbb{L}_1 \doteq \begin{pmatrix} 1 & 0 & 0 & 0 & \dots \\ -\frac{a_{21}}{a_{11}} & 1 & 0 & 0 & \dots \\ -\frac{a_{31}}{a_{11}} & 0 & 1 & 0 & \dots \\ -\frac{a_{41}}{a_{11}} & 0 & 0 & 1 & \dots \\ \vdots & 0 & 0 & 0 & \ddots \end{pmatrix}. \tag{9.53}$$

Now we have $\tilde{\mathbb{A}} = \mathbb{L}_1\mathbb{A}$ with no entries in the first column (except for the diagonal element). Proceeding, we construct $\mathbb{L}_2$ to zero the entries in the second column of $\tilde{\mathbb{A}}$ – the form is the same:

$$\mathbb{L}_2 \doteq \begin{pmatrix} 1 & 0 & 0 & 0 & \dots \\ 0 & 1 & 0 & 0 & \dots \\ 0 & -\frac{\tilde{a}_{32}}{\tilde{a}_{22}} & 1 & 0 & \dots \\ 0 & -\frac{\tilde{a}_{42}}{\tilde{a}_{22}} & 0 & 1 & \dots \\ 0 & -\frac{\tilde{a}_{52}}{\tilde{a}_{22}} & 0 & 0 & \dots \end{pmatrix}. \tag{9.54}$$

Now we have $\mathbb{L}_2\mathbb{L}_1\mathbb{A}$, a matrix with no entries below the diagonal in the first two columns. The construction proceeds until we have no entries below the diagonal for all columns (except the last, whose final entry is on the diagonal).

Algorithmically, given a matrix $\mathbb{A}$ with no entries in the first $k-1$ columns, we construct $\mathbb{L}_k$ as shown in Algorithm 9.2. You can see the advantage to this special

---

**Algorithm 9.2** Constructing $\mathbb{L}_k$

$\mathbb{L}_k = \mathbb{I}$
$\hat{\mathbb{L}}_k = \mathbb{I}$
**for** $j = k+1 \to n$ **do**
$\quad \mathbb{L}_k[j, k] = -\mathbb{A}[j, k]/\mathbb{A}[k, k]$
$\quad \hat{\mathbb{L}}_k[j, k] = -\mathbb{L}_k[j, k]$
**end for**

form – we get the matrix $\mathbb{L}_k$ and its inverse $\hat{\mathbb{L}}_k$ for free. Now it's just a matter of repeated application, and formation of the multiplication that gives the inverse. The full construction is shown in Algorithm 9.3.

---

**Algorithm 9.3** LU decomposition

---

$\mathbb{L} = \mathbb{I}$
$\mathbb{U} = \mathbb{A}$
**for** $k = 1 \rightarrow n - 1$ **do**
    $\mathbb{L}_k = \mathbb{I}$
    $\hat{\mathbb{L}}_k = \mathbb{I}$
    **for** $j = k + 1 \rightarrow n$ **do**
        $\mathbb{L}_k[j, k] = -\mathbb{U}[j, k]/\mathbb{U}[k, k]$
        $\hat{\mathbb{L}}_k[j, k] = -\mathbb{L}_k[j, k]$
    **end for**
    $\mathbb{U} \leftarrow \mathbb{L}_k \mathbb{U}$
    $\mathbb{L} \leftarrow \mathbb{L} \hat{\mathbb{L}}_k$
**end for**

---

### 9.3.7 Pivoting

In the statement of the LU decomposition theorem, I stuffed some extra conditions in a footnote. We now return to this note, and carefully define it by way of an example – it will also become clear that the "fix" is physically irrelevant. The matrix

$$\mathbb{A} = \begin{pmatrix} 0 & 1 \\ 1 & 1 \end{pmatrix} \tag{9.55}$$

is invertible (for any **b**, we can find **x** such that $\mathbb{A}\mathbf{x} = \mathbf{b}$). But our procedure for successively lower-triangularizing $\mathbb{A}$ will fail, since the first step involves generating $\mathbb{L}_1$ that zeroes out the entries below the diagonal in the first column of $\mathbb{A}$ and relies on the existence of $1/a_{11}$ (see (9.53)), which is infinite for the matrix in (9.55).

But for us, this matrix is part of a matrix equation, we are supposed to solve:

$$\begin{pmatrix} 0 & 1 \\ 1 & 1 \end{pmatrix} \begin{pmatrix} x_1 \\ x_2 \end{pmatrix} = \begin{pmatrix} b_1 \\ b_2 \end{pmatrix} \tag{9.56}$$

for $x_1$ and $x_2$ given $b_1$ and $b_2$. That amounts to the pair of equations:

$$x_2 = b_1$$
$$x_1 + x_2 = b_2. \tag{9.57}$$

It is the *order* in which these equations are taken that is causing the problem. After all, if we had started with this pair and transcribed them as:

$$x_1 + x_2 = b_2$$
$$x_2 = b_1 \tag{9.58}$$

then we would write the matrix form:

$$\begin{pmatrix} 1 & 1 \\ 0 & 1 \end{pmatrix} \begin{pmatrix} x_1 \\ x_2 \end{pmatrix} = \begin{pmatrix} b_2 \\ b_1 \end{pmatrix} \tag{9.59}$$

and now the matrix on the left is already upper triangular. The solution (for $x_1$ and $x_2$) to either (9.56) or (9.59) has got to be the same; all we did was re-order the equations, and yet one of them is amenable to LU decomposition, the other is not.

This difference is the crux of the "leading principal submatrices invertible" requirement for LU factorization. A leading principal submatrix of $\mathbb{A}$ is just a square matrix that is formed by taking the $j \times j$ block of $\mathbb{A}$ starting at the upper left-hand corner. For the matrix $\mathbb{A}$ in (9.55), the first leading principal submatrix is 0 (just the $A_{11}$ component), which is not invertible. The first leading principal submatrix of the matrix in (9.59) is 1, which is invertible.

Operationally, then, when we construct the LU decomposition using Algorithm 9.3, we could get a zero as a diagonal entry for one of two reasons – it could be, as above, a "bad" ordering for the equations we want to solve, or it could mean that $\mathbb{A}$ is singular. In the former case, we introduce a "pivoting" (fancy name for re-ordering the equations, as in going from (9.57) to (9.58)) step to ensure that all leading principal submatrices are invertible.

## 9.4 Determinants

Either of our factorizations, $\mathbb{A} = \mathbb{L}\mathbb{U} \in \mathbb{R}^{n \times n}$ or $\mathbb{A} = \mathbb{Q}\mathbb{R}$, allows us to easily calculate the determinant of (invertible) $\mathbb{A}$. For any matrices $\mathbb{A}$ and $\mathbb{B}$, we have $\det(\mathbb{A}\mathbb{B}) = \det(\mathbb{A})\det(\mathbb{B})$ (see [3], for example). So for the LU factorization, we can write:

$$\det(\mathbb{A}) = \det(\mathbb{L})\det(\mathbb{U}). \tag{9.60}$$

The determinant of an upper triangular matrix is just the product of its diagonal elements (as you will show in Problem 9.7), and the same is true for a lower triangular matrix. We can simplify further:

$$\det \mathbb{A} = \prod_{i=1}^{n} L_{ii} U_{ii}. \tag{9.61}$$

Similarly, for the QR decomposition, we have:

$$\det(\mathbb{A}) = \det(\mathbb{Q})\det(\mathbb{R}).\qquad(9.62)$$

The determinant of an orthogonal matrix is $\pm 1$ automatically (for rotations, the determinant is $+1$), and $\mathbb{R}$ is an upper triangular matrix,

$$\det \mathbb{A} = \pm \prod_{i=1}^{n} R_{ii}.\qquad(9.63)$$

## 9.5 Constructing $\mathbb{A}^{-1}$

Let's work out the timing associated with the calculation of, say, the QR decomposition – referring to Algorithm 9.1, each entry of the matrix $\mathbb{R}$ requires $n$ operations (a dot product), and there are $O(n^2)$ entries in $\mathbb{R}$, so the construction of the decomposition has $T(n) = O(n^3)$ (an unavoidable catastrophe). Once you have the decomposition, together with a provided vector $\mathbf{b}$, finding $\mathbf{x}$ in $\mathbb{A}\mathbf{x} = \mathbf{b}$ requires $O(n^2)$ operations (think of the multiplication of $\mathbf{b}$ by $\mathbb{Q}^T$).

It is interesting that nowhere in this chapter, which is on matrix inversion, do we actually construct the matrix inverse $\mathbb{A}^{-1}$. And yet, many programs will provide precisely such a quantity, a matrix in $\mathbb{R}^{n \times n}$ that has the property that $\mathbb{A}^{-1}\mathbb{A} = \mathbb{I}$. In practice, we have no real use for such an object – as you'll see in a moment, the construction of this general matrix inverse requires $n$ separate inversions, while we normally have a concrete vector $\mathbf{b}$ that requires only one "inversion" (one calculation of the QR decomposition, and appropriate construction of $\mathbf{x}$ adapted to this $\mathbf{b}$, rather than $n$ separate versions). So, while the QR (or LU) factorization is itself the most expensive part of the inversion, if we insist on forming the full matrix $\mathbb{A}^{-1}$ we introduce an additional $O(n^3)$ set of operations.

In order to make $\mathbb{A}^{-1} \in \mathbb{R}^{n \times n}$, think of a generic vector $\mathbf{b} = \sum_{i=1}^{n} b_i \mathbf{e}_i$ where $\{\mathbf{e}_i\}_{i=1}^{n}$ is the standard basis set.[5] We know that, as a matter of notation, the solution $\mathbf{x}$ to $\mathbb{A}\mathbf{x} = \mathbf{b}$ can be written as

$$\mathbf{x} = \sum_{i=1}^{n} b_i \mathbb{A}^{-1}\mathbf{e}_i\qquad(9.64)$$

so that we are interested in the individual products $\mathbb{A}^{-1}\mathbf{e}_i$ – let $\mathbf{x}_i$ be the solution to:

$$\mathbb{A}\mathbf{x}_i = \mathbf{e}_i \quad\longrightarrow\quad \mathbf{x}_i = \mathbb{A}^{-1}\mathbf{e}_i.\qquad(9.65)$$

[5] The basis vectors $\mathbf{e}_i$ have one non-zero entry, a 1 at location $i$.

Then we can write (9.64) as:

$$\mathbf{x} = \sum_{i=1}^{n} b_i \mathbf{x}_i, \tag{9.66}$$

and referring back to (9.5), we see that this sum can be thought of as a matrix vector multiplication – the matrix is what we would call $\mathbb{A}^{-1}$, so it is clear that our inverse matrix has columns that are the vectors $\{\mathbf{x}_i\}_{i=1}^{n}$:

$$\mathbb{A}^{-1} \doteq [\mathbf{x}_1 \quad | \quad \mathbf{x}_2 \quad | \quad \mathbf{x}_3 \quad | \quad \dots \quad | \quad \mathbf{x}_n]. \tag{9.67}$$

Back to the timing question, we can see that once we have formed the QR decomposition of $\mathbb{A}$, each vector $\mathbf{x}_i$ requires $O(n^2)$ operations to solve $\mathbb{R}\mathbf{x}_i = \mathbb{Q}^T \mathbf{e}_i$, and there are $n$ of them that we must construct, meaning that in order to make $\mathbb{A}^{-1}$ we have added another $O(n^3)$ to the total number of operations.

## Further reading

1. Arfken, George B. & Hans J. Weber. *Mathematical Methods for Physicists.* Academic Press, 2001.
2. Boas, Mary. *Mathematical Methods in the Physical Sciences.* Wiley, 2005.
3. Demmel, James W. *Applied Numerical Linear Algebra.* Siam, 1997.
4. Golub, Gene H. & Charles F. Van Loan. *Matrix Computations.* The Johns Hopkins University Press, 1996.
5. Isaacson, Eugene & Herbert Bishop Keller. *Analysis of Numerical Methods.* Dover, 1994.
6. Trefethen, Loyd N. & David Bau III. *Numerical Linear Algebra.* Siam, 1997.

## Problems

### Problem 9.1
Find the QR decomposition of the matrix:

$$\mathbb{A} = \begin{pmatrix} 5 & 6 \\ 7 & 8 \end{pmatrix}. \tag{9.68}$$

### Problem 9.2
The QR decomposition is defined for $\mathbb{A} \in \mathbb{R}^{n \times m}$ – it reads: $\mathbb{A} = \mathbb{Q}\mathbb{R}$ for $\mathbb{Q} \in \mathbb{R}^{n \times m}$ and $\mathbb{R} \in \mathbb{R}^{m \times m}$. Generate the decomposition for

$$\mathbb{A} \doteq \begin{pmatrix} 1 & 2 \\ 3 & 4 \\ 5 & 6 \end{pmatrix}. \tag{9.69}$$

### Problem 9.3
Prove that the product of $\mathbb{L}_j$ and $\mathbb{L}_k$ (both in $\mathbb{R}^{n \times n}$, defined in Section 9.3.6) has ones along the diagonal, the $j$th column of $\mathbb{L}_j$, and the $k$th column of $\mathbb{L}_k$ as its $j$ and $k$

columns for $j < k$. Hint – write the elements of the matrices explicitly, so that $(\mathbb{L}_j)_{pq} = \delta_{pq} + \ell_p \delta_{qj}$ where $\boldsymbol{\ell} \in \mathbb{R}^n$ is the $j$th column vector of $\mathbb{L}_j$, and similarly for $\mathbb{L}_k$, then put these into the definition of matrix multiplication and simplify using the sums over Kronecker deltas.

### Problem 9.4
Verify, for $n = 4$, that $\mathbb{L}_1^{-1}$ (for $\mathbb{L}_1$ with generic entries) is given by negating the entries of $\mathbb{L}_1$ below the diagonal.

### Problem 9.5
Given the iterative zeroing that goes on in constructing the LU decomposition: $\mathbb{L}_{N-1}\mathbb{L}_{N-2}\ldots\mathbb{L}_2\mathbb{L}_1\mathbb{A} = \mathbb{U}$, and that we can construct the inverse of the successive $\mathbb{L}_k$ while we are constructing $\mathbb{L}_k$, write the matrix multiplication that you will end up doing to construct $\mathbb{L}$ (in $\mathbb{A} = \mathbb{L}\mathbb{U}$) in terms of the inverses of $\mathbb{L}_k$.

### Problem 9.6
Prove that an orthogonal matrix $\mathbb{Q}$ has $\det(\mathbb{Q}) = \pm 1$.

### Problem 9.7
Prove (by induction) that $\det(\mathbb{L}) = \prod_{j=1}^n L_{jj}$ for a lower triangular matrix $\mathbb{L} \in \mathbb{R}^{n \times n}$ (this property holds for $\mathbb{U}$ upper triangular as well).

### Problem 9.8
Compute the LU factorization "by hand," following Algorithm 9.3, for the matrix:

$$\mathbb{A} = \begin{pmatrix} 1 & 2 & 3 \\ 4 & 5 & 6 \\ 7 & 8 & 10 \end{pmatrix}. \tag{9.70}$$

From your decomposition, find the determinant of $\mathbb{A}$.

### Problem 9.9
Using the matrix inverse definition in Section 9.5, find the inverse of the matrix:

$$\mathbb{A} = \begin{pmatrix} 1 & 2 \\ 0 & 3 \end{pmatrix} \tag{9.71}$$

(exploit the fact that this matrix is upper-triangular to construct the two vectors $\mathbf{x}_1$ and $\mathbf{x}_2$ you need).

## Lab problems

### Problem 9.10
Implement Algorithm 9.3, taking a matrix $\mathbb{A}$ and returning the matrices $\mathbb{L}$ and $\mathbb{U}$ with $\mathbb{A} = \mathbb{L}\mathbb{U}$. Make sure that your function exits, appropriately, if $\mathbb{A}$ is singular (note that

your function will also fail if $\mathbb{A}$ requires pivoting – we will be testing on matrices where pivoting is not necessary). Have the function return the determinant of the matrix $\mathbb{A}$. Test your decomposition on the matrix (9.70).

## Problem 9.11

Using your function from the previous problem, implement a lower-triangular and upper-triangular (provided as part of the chapter notebook) back solver to obtain a full LU-based matrix inversion routine. Use your routine to solve 100 separate $\mathbb{A}\mathbf{x} = \mathbf{b}$ problems, with $\mathbb{A} \in \mathbb{R}^{100 \times 100}$, $\mathbf{b} \in \mathbb{R}^{100}$, both generated randomly out of elements $\in [-1, 1]$. We take these random matrices to reduce the chance that the matrix will be singular, and that pivoting will be required. For each inversion, calculate the residual $\mathbf{r} \equiv \mathbb{A}\mathbf{x} - \mathbf{b}$ and find the maximum (absolute value) element. Make a plot of these elements, and calculate the mean of them over the 100 runs.

## Problem 9.12

Construct a random matrix $\mathbb{A} \in \mathbb{R}^{100 \times 100}$, and using your LU decomposition, find $\mathbb{A}^{-1}$ – verify that $\mathbb{A}^{-1}\mathbb{A} = \mathbb{I}$. Keep in mind, here, that you are providing one hundred different right-hand sides (the $\{\mathbf{e}_i\}_{i=1}^{100}$), but the matrix $\mathbb{A}$ is always the same, so you only need to perform the LU decomposition once.

## Problem 9.13

The classical Gram–Schmidt algorithm is easiest to describe, but it is "unstable." In practice, people use the "modified" Gram-Schmidt form – the ordering is a little different, although the output is identical (in theory). The procedure is shown in Algorithm 9.4 below – it proceeds given an input matrix $\mathbb{A} \in \mathbb{R}^{n \times n}$, the columns of that matrix are referred to as $\{\mathbf{a}_i\}_{i=1}^{n}$ as usual. You should compare this with Algorithm 9.1, they differ in only one line. Replace the `QRDecompose` function

---

**Algorithm 9.4** Modified Gram–Schmidt

> **for** $j = 1 \rightarrow n$ **do**
>  $\mathbf{q}_j \leftarrow \mathbf{a}_j$
>  **for** $i = 1 \rightarrow j - 1$ **do**
>    $R_{ij} \leftarrow \mathbf{Q}_i \cdot \mathbf{q}_j$
>    $\mathbf{q}_j \leftarrow \mathbf{q}_j - R_{ij}\mathbf{Q}_i$
>  **end for**
>  $R_{jj} \leftarrow \sqrt{\mathbf{q}_j \cdot \mathbf{q}_j}$
>  $\mathbf{Q}_j = \frac{\mathbf{q}_j}{R_{jj}}$
> **end for**

---

from the chapter notebook with modified Gram–Schmidt. The rest of the QR solution machinery remains the same. Try solving one hundred inversion problems as in Problem 9.11, again plotting the maximum absolute value of the residual for each run, and providing the mean. This establishes that modified Gram–Schmidt is doing the same thing as classical Gram–Schmidt.

### Problem 9.14

In this problem, we'll look at a matrix that cannot be decomposed using the classical Gram–Schmidt algorithm. The first step is to generate such a matrix (this example is modified from [45]):

(a) One way to "break" the classical algorithm is to use a matrix with a wide range of eigenvalues (so that the ratio of the smallest to the largest is small, making the matrix "look" singular). We can generate such a matrix by taking $\mathbb{A} = \mathbb{V}\mathbb{L}\mathbb{V}^T$ for orthogonal $\mathbb{V}$ and diagonal $\mathbb{L}$, reverse engineering a matrix $\mathbb{A}$ with the desired spectrum. Start with a matrix $\mathbb{B} \in \mathbb{R}^{50 \times 50}$ with random entries (use `RandomReal[{-1,1}]`). Find the QR decomposition of this matrix (using either classical or modified Gram–Schmidt), and take the orthogonal matrix to be $\mathbb{V}$. For $\mathbb{L}$, make a diagonal matrix with entries $L_{jj} = 1.5^{-j}$. Finally, perform the multiplication to define $\mathbb{A}$.

(b) Now compute the QR decomposition of $\mathbb{A}$ using both the classical and modified algorithms. First, ensure that $\mathbb{QR} - \mathbb{A}$ is small (you can find the largest absolute value of the elements of this residual matrix) for both algorithms. The problem with the classical Gram–Schmidt decomposition here is the orthogonality of the $\mathbb{Q}$ – find the maximum absolute value of the elements of $\mathbb{Q}^T\mathbb{Q} - \mathbb{I}$ for both algorithms – note that if $\mathbb{Q}$ is not orthogonal, we lose the ability to solve the inverse problem using the QR decomposition.

### Problem 9.15

For the following simple circuit (Figure 9.5), generate the matrix equation that you need to find the current through each resistor. Using the QR matrix inversion routine (with your modified Gram–Schmidt decomposition from Problem 9.13), find the current through each of the resistors for $R_1 = 5\,\Omega$, $R_2 = 10\,\Omega$, $R_3 = 1\,\Omega$, $R_4 = 20\,\Omega$, $R_5 = 100\,\Omega$, and $R_6 = 200\,\Omega$ if $V_0 = 9$ V. Be careful, not all of the matrix formulations you can generate using Kirchhoff's rules are invertible.

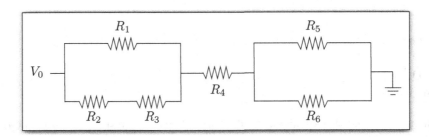

Figure 9.5  Circuit for Problem 9.15.

## Problem 9.16

We can use inversion to generate polynomial interpolating functions as described in Section 9.2.2. Take the discrete data:

$$d_j = \sin(2\pi j \Delta t), \tag{9.72}$$

for $j = 1 \longrightarrow n$, with $\Delta t = 1/n$, and $n = 12$, so that we have sampled twelve points of the sine function. Make the appropriate $12 \times 12$ Vandermonde matrix (the $k$th column of which will be a vector with entries $t_j^{k-1}$ for $j = 1 \longrightarrow n$ as in (9.15)). By solving $\mathbb{A}\mathbf{x} = \mathbf{d}$, you are finding the vector of coefficients in the polynomial interpolating function:

$$f(t) = \sum_{k=1}^{n} x_k t^{k-1} \tag{9.73}$$

that will match the data at the provided points. Using your LU solver, find the coefficients, and make a continuous plot of the function $f(t)$, together with the discrete data points $\{d_j\}_{j=1}^n$.

## Problem 9.17

We used the built-in matrix inverse routine in `Mathematica` back in Chapter 4 to invert the matrices we generated using finite differences. Go back to the chapter notebook, and re-do the one-dimensional example whose numerical solution is shown in Figure 4.1. All you will do is replace the `LinearSolve` command with your own (use either LU or QR with modified Gram–Schmidt). Plot your numerical solution together with the exact one.

## Problem 9.18

As in the last problem, re-do the parallel plate example from Chapter 4 (the setup can be found in the chapter notebook), replacing `LinearSolve` with your LU or QR solver (make it easy on yourself, and set $N_x = N_y = 20$ instead of 40), plot the resulting potential.

# 10

# The eigenvalue problem

The two major problems of numerical linear algebra are: 1. matrix inversion/least squares, and 2. the eigenvalue problem. We covered matrix inversion in Chapter 9, so we are left with least squares and the eigenvalue problem; these are connected through their solution method. The least squares problem makes use of the rectangular QR decomposition (the natural extension of the square QR decomposition we used to invert matrices) and this decomposition can also be used to find the eigenvalues and eigenvectors of a matrix. So while the least squares problem: "Find $\mathbf{x}$ such that $\|\mathbb{A}\mathbf{x} - \mathbf{b}\|$ is minimized for given $\mathbb{A}$ and $\mathbf{b}$" is very different from the eigenvalue problem: "Find all $\mathbf{v}$ and $\lambda$ such that $\mathbb{A}\mathbf{v} = \lambda\mathbf{v}$", the techniques used to solve them (at least, the ones presented here) both rely on the QR decomposition.

## 10.1 Fitting data

We return to the problem of fitting data. In Section 9.2.2, we set up the square version of the data fitting problem. Here, we'll re-state that problem with an eye towards generalizing it: Given $n$ data points $\{d_j\}_{j=1}^n$ (think of a function of time, sampled at equal intervals) we choose a set of functions $\{f_k(t)\}_{k=1}^m$, and we want to find coefficients $\{\alpha_k\}_{k=1}^m$ such that:

$$d_j \approx \sum_{k=1}^m \alpha_k f_k(t_j) \quad \text{for } j = 1 \longrightarrow n \tag{10.1}$$

where $t_j = j\Delta t$. Unlike (9.12), we have no way of enforcing equality here – the difference is that in general, $m \neq n$, so our matrix formulation will not yield a square matrix (and hence our notion of matrix inverse fails).

In order to be quantitative, and more important, concrete, define the problem as follows: Find $\{\alpha_k\}_{k=1}^m$ such that:

$$r_j \equiv d_j - \sum_{k=1}^m \alpha_k f_k(t_j) \quad \text{for } j = 1 \longrightarrow n \tag{10.2}$$

has minimal $\sum_{j=1}^n r_j^2$. Let's phrase this as a matrix-vector equation, similar to (9.13). First define the residual vector $\mathbf{r}$ from the components in (10.2):

$$\mathbf{r} \doteq \underbrace{\begin{pmatrix} d_1 \\ d_2 \\ \vdots \\ d_n \end{pmatrix}}_{\equiv \mathbf{b}} - \underbrace{\begin{pmatrix} f_1(t_1) & f_2(t_1) & \cdots & f_m(t_1) \\ f_1(t_2) & f_2(t_2) & \cdots & f_m(t_2) \\ \vdots & \vdots & \cdots & \vdots \\ f_1(t_n) & f_2(t_n) & \cdots & f_m(t_n) \end{pmatrix}}_{\equiv \mathbb{A}} \underbrace{\begin{pmatrix} \alpha_1 \\ \alpha_2 \\ \vdots \\ \alpha_m \end{pmatrix}}_{\equiv \mathbf{x}}, \tag{10.3}$$

or $\mathbf{r} = \mathbf{b} - \mathbb{A}\mathbf{x}$. This vector is a function of the coefficients $\{\alpha_k\}_{k=1}^m$, and we want to find the value of those coefficients such that $r^2$ is minimized. The problem is called the "least squares" problem for precisely this reason.

As an immediate application, remember the issue of size when we use matrix inverses to fit data – if you have $n$ data points, you must also cook up $n$ functions in order to generate the square matrix in (9.13). That means that if you had data given by the continuous function $d(t) = \sin(16\pi t)$ and you used functions $f_k = \sin(2\pi kt)$, then if $n$ is large, you have to include a bunch of functions in your matrix that will end up providing no contribution. If we could solve the least squares problem, then we could take, say, $m = 10$ regardless of the number of data points, and be assured of capturing the form of $d(t)$.

There is a fundamental difference between the inverse and the least squares form of the data fitting problem. If the matrix $\mathbb{A}$ is square, so that $m = n$, then we know that the vector minimizing the length $r$ of $\mathbf{r}$ in (10.3) is $\mathbf{x} = \mathbb{A}^{-1}\mathbf{b}$, and that means that each data point is matched on the grid (whether the overall match is relevant or not, as in Figure 9.3). Since minimizing $\mathbf{r}$ in general does *not* require equality on the grid (or off), our function, evaluated at the grid point $t_j$:

$$\sum_{k=1}^m \alpha_k f_k(t_j) \tag{10.4}$$

may not equal the value of $d_j$ for any $j$. Yet there is some notion of "closeness" that is being enforced, allowing us to solve for the coefficients $\{\alpha_k\}_{k=1}^m$. Let's see how that goes.

## 10.2 Least squares

We are given a vector $\mathbf{b} \in \mathbb{R}^n$ and a matrix $\mathbb{A} \in \mathbb{R}^{n \times m}$, and we want to find $\mathbf{x} \in \mathbb{R}^m$ such that the (Pythagorean) length[1] of the residual vector $\mathbf{r} \equiv \mathbb{A}\mathbf{x} - \mathbf{b}$ is as small as possible. Proceeding by brute force, we want to minimize the function $M(\mathbf{x}) \equiv r^2$, or:

$$M(\mathbf{x}) = (\mathbb{A}\mathbf{x} - \mathbf{b})^T (\mathbb{A}\mathbf{x} - \mathbf{b}). \tag{10.5}$$

This is a quadratic function in $\mathbf{x}$ (the unknown). Taking the derivative of $M$ with respect to each of the $\{x_k\}_{k=1}^m$ (we'll denote this derivative $\frac{\partial}{\partial \mathbf{x}}$) gives us a linear function, and if we set that to zero, we will find the minimum (or maximum) of $M$:

$$\frac{\partial M(\mathbf{x})}{\partial \mathbf{x}} = 2\mathbb{A}^T (\mathbb{A}\mathbf{x} - \mathbf{b}) = 0. \tag{10.6}$$

Before we address which type of extremum we have, note that this equation is now solvable by inversion, we want $\mathbf{x}$ such that:

$$\boxed{\mathbb{A}^T \mathbb{A}\mathbf{x} = \mathbb{A}^T \mathbf{b}} \tag{10.7}$$

where $\mathbb{A}^T \mathbb{A}$ is a square matrix in $\mathbb{R}^{m \times m}$. That reduces the least squares problem to a matrix inversion. The set of equations expressed by (10.7) is referred to as the "normal equations."

Given a function $M(x) = \sigma x^2$ of a single variable $x$, how do we know that the solution to $\frac{dM}{dx} = 0$ is a maximum versus a minimum? The usual check is the sign of the second derivative at the extremum. For this one-dimensional example, $\frac{dM}{dx} = 0$ gives $x = 0$, and the second derivative, evaluated at $x = 0$, is $\frac{d^2 M}{dx^2}\big|_{x=0} = \sigma$. Our function $M(x)$ is a parabola, and if $\sigma > 0$, the point $x = 0$ represents a minimum. If $\sigma < 0$, $M(x)$ is an upside down parabola, so $x = 0$ is at the "top" and we have a maximum.

---

[1] Again, different norms lead to different minimization problems; least squares refers specifically to the case of $\|\mathbf{w}\| \equiv \sqrt{\mathbf{w} \cdot \mathbf{w}}$ for vector $\mathbf{w}$.

The same logic (and even vaguely geometrical picture) holds for vector-valued $M(\mathbf{x})$. The second derivative of $M(\mathbf{x})$ is, from (10.6),

$$\frac{\partial}{\partial \mathbf{x}}\frac{\partial M}{\partial \mathbf{x}} = \frac{\partial}{\partial \mathbf{x}}\left(2\mathbb{A}^T(\mathbb{A}\mathbf{x} - \mathbf{b})\right) = 2\mathbb{A}^T\mathbb{A}. \tag{10.8}$$

What becomes of the one-dimensional $\sigma > 0$ requirement for a minimum? We have a square matrix, a product of the transpose of $\mathbb{A}$ with itself. The resulting matrix is clearly symmetric and so has real eigenvalues. Then for $\mathbf{x}$ solving (10.7) to be a minimum, we must have *all* eigenvalues positive, so that, in a sense, we have a minimum w.r.t. every "direction." Matrices that are products like $\mathbb{A}^T\mathbb{A}$ *have* positive eigenvalues (basically by construction), so we are guaranteed that the solution to (10.7) is a minimum, and we have solved the least squares problem once we have solved the normal equations.

We can simplify (10.7) using the QR decomposition of $\mathbb{A} \in \mathbb{R}^{n \times m}$ (with $n \geq m$). If $\mathbb{A}$ has full column rank (meaning that it has $m$ linearly independent columns), we can still perform the Gram–Schmidt orthogonalization procedure (via a modification of Algorithm 9.1), leading to an orthogonal matrix $\mathbb{Q} \in \mathbb{R}^{n \times m}$ and a square, invertible, upper triangular matrix $\mathbb{R} \in \mathbb{R}^{m \times m}$ such that $\mathbb{A} = \mathbb{Q}\mathbb{R}$. Then $\mathbb{A}^T\mathbb{A} = \mathbb{R}^T\mathbb{Q}^T\mathbb{Q}\mathbb{R} = \mathbb{R}^T\mathbb{R}$, so the normal equations are:

$$\mathbb{R}^T\mathbb{R}\mathbf{x} = \mathbb{R}^T\mathbb{Q}^T\mathbf{b} \longrightarrow \boxed{\mathbb{R}\mathbf{x} = \mathbb{Q}^T\mathbf{b}} \tag{10.9}$$

and this is relatively easy to solve, since $\mathbb{R}$ is upper triangular.

---

**Example**
We'll work out the procedure for a case that is opposite the example presented in Section 9.2.2. Our artificial data will be constructed by discretizing:

$$d(t) = t\left(t - \frac{1}{5}\right)\left(t - \frac{2}{5}\right)\left(t - \frac{3}{5}\right)\left(t - \frac{4}{5}\right)(t - 1) \tag{10.10}$$

so it is polynomial data. And we'll use a set of sine functions to approximate this data, let $f_k(t) = \sin(k\pi t)$. For our discretization, we'll take $n = 100$, with $t_n = 1$ (a temporal grid with spacing $\Delta t = 1/100$). Finally, let's take $m = 20$, so we have $\sin(\pi t), \sin(2\pi t) \ldots \sin(20\pi t)$ in our set of functions. We're using a sine series to fit a polynomial, whereas in Section 9.2.2 we used a polynomial series to fit a sine. Remember that the problem with that approach, shown in Figure 9.3, was that while the interpolating polynomial matched the data at the grid points, it did not do a good job of predicting the functional values at other locations.

Here, as shown in Figure 10.1, we see the opposite behavior – the fit gives a good general description of the data, while matching the data exactly at no grid point.

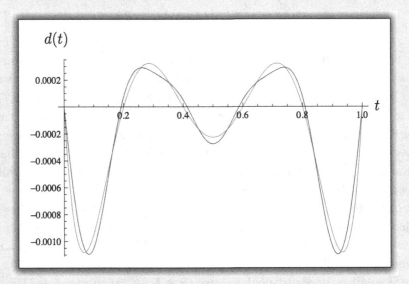

Figure 10.1 The function $d(t)$ from (10.10) is shown in gray, the continuous function $\sum_{k=1}^{m} \alpha_k f_k(t)$, that comes from solving for the least squares coefficients is shown in black.

### $R^2$ *value*

The least squares procedure is guaranteed to minimize (in the Pythagorean sense) the residual $\mathbf{r} \equiv \mathbb{A}\mathbf{x} - \mathbf{b}$, by constructing $\mathbf{x}$ appropriately. But how good a job does it do? The residual itself should "be small," but small compared to what? The most natural choice is to compare $r^2 \equiv \mathbf{r} \cdot \mathbf{r}$ to the (sample) variance of the data itself:

$$\sigma^2 \equiv \frac{1}{n} \sum_{j=1}^{n} \left( b_j - \langle b \rangle \right)^2 \quad \text{with } \langle b \rangle \equiv \frac{1}{n} \sum_{j=1}^{n} b_j. \tag{10.11}$$

The variance is a measure of the "scatter" of the provided data $\{b_j\}_{j=1}^{n}$ – the deviation from the mean, $\langle b \rangle$. We can then form the dimensionless ratio: $r^2/\sigma^2$, that quantity will be small provided it is much less than one. In statistics, the relevant quantity is:

$$R^2 \equiv 1 - \frac{r^2}{\sigma^2} \tag{10.12}$$

so that a good "fit" (small $r^2/\sigma^2$) means that $R^2 \approx 1$.

## 10.3 The eigenvalue problem

For any matrix $\mathbb{A} \in \mathbb{R}^{n \times n}$, there is a set of complex vectors $\{v_i\}_{i=1}^n$ and numbers $\{\lambda_i\}_{i=1}^n$ for which:

$$\mathbb{A}v_i = \lambda_i v_i. \tag{10.13}$$

The vectors $v_i$ are special – when multiplied by $\mathbb{A}$ they are scaled but do not change direction, so that $\mathbb{A}v_i \parallel v_i$. The problem is: Given $\mathbb{A}$, find the eigenvectors $\{v_i\}_{i=1}^n$ and their (associated) eigenvalues $\{\lambda_i\}_{i=1}^n$.

Once we have all the eigenvectors, we can package them into a matrix $\mathbb{V}$ whose columns are the $v_i$, and then the defining equation (10.13) can be written for all vectors at once:

$$\mathbb{A}\mathbb{V} = \mathbb{V}\mathbb{L} \tag{10.14}$$

where $\mathbb{L}$ is a diagonal matrix with $L_{ii} = \lambda_i$.

In physical problems, certainly the ones we have seen so far, the matrices of interest are real, square and *symmetric*: $\mathbb{A}^T = \mathbb{A}$. Symmetric matrices have further, desirable, eigenvector/value properties. First, the eigenvalues are all real, and second, the eigenvectors form a basis for $\mathbb{R}^n$ – they are orthogonal and normalizable so that $\mathbb{V}^T\mathbb{V} = \mathbb{I}$.

## 10.4 Physical motivation

Let's review some of the fundamental physical problems that have explicitly relied on eigenvectors for their solution, and then we'll look at some further applications.

### 10.4.1 Linear springs

Remember (from Chapter 8) that if we have a generic set of springs connecting $n$ masses together in some manner, we can write Newton's second law for $\mathbf{X}(t) \in \mathbb{R}^{3n}$, the position vector for the entire system as:

$$\ddot{\mathbf{X}}(t) = \mathbb{Q}\mathbf{X}(t) + \mathbf{A}, \tag{10.15}$$

with $\mathbb{Q}$ symmetric. The solution to the equations of motion relied on finding $\mathbb{V}$ (whose columns are the eigenvectors of $\mathbb{Q}$), and then defining $\mathbf{\Psi} = \mathbb{V}^T\mathbf{X}(t)$ decoupled the problem into $3n$ independent linear ODEs that are easy to solve.

### 10.4.2 Quantum mechanics

Recall from Chapter 4 that it is possible to discretize differential operators, and represent differential equations as matrix equations (finite difference). The Schrödinger equation governs the non-relativistic quantum mechanical description of particles, and it reads, in one dimension, for an arbitrary potential energy $V(x)$:

$$-\frac{\hbar^2}{2m}\frac{\partial^2\Psi(x,t)}{\partial x^2} + V(x)\Psi(x,t) = i\hbar\frac{\partial\Psi(x,t)}{\partial t}. \tag{10.16}$$

If we use multiplicative separation of variables: $\Psi(x,t) = \psi(x)\phi(t)$, then the above becomes,

$$-\frac{\hbar^2}{2m}\psi''(x)\phi(t) + V(x)\psi(x)\phi(t) = i\hbar\psi(x)\dot{\phi}(t) \tag{10.17}$$

where primes refer to $x$-derivatives and dots denote $t$-derivatives. Dividing this equation by $\psi(x)\phi(t)$ gives:

$$\frac{1}{\psi(x)}\left(-\frac{\hbar^2}{2m}\psi''(x) + V(x)\psi(x)\right) = i\hbar\frac{\dot{\phi}(t)}{\phi(t)}. \tag{10.18}$$

Now, the usual argument: "The left-hand side depends only on $x$, while the right depends only on $t$, so both sides must be (the same) constant." This "separation constant" is called $E$ (and represents the energy of a particle in the state $\psi$), so we have:

$$-\frac{\hbar^2}{2m}\psi''(x) + V(x)\psi(x) = E\psi(x) \qquad i\hbar\frac{\dot{\phi}(t)}{\phi(t)} = E. \tag{10.19}$$

The equation involving $\phi(t)$ can be solved directly, $\phi(t) = e^{-i\frac{E}{\hbar}t}$. The equation for $\psi(x)$ cannot be solved until a potential is specified. Notice that this equation defines a sort of differential eigenvalue problem – given $V(x)$, we want to find $\psi(x)$ and $E$ satisfying

$$\left[-\frac{\hbar^2}{2m}\frac{d^2}{dq^2} + V(x)\right]\psi(x) = E\psi(x). \tag{10.20}$$

We'll turn this into an explicit matrix problem with a concrete example.

#### Infinite square well

Remember, from Section 3.1.3, the "infinite square well" potential

$$V(x) = \begin{cases} 0 & 0 < x < a \\ \infty & x < 0 \text{ or } x > a \end{cases}. \tag{10.21}$$

The function $\psi(x)$ must vanish at both $x = 0$ and $x = a$, so we have two boundaries. The goal is to find the eigenvalues $E$ and eigenfunctions $\psi(x)$, and we did this numerically using the shooting method in Section 3.1.3. Here, we will turn the continuous eigenvalue problem presented by Schrödinger's equation into a discrete eigenvalue problem, one with an actual matrix representing the discretized physical content of the time-independent Schrödinger equation.

We can render the eigenvalue (an energy, in this setting) dimensionless using the combination $\tilde{E} \equiv \frac{2ma^2}{\hbar^2} E$ – the relevant equation from (10.19) reads (with the potential in place)

$$-a^2 \psi''(x) = \tilde{E}\psi(x) \quad \tilde{E} = \frac{2ma^2}{\hbar^2} E. \tag{10.22}$$

Finally, absorb the $a^2$ out front by letting $x \equiv aq$ for dimensionless $q$. Now we have the ODE and boundary conditions:

$$-\psi''(q) = \tilde{E}\psi(q) \quad \psi(0) = \psi(1) = 0. \tag{10.23}$$

By discretizing the differential equation, we will obtain an eigenvalue problem in which the eigenvectors are discrete approximations of the wave functions, and the eigenvalues are the associated (approximate) energies. Take a grid with spacing $\Delta q$ and $q_j = j\Delta q$ for $j = 1 \ldots n$, with $q_{n+1} = 1$, the boundary point. Then the function $\psi(q_j) \equiv \psi_j$ as usual, and the vector $\boldsymbol{\psi}$ has the $\psi_j$ as entries. We can approximate the second derivative appearing in (10.23) using:

$$\psi''(q_j) \approx \frac{\psi_{j+1} - 2\psi_j + \psi_{j-1}}{\Delta q^2}. \tag{10.24}$$

Inserting this on the left of (10.23), we have

$$-\frac{\psi_{j+1} - 2\psi_j + \psi_{j-1}}{\Delta q^2} = \tilde{E}\psi_j \tag{10.25}$$

for $j = 1 \ldots n$.

At $j = 1$ and $n$, the equation (10.25) depends on the boundary points $\psi_0$ and $\psi_{n+1}$ (similar to the boundary points in Section 4.2). From the boundary conditions in (10.23), we know that $\psi_0 = \psi_{n+1} = 0$, and the equations simplify:

$$-\frac{1}{\Delta q^2}(\psi_2 - 2\psi_1) = \tilde{E}\psi_1$$
$$-\frac{1}{\Delta q^2}(-2\psi_n + \psi_{n-1}) = \tilde{E}\psi_n. \tag{10.26}$$

The set of equations (10.25) and (10.26) defines an eigenvalue problem:

$$\mathbb{D}\boldsymbol{\psi} = \tilde{E}\boldsymbol{\psi} \tag{10.27}$$

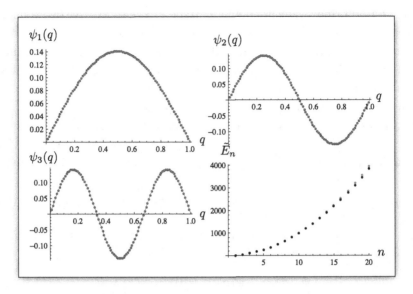

Figure 10.2 The first three eigenvectors for the infinite square well. The exact solution (continuous) and eigenvector approximation (points) are shown. The bottom right-hand corner plot shows the first 20 eigenvalues calculated numerically (black) and exactly (gray, these are just $\tilde{E}_n = n^2\pi^2$).

for the (symmetric) matrix $\mathbb{D}$:

$$
\mathbb{D} \doteq -\left(\frac{1}{\Delta q^2}\right)
\begin{pmatrix}
-2 & 1 & 0 & 0 & \cdots & 0 \\
1 & -2 & 1 & 0 & \cdots & 0 \\
0 & 1 & -2 & 1 & \cdots & 0 \\
\vdots & \vdots & & \ddots & \ddots & \vdots \\
0 & 0 & \cdots & 1 & -2 & 1 \\
0 & 0 & \cdots & 0 & 1 & -2
\end{pmatrix}.
\tag{10.28}
$$

In this particular case, the solutions to (10.23) are well known – there is an infinite family of sine solutions, indexed by the integer $n$:

$$
\psi_n(q) = \sqrt{2}\sin(n\pi q) \qquad \tilde{E}_n = n^2\pi^2,
\tag{10.29}
$$

so that $E_n = \frac{\hbar^2}{2ma^2}\tilde{E}_n = \frac{\hbar^2 n^2\pi^2}{2ma^2}$. The solutions to (10.25) are also available in exact form, as you will establish in Problem 10.8.

For now, we will find the eigenvalues numerically, and then compare the numerical solution with the exact solution to the continuous problem. In Figure 10.2, we see the first three eigenvectors, computed with 100 grid points, and $a = 1$ using both (10.29) (solid curve) and the eigenvectors from (10.27) (points).

**The hydrogen spectrum**

Hydrogen, from a physics point of view, consists of an electron (carrying charge $-e$) moving in the central potential set up by a proton (with charge $e$). The potential is just the usual Coulomb one:

$$V(x) = -\frac{e^2}{4\pi\epsilon_0 x}, \tag{10.30}$$

and we can put this in our nondimensionalized form:

$$\tilde{V}(q) = -\frac{2ma^2e^2}{4\pi\epsilon_0\hbar^2 aq} = -2\left[\frac{me^2}{4\pi\epsilon_0\hbar^2}\right]\frac{a}{q}. \tag{10.31}$$

The factor in brackets on the right defines a length, called the "Bohr radius," and usually denoted $a \equiv \frac{4\pi\epsilon_0\hbar^2}{me^2}$ (lucky coincidence) – let this be our fundamental scaling length $a$, and then the final form of the potential is:

$$\tilde{V}(q) = -\frac{2}{q}. \tag{10.32}$$

The nondimensionalized Schrödinger equation for this specific potential reads (in one dimension)

$$-\frac{d^2\psi(q)}{dq^2} - \frac{2}{q}\psi(q) = \tilde{E}\psi(q). \tag{10.33}$$

Now for the boundary conditions – we will assume (somewhat plausibly) that the wave function vanishes at $q = 0$, and at some numerical infinity $q_\infty$ – those will represent the endpoints of our discrete grid.

If we solve this equation by discretizing it, we can get accurate approximations to the "bound-state" energies of hydrogen. There are two fundamentally different behaviors we can study in both classical and quantum mechanics – bound states are, here, states of negative energy, and are something like the classical orbits associated with central potentials. There are also "scattering states." In the case of hydrogen, these have positive energy, and are a quantum mechanical analogue of unbound classical motion (shoot an electron past a central charge).

### 10.4.3 Image compression

Suppose we have a grayscale (no color), two-dimensional, image. We can view the image as a matrix by associating a grid of points on the page with the matrix entries and assigning at each point a value of 0 for white, 1 for black (there is a continuum of gray from white to black, and a continuum of numbers from 0 to 1 – in the end, we typically pick some number of shades and assign them appropriate values $\in [0, 1]$). An image on a computer screen is a good example – the image has a width and a height in pixels, and each pixel stores (in some manner) the "color" it will display. Suppose we have an image that is $301 \times 301$ pixels – that requires $301^2 = 90\,601$ entries. The solution to the eigenvalue problem gives us a way to reduce the storage of a picture (while increasing the computational cost of displaying it).

Assume that our picture is representable by a square, symmetric matrix (you can always take a picture of interest and generate a square symmetric matrix from it) $\mathbb{A} \in \mathbb{R}^{n \times n}$. Then if we can solve for all the eigenvectors and all the eigenvalues, we have:

$$\mathbb{A} = \mathbb{V} \mathbb{L} \mathbb{V}^T \tag{10.34}$$

where $\mathbb{V}^{-1} = \mathbb{V}^T$ since a symmetric matrix has orthogonal eigenvectors. Now the claim, to be proven below, is that if we take the eigenvalues to be in decreasing (magnitude) order in the matrix $\mathbb{L}$, then if we take just the first $W$ eigenvalues (the largest $W$ ones, in absolute value), make a diagonal matrix $\tilde{\mathbb{L}} \in \mathbb{R}^{W \times W}$, and take the associated eigenvectors as the columns of $\tilde{\mathbb{V}}$, then we have an "approximation" to the matrix $\mathbb{A}$ given by:

$$\tilde{\mathbb{A}} = \tilde{\mathbb{V}} \tilde{\mathbb{L}} \tilde{\mathbb{V}}^T. \tag{10.35}$$

It should be clear that this factorization implies a compression, since we now need to store only $W + W \times n$ pieces of data, the $W$ eigenvalues we chose, and the corresponding $W$ eigenvectors (each of length $n$) that are the columns of $\tilde{\mathbb{V}}$.[2]

In Figure 10.3, we can see the result of this type of truncated compression for a few different choices of $W$ – notice that the quality increases dramatically at first, the difference between $W = 4$ and $W = 16$ is large, but for $W > 64$, the increase in "recognizability" gets smaller and smaller. The reason for this is the decay of the (magnitude) of the eigenvalues, shown in Figure 10.4. The smaller the eigenvalue, the less its eigenvector "contributes" to the overall picture.

---

**Matrix norms**

In order to understand the association of $\tilde{\mathbb{A}}$ with $\mathbb{A}$, we need the notion of a matrix "norm." We know how to take the norm of vectors – given $\mathbf{x} \in \mathbb{R}^n$, we have the Pythagorean length (squared):

$$x^2 \equiv \mathbf{x} \cdot \mathbf{x} = x_1^2 + x_2^2 + \cdots x_n^2 \tag{10.36}$$

and the norm is just the square root of this: $\|x\| \equiv x$. Actually, this is just an example of a norm; there are any number of "lengths" one can define for a vector. In general,

---

[2] The decomposition and truncation described here has a generalization to non-symmetric matrix "pictures," but it is easier to put bounds on these symmetric matrices.

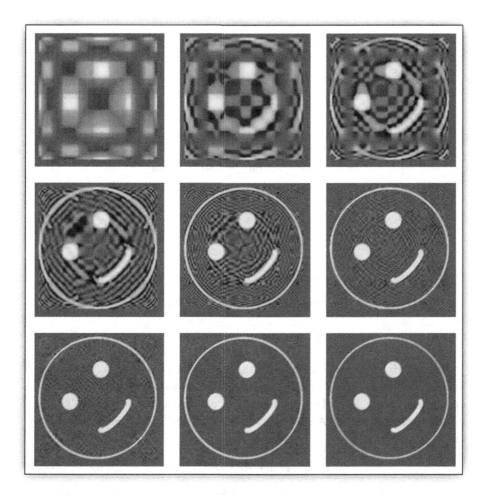

Figure 10.3 Top left is the matrix $\tilde{A}$ with $W = 2$, then going over in the top row, $W = 4$, $W = 8$, in the next row $W = 16$, $W = 32$, $W = 64$ and in the bottom row, $W = 128$, 256, and 301 (the full size of the matrix, so the bottom right is the original image with no approximation).

for $\mathbf{x} \in \mathbb{R}^n$, a norm $\|\mathbf{x}\|$ must satisfy the following three properties (see, for example, [3, 9]):

$$\|\mathbf{x}\| \geq 0 \quad \|\mathbf{x}\| = 0 \longrightarrow \mathbf{x} = 0$$
$$\|\mathbf{x} + \mathbf{y}\| \leq \|\mathbf{x}\| + \|\mathbf{y}\| \quad \mathbf{y} \in \mathbb{R}^n \quad (10.37)$$
$$\|\alpha\mathbf{x}\| = |\alpha| \|\mathbf{x}\| \quad \alpha \in \mathbb{R}.$$

The first of these requires that the norm be greater than or equal to zero, and if $\|\mathbf{x}\| = 0$, then $\mathbf{x} = 0$. The second is the "triangle inequality," and the third defines the

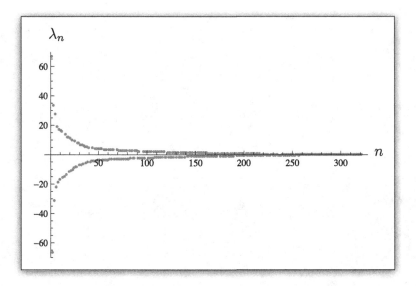

Figure 10.4 The eigenvalues associated with the smiley matrix $\mathbb{A}$.

scaling relation. Any operation on the vector **x** satisfying these three properties is a norm.

For a matrix, the definition of norm is the same, all three properties (10.37) must hold for a matrix $\mathbb{A} \in \mathbb{R}^{n \times n}$. There are many useful definitions (and for finite-dimensional spaces, all of these definitions are "equivalent," so that it doesn't much matter which one we use). A class of natural choices works from the vector definition (an "induced" norm),

$$\|\mathbb{A}\| = \max_{\mathbf{x}:x=1} \|\mathbb{A}\mathbf{x}\| \tag{10.38}$$

that is: "The norm of $\mathbb{A}$ is the largest norm of the vector $\mathbb{A}\mathbf{x}$ with $\|\mathbf{x}\| = 1$." We can take any vector norm we like to build the induced norm, but we'll stick with the Pythagorean case (10.36).

As a two-by-two example let

$$\mathbb{A} \doteq \begin{pmatrix} 1 & 2 \\ 2 & 1 \end{pmatrix}. \tag{10.39}$$

Now take a generic vector **x** with unit length – this can be parametrized by an arbitrary angle $\theta$:

$$\mathbf{x} \doteq \begin{pmatrix} \cos\theta \\ \sin\theta \end{pmatrix}. \tag{10.40}$$

We construct the vector **z** via:

$$\mathbf{z} \equiv \mathbb{A}\mathbf{x} \doteq \begin{pmatrix} \cos\theta + 2\sin\theta \\ 2\cos\theta + \sin\theta \end{pmatrix}. \tag{10.41}$$

To find the value of $\theta$ (and hence the unit vector $\mathbf{x}$) that maximizes $\|\mathbf{z}\|$, we'll take $\frac{d\|\mathbf{z}\|}{d\theta} = 0$ and solve for $\theta$

$$\frac{d}{d\theta}\sqrt{5 + 8\cos\theta\sin\theta} = \frac{1}{2}\frac{\left(-8\sin^2\theta + 8\cos^2\theta\right)}{\sqrt{5 + 8\cos\theta\sin\theta}}, \tag{10.42}$$

and the numerator can be made zero by taking:

$$\cos^2\theta - \sin^2\theta = 0 \longrightarrow \theta = \frac{1}{4}\pi \tag{10.43}$$

(you should check that this is a maximum, rather than a minimum). The vector $\mathbf{x}$ is then:

$$\mathbf{x} = \frac{1}{\sqrt{2}}\begin{pmatrix} 1 \\ 1 \end{pmatrix}, \tag{10.44}$$

and we learn that the norm of $\mathbb{A}$ is $\|\mathbb{A}\| = 3$.

Now we can return to the question of comparing $\mathbb{A} \in \mathbb{R}^{n \times n}$ and $\tilde{\mathbb{A}}$ from (10.35) – what we'd ultimately like is to put a bound on the norm of the difference: $\|\mathbb{A} - \tilde{\mathbb{A}}\|$. To do this, take a generic vector $\mathbf{x}$ decomposed into the basis $\{\mathbf{v}_j\}_{j=1}^n$ (the eigenvectors of the symmetric $\mathbb{A}$):

$$\mathbf{x} = \sum_{j=1}^{n}\alpha_j\mathbf{v}_j. \tag{10.45}$$

If we multiply $\mathbf{x}$ by $\mathbb{A}$, then each eigenvector picks up its eigenvalue,

$$\mathbb{A}\mathbf{x} = \sum_{j=1}^{n}\alpha_j\lambda_j\mathbf{v}_j. \tag{10.46}$$

For the $W$-truncated $\tilde{\mathbb{A}} = \tilde{\mathbb{V}}\tilde{\mathbb{L}}\tilde{\mathbb{V}}^T$, we have

$$\tilde{\mathbb{A}}\mathbf{x} = \sum_{j=1}^{W}\alpha_j\lambda_j\mathbf{v}_j. \tag{10.47}$$

The norm of the difference is:

$$\|\mathbb{A} - \tilde{\mathbb{A}}\| = \max_{\mathbf{x}:x=1}\left\|\left(\mathbb{A} - \tilde{\mathbb{A}}\right)\mathbf{x}\right\|. \tag{10.48}$$

The vector norm on the right, for any $\mathbf{x}$ (so eliminating the "max" and normalization requirement), is

$$\left\|\left(\mathbb{A} - \tilde{\mathbb{A}}\right)\mathbf{x}\right\| = \left\|\sum_{j=W+1}^{n}\alpha_j\lambda_j\mathbf{v}_j\right\| = \sqrt{\sum_{j=W+1}^{n}\alpha_j^2\lambda_j^2}. \tag{10.49}$$

When we take the maximum and require that $\mathbf{x}$ be normalized, we can develop the inequality

$$\max_{\mathbf{x}:x=1} \left\| (\mathbb{A} - \tilde{\mathbb{A}}) \mathbf{x} \right\| = \max_{\mathbf{x}:x=1} \left[ \sqrt{\sum_{j=W+1}^{n} \alpha_j^2 \lambda_j^2} \right] \leq \sqrt{\sum_{j=W+1}^{n} \lambda_j^2} \qquad (10.50)$$

by noting that since every element in $\mathbf{x}$ must have absolute value less than or equal to one (if $x = 1$), setting all $\alpha_j = 1$ will over-estimate. The bound on the norm of the difference is

$$\| \mathbb{A} - \tilde{\mathbb{A}} \| \leq \sqrt{\sum_{j=W+1}^{n} \lambda_j^2}. \qquad (10.51)$$

Referring to Figure 10.3, at $W = 128$, we capture the primary features of the image, and could average or smooth to get rid of the remaining detritus, so that's not a bad place to stop. The square root of the sum of the squares of the eigenvalues above $W = 128$, appearing on the right-hand side of the inequality in (10.51), is $\approx 10$, while the norm of the matrix itself, in this case, is $\| \mathbb{A} \| \approx 67$ (and 10 is small compared to this).

## 10.5 The power method

How do we calculate the eigenvalues and eigenvectors of a matrix? The answer is surprisingly simple, not to mention cute. We have a symmetric matrix $\mathbb{A} \in \mathbb{R}^{n \times n}$, and suppose that its spectrum is non-degenerate and ordered, so that we'll take $\lambda_1 > \lambda_2 > \ldots > \lambda_n$ – in addition, we'll assume $|\lambda_n| > 1$ (we can always ensure this is the case by multiplying $\mathbb{A}$ by a constant). We know that any random vector in $\mathbb{R}^n$ can be decomposed in the eigenvectors of $\mathbb{A}$, whatever they may be, using coefficients $\{\beta_j\}_{j=1}^n$:

$$\mathbf{x} = \sum_{j=1}^{n} \beta_j \mathbf{v}_j \qquad (10.52)$$

for $\mathbf{v}_j$ an eigenvector of $\mathbb{A}$ with eigenvalue $\lambda_j$. If we multiply the vector $\mathbf{x}$ by $\mathbb{A}^q$, then

$$\mathbb{A}^q \mathbf{x} = \sum_{j=1}^{n} \beta_j \lambda_j^q \mathbf{v}_j, \qquad (10.53)$$

and now the point: For the largest (in absolute value) eigenvalue, with $q$ large enough, we will have $\lambda_1^q$ large compared to all the other $\lambda_j^q$ in the sum, so the term associated with the largest eigenvalue will dominate. That is, for large $q$:

$$\mathbb{A}^q \mathbf{x} = \beta_1 \lambda_1^q \mathbf{v}_1 + \sum_{j=2}^{n} \beta_j \lambda_j^q \mathbf{v}_j \approx \beta_1 \lambda_1^q \mathbf{v}_1. \tag{10.54}$$

The product $\mathbb{A}^q \mathbf{x}$ is parallel to $\mathbf{v}_1$, and then the eigenvector associated with the maximum eigenvalue can be obtained by normalizing

$$\mathbf{v}_1 \approx \frac{\mathbb{A}^q \mathbf{x}}{\|\mathbb{A}^q \mathbf{x}\|}. \tag{10.55}$$

The eigenvalue can be extracted from the defining equation once we have $\mathbf{v}_1$, take:

$$\mathbb{A}\mathbf{v}_1 = \lambda_1 \mathbf{v}_1 \longrightarrow \mathbf{v}_1^T \mathbb{A}\mathbf{v}_1 = \lambda_1, \tag{10.56}$$

by dotting $\mathbf{v}_1$ on both sides. The procedure for finding the eigenvector associated with the largest eigenvalue is straightforward – just multiply by $\mathbb{A}$ over and over again, this defines the "power method."

---

**Example**
Take the simple matrix,

$$\mathbb{A} = \begin{pmatrix} 33 & 39 & 45 \\ 39 & \frac{93}{2} & 54 \\ 45 & 54 & 63 \end{pmatrix} \tag{10.57}$$

so that $\mathbb{A}$ has eigenvectors that span $\mathbb{R}^3$ (since $\mathbb{A}$ is symmetric), and generate a random vector $\mathbf{x}$, then

$$\mathbf{y} = \mathbb{A}^{10} \mathbf{x} \quad \hat{\mathbf{y}} \sim \begin{pmatrix} 0.479671 \\ 0.572368 \\ 0.665064 \end{pmatrix}. \tag{10.58}$$

The normalization doesn't matter, making a unit vector $\hat{\mathbf{y}}$ (taking away the $\beta_1 \lambda_1^{10}$ factor) doesn't change its scaling under multiplication. For the approximate eigenvector $\hat{\mathbf{y}}$, the associated eigenvalue is $\lambda \sim 141.929293290185$. From its characteristic polynomial, the exact eigenvalues of this matrix are:

$$\lambda_{1,2} = \frac{3}{4}(95 \pm \sqrt{8881}) \quad \lambda_3 = 0 \tag{10.59}$$

and abusing the notion of precision slightly, this puts our numerically determined eigenvalue within $10^{-13}$ of a numerical representation of the exact value.

---

Of course, we are not done, but the basic idea persists. To get the next eigenvector (the one associated with $\lambda_2$), we multiply by $\mathbb{A}$, and project out any dependence on the existing eigenvector – it's also good to normalize at each step.

If you have $\mathbf{v}_1$, then start with a random $\mathbf{z}_0$, and iterate:

$$\mathbf{z}_{i+1} = \mathbb{A}\mathbf{z}_i - [(\mathbb{A}\mathbf{z}_i) \cdot \mathbf{v}_1]\,\mathbf{v}_1$$

$$\mathbf{z}_{i+1} = \frac{\mathbf{z}_{i+1}}{\sqrt{\mathbf{z}_{i+1} \cdot \mathbf{z}_{i+1}}},  \tag{10.60}$$

(the second line represents normalizing after each step). After some number of iterations, $\|\mathbf{z}_{i+1} - \mathbf{z}_i\|$ will be acceptably small, and you'll have the eigenvector $\mathbf{v}_2$, associated with the second largest eigenvalue. As we obtain more and more eigenvectors, we need to do more and more projection to stay on target.

There are mathematical difficulties – one needs to be careful with degenerate eigenvalues. In addition, there are numerical difficulties – this is not a very efficient way to find eigenvectors, and for eigenvalues close to degenerate, the number of iterations required to separate two vectors can be prohibitive. Regardless, more sophisticated methods proceed from this idea. We'll look at a complementary pair, simultaneous iteration and QR.

## 10.6 Simultaneous iteration and QR iteration

In simultaneous iteration, we start with a set of basis vectors, and successively multiply by $\mathbb{A}$ in order to find *all* the eigenvectors of the matrix. It is a simultaneous version of the procedure sketched above (typified by, for example, (10.60)). We start with an orthogonal matrix, and multiply by $\mathbb{A}$ (so each column of the matrix is multiplied by $\mathbb{A}$), then orthogonalize the columns (the generalization of projecting out "known" eigenvectors, as in (10.60)) and normalize them. The algorithm for simultaneous iteration (SI) is shown in Algorithm 10.1 (the final line of the iteration, computing $\bar{\mathbb{A}}_j$ is unnecessary, but useful when comparing with other methods). When we're done, the matrix $\bar{\mathbb{Q}}_M$ should have columns that are the eigenvectors of $\mathbb{A}$, and then $\bar{\mathbb{A}}_M = \bar{\mathbb{Q}}_M^T \mathbb{A} \bar{\mathbb{Q}}_M$ is the diagonal matrix with the eigenvalues of $\mathbb{A}$ on the diagonal.

---

**Algorithm 10.1** Simultaneous iteration

---

Pick an orthonormal matrix $\bar{\mathbb{Q}}_0$
**for** $j = 1 \to M$ **do**
  $\mathbb{Z} \leftarrow \mathbb{A}\bar{\mathbb{Q}}_{j-1}$ (multiply the columns of $\bar{\mathbb{Q}}_{j-1}$ by $\mathbb{A}$)
  $\bar{\mathbb{Q}}_j \bar{\mathbb{R}}_j = \mathbb{Z}$ (calculate the QR decomposition of $\mathbb{Z}$)
  $\bar{\mathbb{A}}_j \leftarrow \bar{\mathbb{Q}}_j^T \mathbb{A} \bar{\mathbb{Q}}_j$
**end for**

---

We are effectively multiplying a matrix $\bar{\mathbb{Q}}_0$ by $\mathbb{A}$ over and over, so that you can think of $\bar{\mathbb{Q}}_j$ as $\mathbb{A}^j \bar{\mathbb{Q}}_0$. To avoid the increase in magnitude (for the columns of $\bar{\mathbb{Q}}_j$) associated with that multiplication, we generate a set of orthonormal columns that spans the same space, that is the role played by the QR decomposition of $\mathbb{Z}$ – at any step, $\mathbb{Z}$ and $\bar{\mathbb{Q}}_j$ have columns spanning the same space. Basically, we are normalizing all of the iterates of the eigenvectors while projecting out their dependence on each other, one at a time, starting with the first column, which will eventually become the eigenvector associated with the maximum eigenvalue. We multiply and orthogonalize, multiply and orthogonalize, but when do we stop? In Algorithm 10.1, the loop ends at $M$ – that is artificial, $M$ is only introduced for counting and timing – how big does $M$ have to be to achieve convergence? That depends on the matrix – we would typically run until $\|\bar{\mathbb{A}}_j - \bar{\mathbb{A}}_{j-1}\|$ is small.

Another method for solving the eigenvalue problem is called the "QR algorithm" – it uses the QR decomposition directly to perform the multiplication by $\mathbb{A}$ found in the first line of Algorithm 10.1. The QR algorithm for finding the eigenvalues and eigenvectors of a matrix is given in Algorithm 10.2. At the

---

**Algorithm 10.2** QR Algorithm

$\mathbb{A}_0 \leftarrow \mathbb{A}$
$\mathbb{V}_0 \leftarrow \mathbb{I}$
**for** $j = 1 \rightarrow M$ **do**
$\quad \mathbb{Q}_j \mathbb{R}_j = \mathbb{A}_{j-1}$ (compute the QR decomposition of $\mathbb{A}_{j-1}$)
$\quad \mathbb{A}_j \leftarrow \mathbb{R}_j \mathbb{Q}_j$
$\quad \mathbb{V}_j \leftarrow \mathbb{V}_{j-1} \mathbb{Q}_j$
**end for**

---

end of this process (whenever you decide to stop), you will have the approximate factorization:

$$\mathbb{A} = \mathbb{V}_M \mathbb{A}_M \mathbb{V}_M^T \qquad (10.61)$$

for $\mathbb{V}_M$, $\mathbb{A}_M$ the final iterates of your implementation. $\mathbb{A}_M$ will be (upon "completion") a diagonal matrix with the eigenvalues of $\mathbb{A}$, and $\mathbb{V}_M$ is orthogonal.

While SI is a plausible generalization of the power method (via (10.60)), applied to a set of vectors, it is not clear how QR relates to the power method – how do we know the two procedures, SI and QR (ideally) produce the same decomposition? Both methods are basically factoring the $k$th power of $\mathbb{A}$ into $\mathbb{Q}$ and $\mathbb{R}$, so we have an orthogonal basis for $\mathbb{A}^k$ (relevant for eigenvectors by the discussion at the end of Section 10.5). But we can actually show that QR and simultaneous iteration are the same. That proof (from [45]) gives a sense of the inductive procedure applied to these algorithms. For the proof, take $\bar{\mathbb{Q}}_0 = \mathbb{I}$ for simultaneous iteration. Then we will show that QR and simultaneous iteration generate, at each step, identical $A_j = \bar{A}_j$ and $V_j = \bar{Q}_j$, hence the computed eigenvectors and eigenvalues are the same for the two methods. Assume this is true for the $j$th step; then we have, at the next step:

$$
\begin{array}{cc}
\text{QR} & \text{SI} \\[4pt]
Q_{j+1}R_{j+1} = A_j & \bar{Q}_{j+1}\bar{R}_{j+1} = A\bar{Q}_j \\[4pt]
A_{j+1} = R_{j+1}Q_{j+1} & \bar{A}_{j+1} = \bar{Q}_{j+1}^T A\bar{Q}_{j+1} \\[4pt]
V_{j+1} = V_j Q_{j+1} &
\end{array}
\tag{10.62}
$$

Now, from the previous step, we have $\bar{Q}_j^T A \bar{Q}_j = A_j$ where $A_j = \bar{A}_j$ by our inductive assumption. Then the QR decomposition on the SI side gives

$$
\bar{Q}_{j+1}\bar{R}_{j+1} = \bar{Q}_j A_j \tag{10.63}
$$

and we can now use the QR decomposition on the QR side to relate the $\bar{Q}$ to the $Q$

$$
Q_{j+1}R_{j+1} = \bar{Q}_j^T \bar{Q}_{j+1}\bar{R}_{j+1}. \tag{10.64}
$$

Both the left and right-hand sides are QR decompositions of the same matrix $A_j$, and the QR decomposition is unique (for $R$ with positive diagonal entries), so we must have

$$
Q_{j+1} = \bar{Q}_j^T \bar{Q}_{j+1} \qquad R_{j+1} = \bar{R}_{j+1}. \tag{10.65}
$$

Constructing the updated $V_{j+1} = V_j Q_{j+1}$ using the above and $V_j = \bar{Q}_j$ (by inductive assumption), we have

$$
V_{j+1} = \bar{Q}_{j+1} \tag{10.66}
$$

as desired. For the matrix $A_{j+1}$ equality, take the QR update written in terms of the barred matrices (use $\bar{R}_{j+1}\bar{Q}_j^T = \bar{Q}_{j+1}^T A$ from the first line of the SI step):

$$
A_{j+1} = R_{j+1}Q_{j+1} = \bar{R}_{j+1}\bar{Q}_j^T \bar{Q}_{j+1} = \bar{Q}_{j+1}^T A\bar{Q}_{j+1} = \bar{A}_{j+1}. \tag{10.67}
$$

So given $A_j = \bar{A}_j$ and $V_j = \bar{Q}_j$, we have $A_{j+1} = \bar{A}_{j+1}$ and $V_{j+1} = \bar{Q}_{j+1}$ as desired. The result is clearly true for $j = 0$ (since we took $\bar{Q}_0 = \mathbb{I}$), so true for all $j$.

### 10.6.1 Timing

Let's think about the timing of QR iteration for finding eigenvalues and eigenvectors. Referring to Algorithm 10.2, we compute the QR decomposition of a matrix $M$ times. The QR decomposition itself has timing (see Section 9.5) $T(n) = O(n^3)$ for matrices in $\mathbb{R}^{n \times n}$. Since we perform $M$ decompositions, the timing of the QR method is $T(n) = O(Mn^3)$. But what is $M$? That depends on how accurately we want the eigenvalues and eigenvectors. If we require that $\|\mathbb{A}_{M+1} - \mathbb{A}_M\| \leq \epsilon$ for small $\epsilon$, $M$ could be very large. It is difficult to predict how many iterations of the method are required to achieve a certain tolerance, since the structure of the eigenvalues themselves determines the rate of convergence (if the eigenvalues are well-separated, $M$ will be smaller than if the eigenvalues are very close together).

We may not be interested in the entire spectrum, and our QR method gives us everything all at once – we'll see in Chapter 11 a way to limit the number of eigenvalues we obtain accurately, reducing the cost by truncating the problem.

## 10.7 Quantum mechanics and perturbation

Our numerical (finite difference) eigenvalue/eigenvector solution for the time-independent Schrödinger equation can handle any potential $V(x)$ we choose to provide. This gives us a great deal of freedom in our numerical solutions, freedom that is not available analytically (there are only a small number of potentials for which (10.20) has eigenvalue/vector combinations that can be solved exactly using familiar functions). We have focused on physical problems with known analytic solutions to test the numerical methods developed, but these methods are most powerful when used to predict the behavior of a physical system whose solution is not known. It is difficult to tell, absent an explicit solution, how well our numerical method is doing. So in this section, we will set up some simple perturbation machinery to calculate the corrections to the energy spectrum of a quantum mechanical system when a small, but analytically intractable, potential is added. Then we'll solve the same problem numerically – that solution will be "exact" up to numerical errors. We can then compare the perturbative result with the exact numerical result.

### *10.7.1 Perturbation setup*

Suppose we have a quantum mechanical problem, of the form:

$$-\frac{\hbar^2}{2m}\psi''(x) + V(x)\psi(x) = E\psi(x) \quad \psi(0) = \psi(a) = 0, \quad (10.68)$$

where we will take $V(x)$ to be small. If $V(x)$ were zero, then we have an infinite square well of length $a$ – that's a reasonable "background" problem for us – numerically, it is natural to set the wave function to zero at two spatial locations, and that defines an infinite square well implicitly.

What we want is a partial description of what happens when $V(x)$ is close to zero – to make this quantitative, we start by moving to dimensionless variables as always. Let $x = aq$, $\tilde{E} \equiv \frac{2ma^2}{\hbar^2} E$, and $\tilde{V}(q) = \frac{2ma^2}{\hbar^2} V(aq)$, then (10.68) becomes

$$-\psi''(q) + \tilde{V}(q)\psi(q) = \tilde{E}\psi(q) \quad \psi(0) = \psi(1) = 0. \tag{10.69}$$

In this dimensionless setting, if $\tilde{V}(q) = 0$, our solutions are $\psi_n(q) = \sqrt{2}\sin(n\pi q)$ with $\tilde{E}_n = n^2\pi^2$. Now "$\tilde{V}(q)$ is small" means that, for all $q$, $\tilde{V}(q) = \epsilon\hat{V}(q)$ with $\hat{V}(q) \sim 1$ and $\epsilon$ small: $\epsilon \ll 1$.

The perturbative approach to this problem is to assume that $\psi(q)$ and $\tilde{E}$ can be expanded in powers of $\epsilon$ as (see [29] for details of this procedure in more general settings)

$$\psi(q) = \sum_{j=0}^{\infty} \epsilon^j \psi^{(j)}(q) \quad \tilde{E} = \sum_{j=0}^{\infty} \epsilon^j \tilde{E}^{(j)} \tag{10.70}$$

for a set of independent functions $\{\psi^{(j)}(q)\}_{j=0}^{\infty}$ and constants $\{\tilde{E}^{(j)}\}_{j=0}^{\infty}$ that are the targets, those are what we will try to find. Our assumption is that these functions and constants are all of order unity, and therefore the different terms in the sums do not talk to one another – terms of size $\epsilon$ cannot be combined with terms of size $\epsilon^2$, for example.

We insert the expressions (10.70) into (10.69):

$$\sum_{j=0}^{\infty} \epsilon^j \left[ -\frac{d^2\psi^{(j)}(q)}{dq^2} + \epsilon\psi^{(j)}(q)\hat{V}(q) \right] = \sum_{j=0}^{\infty} \epsilon^j \tilde{E}^{(j)} \sum_{k=0}^{\infty} \epsilon^k \psi^{(k)}(q). \tag{10.71}$$

And we collect in powers of $\epsilon$ – again, terms of different size cannot combine, so we must have equality at each power of $\epsilon$. The first two powers of $\epsilon$ then give us the pair of equations:

$$-\frac{d^2\psi^{(0)}}{dq^2} = \tilde{E}^{(0)}\psi^{(0)}$$

$$-\frac{d^2\psi^{(1)}}{dq^2} + \hat{V}\psi^{(0)} = \tilde{E}^{(0)}\psi^{(1)} + \tilde{E}^{(1)}\psi^{(0)}. \tag{10.72}$$

From the first of this pair, we learn that the $\epsilon^0$, i.e. unperturbed, contribution is precisely the wave function and energy of the original infinite square well – that is, $\psi^{(0)}(q) = \psi_n(q)$ for integer $n$, with $\tilde{E}^{(0)} = \tilde{E}_n \equiv n^2\pi^2$. Then inserting this information in the second equation, we have

$$-\frac{d^2\psi^{(1)}}{dq^2} + \hat{V}\psi_n = \tilde{E}_n\psi^{(1)} + \tilde{E}^{(1)}\psi_n. \tag{10.73}$$

The wave functions of the infinite square well satisfy the orthonormality condition:

$$\int_0^1 \psi_n(q)\psi_m(q)dq = \delta_{mn}. \tag{10.74}$$

We can use this to make progress in (10.73): multiply both sides by $\psi_n(q)$ and integrate,

$$-\int_0^1 \frac{d^2\psi^{(1)}}{dq^2}\psi_n(q)dq + \int_0^1 \hat{V}\psi_n^2 dq = \tilde{E}_n \int_0^1 \psi^{(1)}\psi_n(q)dq + \tilde{E}^{(1)}. \tag{10.75}$$

The first term can be integrated by parts (assuming $\psi^{(1)} = 0$ at $q = 0$ and 1) twice – and once both derivatives have been flipped onto the $\psi_n(q)$, we can use the fact that $\psi_n''(q) = -\tilde{E}_n\psi_n(q)$ to simplify – the first term on the left is then identical to the first term on the right and they cancel, leaving the simple form:

$$\tilde{E}^{(1)} = \int_0^1 \hat{V}(q)\psi_n(q)^2 dq. \tag{10.76}$$

This is the result of "first-order" perturbation theory applied to a small potential $V(x)$ introduced in an infinite square well "background." It tells us that, to first order in the small parameter of the potential, the $n$th energy of the system is:

$$\tilde{E} = n^2\pi^2 + \epsilon \int_0^1 \hat{V}(q)\psi_n(q)^2 dq. \tag{10.77}$$

---

**Example**

Take, as the perturbing potential, a small step extending half-way across the well:

$$\tilde{V}(q) = \begin{cases} 0 & 0 < q < \frac{1}{2} \\ \epsilon & \frac{1}{2} \le q < 1 \end{cases}. \tag{10.78}$$

To use (10.77), we need the integral:

$$\int_0^1 \hat{V}(q)\psi_n(q)^2 dq = \int_0^{\frac{1}{2}} \psi_n(q)^2 dq = \frac{1}{2} \tag{10.79}$$

(since $\tilde{V} = \epsilon$ over half the step, $\hat{V} = 1$ over that interval) so the spectrum has shifted by a constant amount:

$$\tilde{E} = n^2\pi^2 + \frac{1}{2}\epsilon. \tag{10.80}$$

That's the analytic, but perturbative result. For the numerical calculation of the spectrum, we just augment our $\mathbb{D}$ from (10.28) with a diagonal matrix that has ones in the lower half (the step function is, remember, a function of $q$) – call this matrix $\hat{\mathbb{V}}$ – our full matrix operator is, then:

$$\mathbb{D} + \epsilon\hat{\mathbb{V}}. \tag{10.81}$$

If we calculate the numerical spectrum for $\epsilon = 1$, we would expect all energies to be shifted up by a half (based on the perturbative result) $\tilde{E} = n^2\pi^2 + \frac{1}{2}$. For the ground state energy, then, we should have: $\pi^2 + \frac{1}{2} \approx 10.3696$. From the determination of the eigenvalues of the matrix in (10.81), we have (for the smallest eigenvalue) $\lambda_{\text{min}} \approx 10.3643$ using $N = 2000$ (with $\Delta q = 1/(2001)$). There are two sources of error here – first we are calculating the *exact* numerical solution to the problem, and that introduces error from the approximation to the differential operator. Second, we only took the first term in the perturbation expansion of the analytical solution as our "known" value, and we could do better by going out further in perturbative order.

On the numerical side, we still get relevant results for large $\epsilon$ – there is no validity constraint on the size of $\epsilon$, as there is for the perturbative approach. Suppose, for example, we make the potential with $\epsilon = 475$. Then the energy shift is dramatic, well beyond what we can predict with first order perturbation theory. The discrete problem is simple, and the first four eigenfunctions are shown in Figure 10.5.

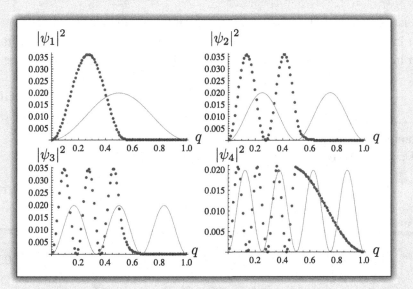

Figure 10.5 The probability density for the first four eigenfunctions of the infinite well potential with a step extending half-way across the bottom – the strength of the step is set large, at $\epsilon = 475$. The unperturbed probabilities associated with the infinite square well bound states are shown in gray for comparison.

# Further reading

1. Arfken, George B. & Hans J. Weber. *Mathematical Methods for Physicists*. Academic Press, 2001.
2. Boas, Mary. *Mathematical Methods in the Physical Sciences*. Wiley, 2005.

3. Demmel, James W. *Applied Numerical Linear Algebra*. Siam, 1997.
4. Golub, Gene H. & Charles F. Van Loan. *Matrix Computations*. The Johns Hopkins University Press, 1996.
5. Griffiths, David J. *Introduction to Quantum Mechanics*. Pearson Prentice Hall, 2005.
6. Hinch, E. J. *Perturbation Methods*. Cambridge University Press, 1991.
7. Trefethen, Loyd N. & David Bau III. *Numerical Linear Algebra*. Siam, 1997.

## Problems

### Problem 10.1

A damped harmonic oscillator has equation of motion (from (4.10)):

$$\ddot{x}(t) = -\omega^2 x(t) - 2b\dot{x}(t). \tag{10.82}$$

Solve this equation for initial conditions $x(0) = 0$ and $\dot{x}(0) = v_0$. The solution of interest looks as shown in Figure 10.6.

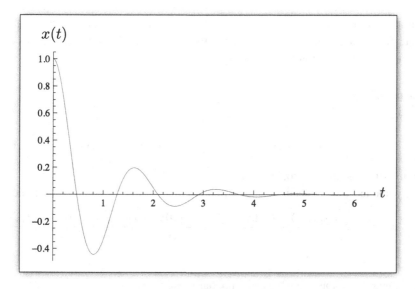

Figure 10.6 Damped harmonic oscillator solution.

### Problem 10.2

Show that if a matrix $\mathbb{A} \in \mathbb{R}^{n \times n}$ is symmetric, an eigenvector $\mathbf{v}$ with eigenvalue $\lambda$ has $\lambda \in \mathbb{R}$.

### Problem 10.3

Prove that the matrix $\mathbb{A}^T \mathbb{A}$ for symmetric $\mathbb{A}$ has positive eigenvalues.

### Problem 10.4

Two eigenvectors, $\mathbf{v}_1$ and $\mathbf{v}_2$ of a matrix $\mathbb{A} \in \mathbb{R}^{n \times n}$ share an eigenvalue, $\lambda$. Show that any linear combination of these two eigenvectors is also an eigenvector with eigenvalue $\lambda$. The eigenvectors form a subspace – when a method finds such eigenvectors, it can end up with any pair, and most routines will return an orthonormalized basis for the subspace.

## Problem 10.5
Prove that the following is a norm:

$$\|\mathbb{A}\| = \max_{ij}|A_{ij}|, \tag{10.83}$$

i.e. $\|\mathbb{A}\|$ is the largest element of $\mathbb{A}$ (in absolute value).

## Problem 10.6
The Schrödinger equation for a harmonic oscillator can be written in the form:

$$-\psi''(q) + q^2\psi(q) = \tilde{E}\psi(q). \tag{10.84}$$

Write the normalized wave function and eigenvalues for this choice of units (use the known solution from, for example [23]). Write a `Mathematica` function that will generate the solution and eigenvalue given $n$, the integer indexing the bound states (you can use the built-in function `HermiteH`).

## Problem 10.7
(a) Find $p$ in

$$\psi''(x_j) = \frac{\psi(x_{j+1}) - 2\psi(x_j) + \psi(x_{j-1})}{\Delta x^2} + O(\Delta x^p), \tag{10.85}$$

from (10.24) with the error in place.
(b) We could use a more accurate approximation to $\psi''$: Find the coefficients $a$ and $b$ in

$$\psi''(x_j) = \frac{a\psi(x_{j+2}) + b\psi(x_{j+1}) - 2(a+b)\psi(x_j) + b\psi(x_{j-1}) + a\psi(x_{j-2})}{\Delta x^2}$$
$$+ O(\Delta x^4). \tag{10.86}$$

This approximation requires more boundary data.

## Problem 10.8
In the case of the infinite square well, with $-\psi''(q) = \tilde{E}\psi(q)$, we can find the discretized eigenvalues exactly, and compare those with the known energy spectrum $\tilde{E} = n^2\pi^2$. Take $\psi(q) = Ae^{i\pi kq}$ for integer $k$. Now discretize using $q_j = j\Delta q$, so that: $\psi_j = Ae^{i\pi kj\Delta q}$. Insert this directly in the discretized form:

$$\frac{-\psi_{j+1} + 2\psi_j - \psi_{j-1}}{\Delta q^2} = \lambda_k\psi_j \tag{10.87}$$

and find the eigenvalues (indexed by $k$). Subtract the continuous solution, $k^2\pi^2$ and Taylor expand this residual for $\Delta q \longrightarrow 0$. What is the size of the leading error in your discrete approximation to the energy spectrum for a particle in a box? For what value $k$ is the size of this error of the same order as the "dominant" $k^2\pi^2$ term (write in terms of $\Delta q$)? For what $k$ is the error order one?

## Problem 10.9

Using (10.77), find the perturbed energies for an infinite square well (extending from $q = 0$ to 1) with a perturbed bottom, $\tilde{V}(q) = -\epsilon q$. This perturbation comes from a constant force acting on a particle confined to the well. An example would be gravity near the surface of the Earth.

## Problem 10.10

Work out the energy spectrum to first order for an infinite square well (from $0 \longrightarrow 1$) with perturbing potential: $\tilde{V}(q) = \epsilon \sin(\pi q)$.

## Lab problems

## Problem 10.11

Download the data file "ch10.dat" from the book's website. This has data appropriate to a mass on a spring oscillating in a viscous medium (the damped harmonic oscillator from Problem 10.1) and consists of pairs $\{t, x\}$ giving the time $t$ and location of the mass $x$ at that time. The data has been sampled at equal temporal intervals (so that $\Delta t$ is a constant). Read in the data, and use your FFT routine to find the frequency of oscillation for the mass. Generate a set of functions to fit the data using this frequency and appropriate decaying functions (assume $b$ in (10.82) is an integer and use your solution to Problem 10.1). Solve the normal equations (10.9) using the built-in function QRDecomposition, but your own back solver (or use UDBacksolve from the Chapter 9 notebook). You should be able to refine your functions to get "perfect" agreement (an $R^2$ value of one). Using that case, find $\omega$ and $b$, the physical parameters governing the system.

## Problem 10.12

We can use least squares to fit noisy data.
(a) We'll generate fake noisy data from

$$d(t) = \sin(2\pi t) - 4 \sin(2\pi 4t) + 2 \sin(2\pi 5t) + F \qquad (10.88)$$

where $F$ is a random number drawn from a flat distribution $[-5, 5]$ (accessible in Mathematica by the command RandomReal[{-5,5}]). Write a function that outputs the vector **d** with entries $d_j = d(t_j)$ for $t_j = j\Delta t$, $j = 1 \rightarrow 100$ and $\Delta t = 0.01$.
(b) We'll use the sinusoidal functions $f_k(t) = \sin(2\pi kt)$ for $k = 1 \rightarrow 10$ in the decomposition (10.1). Generate the matrix $\mathbb{A}$ from (10.3) and solve the normal equations for 100 different instantiations of the random noise (i.e. one hundred different vectors of data made by your function from part a) – what are the average

coefficients $\{\alpha_k\}_{k=1}^{10}$ for the 100 data sets and how do they compare to their target values from (10.88)?

## Problem 10.13

Implement the QR and SI eigensolvers. For your initial, orthogonal matrix, in SI, take the identity. In both cases, make $M$ (the number of iterations) an input in your function, and use your modified Gram–Schmidt function from Problem 9.13 to perform the decompositions. Using each method, find the eigenvectors and eigenvalues of the matrix:

$$
\mathbb{A} = \begin{pmatrix} 1 & 2 & 3 & 4 \\ 2 & 5 & 6 & 7 \\ 3 & 6 & 8 & 9 \\ 4 & 7 & 9 & 10 \end{pmatrix}. \tag{10.89}
$$

How big does $M$ have to be to get, for example, the off-diagonal elements of $\bar{\mathbb{A}}_M$ (for SI) and $\mathbb{A}_M$ (for QR) to be $\leq 10^{-13}$?

## Problem 10.14

Discretize the Schrödinger equation with the harmonic oscillator potential as written in (10.84). Take $q_\infty = 10$, and discretize using $q_j = -q_\infty + j\Delta q$ with $\Delta q = \frac{2q_\infty}{N+1}$ and $N = 100$.

(a) Find the eigenvalues and eigenvectors of the discretized matrix on the left of (10.84) using the QR method you implemented in the previous problem with $M = 500$ (i.e. set a maximum of five hundred iterations). Plot the exact energies from Problem 10.6 for $n = 1$ to 50 together with the first fifty (smallest to largest) eigenvalues determined by the QR method. At what $n$, roughly, do you begin to diverge from the exact energies (i.e. where do you lose the linear form of the exact solution)?

(b) Plot the first and tenth $n = 0$, $n = 9$ eigenvectors, determined numerically, as points, with the exact solutions on top. You must normalize your numerical eigenvectors to match the quantum mechanical normalization, which in discrete (approximate) form, reads: $\sum_{j=1}^{N} |\psi_j|^2 \Delta q = 1$ (i.e. divide your eigenvectors by $\sqrt{\Delta q}$). Given that you have implicitly put the oscillator potential inside an infinite square well, can you see why, from the wave functions, the eigenvalues begin to diverge from the linear exact form?

## Problem 10.15

You obtained the spectrum for hydrogen by shooting back in Chapter 3. Try it again using the finite difference approach – make a matrix by approximating (10.33) using the discretization scheme from (10.24) with $q_j = j\Delta q$ for $j = 1 \longrightarrow N$ and $\Delta q = \frac{q_\infty}{N+1}$ (notice that we omit the origin). Using $N = 10\,000$ and $q_\infty = 100$, find the first six negative (bound-state) energies (use the built-in function `Eigenvalues` this time; the matrices are large). Fit this data using least squares with a reasonable set of functions of $n$ (an integer in this case) and give the $R^2$ value for your fit (use, as in Problem 10.11, the built-in QRDecomposition to handle the non-square QR decomposition, but then solve the least squares problem using UDBacksolve).

## Problem 10.16

Download the mystery picture in matrix form "ch10.mystery.mat" from the book's website. Use the built-in Eigensystem command to obtain the eigenvectors and eigenvalues associated with this matrix, and make plots taking the first few eigenvectors as approximations to the final picture as in (10.35) (see the chapter notebook for an example of this process). Roughly how many eigenvectors do you need to recognize the person in the picture? (Do *not* simply look at the full matrix as a picture; that spoils the "fun.")

## Problem 10.17

Here, we'll generate the discrete matrix appropriate to the perturbation problem from Problem 10.9. Write a function that returns the appropriate matrix $\mathbb{D} + \epsilon \hat{\mathbb{V}}$ for $\hat{V}(q) = -q$ given $\epsilon$, and the number of grid points $N$.

(a) First find the eigenvalues for the matrix with $\epsilon = 0$ and $N = 1000$ with $\Delta q = \frac{1}{N+1}$, and verify that errors of order one emerge at the appropriate value of $k$ (the index of the sorted eigenvalues) from Problem 10.8.

(b) Now generate the matrix with $\epsilon = 0.1$, $N = 1000$, $\Delta q = \frac{1}{N+1}$ – find the eigenvalues (use Eigenvalues, the built-in function, to save time), sort them, and compare with the perturbation result you computed in Problem 10.9 – where do the finite difference and perturbative results begin to diverge?

## Problem 10.18

Generate the finite difference approximation to the perturbing potential from Problem 10.10, using $N = 1000$, $\Delta q = \frac{1}{N+1}$, and $\epsilon = .1$, find the eigenvalues, and sort them. Subtract from each eigenvalue the result of the first-order perturbation analysis you did in Problem 10.10 and plot the result.

# 11

# Iterative methods

We have seen some direct methods for inverting, and calculating the eigenvalues and eigenvectors of matrices. In many cases, however, these direct methods will take too long to solve a particular problem. In addition, we may only have interest in approximate solutions. When solving the Poisson problem, for example, we may already be on a coarse grid, and require only a qualitative description of the solution – high precision and "exact" matrix inverses are unnecessary. On the eigenvalue side, we may only need part of the spectrum of a matrix, maybe we just want a few bound states for a potential in quantum mechanics, for example. In these cases, what we would like is a process that ultimately would give the full inverse, or the complete set of eigenvalues, but a process that can be truncated "along the way" and still provide partial information.

There are two broad schemes for these approximate methods, and we'll describe and see examples of both. The first approach is relevant to matrix inversion, and involves decomposing a matrix into a simple (to invert) part, and a (hopefully small) "other" part. We proceed to invert the simple part and use that inversion to drive an iteration that will converge to the exact numerical solution (computed using QR factorization, for example). The second approach involves constructing a particular subspace of $\mathbb{R}^n$, called a Krylov subspace, and we invert matrices and find eigenvalues within that subspace. The Krylov subspace is smaller than $\mathbb{R}^n$, and so our linear algebra problems require fewer operations to solve in it.

## 11.1 Physical motivation

We have examples of problems that produce large matrices, the finite difference approach to PDE solving has matrix size that depends on our grid resolution. In three dimensions, for example, if we take a grid of $N$ points in each of the $x$, $y$ and $z$ directions, the matrix representing the Laplacian would be in $\mathbb{R}^{N^3 \times N^3}$. For all but the most modest $N$, these matrices will be too large to invert efficiently (or at all).

274

For the eigenvalue problem, large matrices can arise as an attempt to find accurate quantum mechanical energies. In the case of hydrogen (described in Section 10.4.2), we had two parameters to tune in order to find bound states numerically: the value of the right-hand endpoint (the numerical "infinity" in the problem), and the number of points in our grid. By setting the endpoint large, we obtain more bound states, but then we must make $N$ large to get good approximations for the energies. If the goal is a lot of bound states at high accuracy, we need a large $N$, and the QR iteration method is expensive (see Section 10.6.1).

### 11.1.1 Particle motion

Aside from the timing of an individual inverse, there are physical problems in which we have many repeated inversions, so that even a relatively small matrix needs to be inverted many times. Consider, for example, the problem of the motion of a set of $N$ particles in our usual two-dimensional box with grounded walls. We now have a coupled PDE-particle system – we are calculating the potential for the set of particles, and letting the Lorentz force (electrostatic) govern the time-evolution of the particles, which then changes the potential, changing the motion, etc.

The full particle-field system is:

$$\nabla^2 V = -\frac{1}{\epsilon_0} \sum_{j=1}^{N} q_j \delta^2(\mathbf{r} - \mathbf{r}_j), \quad V(\partial \Omega) = 0$$

$$m_j \ddot{\mathbf{r}}_j = -q_j \nabla V \quad \text{for } j = 1 \to N$$

(11.1)

where we are in two dimensions (hence the $\delta^2$), and the particles are confined to some domain $\Omega$ with boundary $\partial\Omega$ (on which the potential is given). This is the full problem of electrostatics where we are not just treating particle motion in the presence of a fixed field. Incidentally, one might ask whether such a complicated field-particle approach is really necessary. After all, we know that the conducting boundaries can often be handled effectively by the image method (for example). And indeed, it is in theory possible to simply calculate the superposition of all real and image charges at each time step, use that to develop the potential and then update as a simple $2N$-dimensional ODE. The problem is the number of image charges – we need an efficient method for calculating their location (and more important, a way to sum the infinite set of their force contributions). Rather than do that, we will combine our Poisson solver with a simple finite difference.

The basic approach is clear, we must solve Poisson's equation in $\Omega$ (our grounded box, say), and use the resulting potential to update the particle locations via Newton's second law. If we take steps using the Verlet update from Section 2.2, and let

$\mathbf{r}^n_j$ refer to the position of the $j$th particle at time $t_n$, then:

$$\mathbf{r}^{n+1}_j = 2\mathbf{r}^n_j - \mathbf{r}^n_{j-1} + \Delta t^2 \frac{\mathbf{F}^n_j}{m_j} \tag{11.2}$$

where the force comes from $-q\nabla V$ at time level $n$ evaluated at the particle location: $\mathbf{F}^n_j \equiv -q_j \nabla V^n|_{\mathbf{r}=\mathbf{r}_j}$ where $V^n$ is the potential at time $t_n$. For each time step, then, we must know the potential $V$ everywhere in $\Omega$ – that means that we have to solve the Poisson problem at time $t_n$, and then again at time $t_{n+1}$ (since the particles generating the potential have now moved), etc. Each step of the dynamics update requires a solution of Poisson's problem, and hence a matrix inverse.

If we want to run the dynamics for a long time, then we have a lot of matrix inversions to perform, and again, that may be prohibitive. What we'd like is a faster way to solve the inverse problem given the highly structured, and sparse, matrix associated with the discretization of the Laplace operator.

### 11.1.2 Advertisement: sparse matrices

Many of the physical problems we have encountered lead to large matrices, but ones with relatively few non-zero entries. Think of the discrete (generalized) Laplacian matrix in one dimension (4.45) – that only has non-zero entries on the diagonal, super and sub-diagonals. So each row of the matrix has at most three entries, regardless of the number of grid points we use. We call a matrix in $\mathbb{R}^{n\times n}$ "sparse" if it has $s \ll n$ non-zero entries per row.

Sparsity can be used to speed up matrix-vector operations. First, let's think of a sparse vector $\mathbf{a} \in \mathbb{R}^n$, with only $s \ll n$ non-zero entries. How long would the dot product of $\mathbf{a}$ with $\mathbf{b} \in \mathbb{R}^n$ take? For generic vectors, the dot product has timing $T(n) = O(n)$ (multiplications and additions), but it is clear that if $\mathbf{a}$ has only $s$ entries, then the dot product with any other vector should take time $O(s)$. We can accomplish this by storing only the location and value of the non-zero entries of $\mathbf{a}$, and then picking out the relevant entries from $\mathbf{b}$. Suppose, to be concrete, that $\mathbf{a}$ only has two non-zero entries, $a_1$ and $a_n$. Then

$$\mathbf{a} \cdot \mathbf{b} = a_1 b_1 + a_n b_n, \tag{11.3}$$

we need to perform only two multiplications to obtain the dot product, regardless of the length of the vector.

If we have a sparse matrix, $\mathbb{A} \in \mathbb{R}^{n\times n}$, where each row only has $s$ entries, then matrix-vector multiplication can be done quickly. Normally, multiplying $\mathbb{A}$ by $\mathbf{b} \in \mathbb{R}^n$ would require $O(n^2)$ operations (that's $n$ dot-products), but again, by storing only the non-zero entries of $\mathbb{A}$, it should be clear that we can accomplish the full multiplication with just $O(ns)$ multiplications – we take the dot product of a row of $\mathbb{A}$, with only $s$ entries, with $\mathbf{b}$. That requires $O(s)$ operations, and we do this $n$ times.

A common theme for the approximate methods we will discuss in this chapter is that they replace nonlinear operations (like those appearing in the Gram–Schmidt procedure, Algorithm 9.1) with many (iterative) *linear* operations – successive multiplication by a matrix, for example. Physically, many of the matrices we study are sparse, and so matrix-vector multiplication is cheap.

## 11.2 Iteration and decomposition

We will focus on matrix inverse problems associated with PDE solution, since that is an obvious place where size is an issue. There is a lot of room for creativity in setting up and executing the iterative approach here, and this is where numerical work can become very problem-specific – matrices associated with PDEs will be sparse, and that is an exploitable property, they also have predictable spectral structure which has specially tailored decompositions that work well for accuracy and convergence.

Remaining as general as possible (by referring to the generic $\mathbb{A}\mathbf{x} = \mathbf{b}$ problem), we will look at some popular decomposition schemes, and apply them to some of the problems set up in previous chapters. If we work in low enough dimension (coarse grids for PDE operators), we can even compare with "exact" results. At the heart of these iterative schemes is a matrix decomposition. Suppose we have $\mathbb{A} \in \mathbb{R}^{n \times n}$ (and assume invertible). We can split $\mathbb{A} = \mathbb{P} - \mathbb{Q}$ for two matrices $\mathbb{P}$ and $\mathbb{Q}$, where $\mathbb{P}^{-1}$ exists (we assume $\mathbb{A}$ is itself invertible).

Our problem can then be written

$$\mathbb{A}\mathbf{x} = \mathbf{b} \longrightarrow (\mathbb{P} - \mathbb{Q})\mathbf{x} = \mathbf{b}. \tag{11.4}$$

Now for the iteration: We can solve for $\mathbf{x}$ in terms of itself as

$$\mathbf{x} = \underbrace{\mathbb{P}^{-1}\mathbb{Q}}_{\equiv \mathbb{S}}\mathbf{x} + \underbrace{\mathbb{P}^{-1}\mathbf{b}}_{\equiv \mathbf{c}}. \tag{11.5}$$

It is easy to cook up a $\mathbb{P}$ and then take $\mathbb{Q} = \mathbb{P} - \mathbb{A}$ satisfying the decomposition form, but does the implied iteration:

$$\mathbf{x}_{m+1} = \mathbb{S}\mathbf{x}_m + \mathbf{c} \tag{11.6}$$

actually converge to the true solution? To answer this question, define $\mathbf{x} \equiv \mathbb{A}^{-1}\mathbf{b}$. We are interested in the difference between the $(m+1)$st iterate and $\mathbf{x}$ itself. The difference can be written as:

$$\mathbf{x}_{m+1} - \mathbf{x} = \mathbb{S}(\mathbf{x}_m - \mathbf{x}) + \mathbf{c} + \mathbb{S}\mathbf{x} - \mathbf{x}. \tag{11.7}$$

The expression on the right can be simplified by noting that

$$\mathbf{c} + (\mathbb{S} - \mathbb{I})\mathbf{x} = \mathbb{P}^{-1}\mathbf{b} + \mathbb{P}^{-1}(\mathbb{Q} - \mathbb{P})\mathbf{x}$$
$$= \mathbb{P}^{-1}(\mathbf{b} - \mathbb{A}x) = 0. \tag{11.8}$$

The difference between the $m$th and $(m + 1)$st iterate and the true solution $\mathbf{x}$ is then

$$\mathbf{x}_{m+1} - \mathbf{x} = \mathbb{S}(\mathbf{x}_m - \mathbf{x})$$
$$= \mathbb{S}\mathbb{S}(\mathbf{x}_{m-1} - \mathbf{x})$$
$$\vdots$$
$$= \mathbb{S}^{j+1}\left(\mathbf{x}_{m-j} - \mathbf{x}\right) \tag{11.9}$$
$$\vdots$$
$$= \mathbb{S}^{m+1}(\mathbf{x}_0 - \mathbf{x})$$

where $\mathbf{x}_0$ is our initial guess. We have seen this type of matrix-power product before in Section 10.5, and assuming the eigenvectors of $\mathbb{S}$ are complete and orthonormal (easy to accomplish by taking $\mathbb{S}$ symmetric), we know that the arbitrary vector $\mathbf{x}_0 - \mathbf{x}$ has a decomposition in terms of these eigenvectors. Then repeated multiplication by $\mathbb{S}$ will yield:

$$\mathbb{S}^{m+1}(\mathbf{x}_0 - \mathbf{x}) \propto |\lambda_{\max}|^{m+1}\mathbf{v}_{\max} \tag{11.10}$$

for the largest eigenvalue (in absolute value) $\lambda_{\max}$ and its associated eigenvector $\mathbf{v}_{\max}$. If we consider the length of $\mathbf{x}_{m+1} - \mathbf{x}$, then, we have

$$\|\mathbf{x}_{m+1} - \mathbf{x}\| \propto |\lambda_{\max}|^{m+1}, \tag{11.11}$$

and our method will converge provided that the largest eigenvalue of $\mathbb{S}$ has magnitude less than one.

So one requirement we need to satisfy, in developing a useful decomposition, is that the spectrum of the matrix $\mathbb{S}$ has maximum magnitude less than one. Clearly, another requirement is that $\mathbb{P}^{-1}$ is easy to compute, or else we have just replaced one hard matrix inversion with another. Finally, we note that the update itself requires a matrix-vector multiplication – assuming a dense matrix (few non-zero entries), each step requires $O(n^2)$ operations – so as we tune the splittings, we can allow "easy" matrix inverses. Remember from Section 9.5 that to invert an upper or lower triangular matrix only requires $O(n^2)$ operations, the same order as the matrix-vector multiplication that drives the iteration, so we are not significantly increasing the timing by introducing an inverse of this type.

### 11.2.1 Jacobi's method

Jacobi's method is defined by the following splitting: take $\mathbb{P}$ to be the diagonal of $\mathbb{A}$, and $\mathbb{Q} = \mathbb{P} - \mathbb{A}$. Let's work through a simple example – let

$$\mathbb{A} \doteq \begin{pmatrix} 1 & 2 \\ 0 & 3 \end{pmatrix} \quad \mathbf{b} = \begin{pmatrix} 1 \\ -1 \end{pmatrix}. \tag{11.12}$$

Take the easy-to-invert diagonal and call it $\mathbb{P}$, then $\mathbb{Q} = \mathbb{P} - \mathbb{A}$ and we can construct $\mathbb{S}$ and $\mathbf{c}$:

$$\mathbb{S} \doteq \begin{pmatrix} 0 & -2 \\ 0 & 0 \end{pmatrix} \quad \mathbf{c} \doteq \begin{pmatrix} 1 \\ -\frac{1}{3} \end{pmatrix}. \tag{11.13}$$

If we define the iteration by (11.6) using our current decomposition, then we just need an initial vector to start the process – take $\mathbf{x}_0 \doteq (1 \quad 1)^T$. The iteration gives

$$\mathbf{x} \doteq \begin{pmatrix} 1 \\ 1 \end{pmatrix} \longrightarrow \begin{pmatrix} -1 \\ -\frac{1}{3} \end{pmatrix} \longrightarrow \begin{pmatrix} \frac{5}{3} \\ -\frac{1}{3} \end{pmatrix}, \tag{11.14}$$

where we have the exact solution obtained in three steps. The fact that we end up with the exact answer is engineered in this example (it won't be true in general), and is associated with the eigenvalues of $\mathbb{S}$.

### 11.2.2 Successive over-relaxation (SOR)

A class of method can be defined by taking the decomposition

$$\mathbb{A} = \mathbb{D} - \mathbb{L} - \mathbb{U}, \tag{11.15}$$

where $\mathbb{D}$ is the diagonal portion of $\mathbb{A}$ and is invertible (assuming no zero diagonal elements in $\mathbb{A}$), $-\mathbb{L}$ and $-\mathbb{U}$ are the strictly (meaning excluding the diagonal) lower and upper triangular portions of $\mathbb{A}$, respectively.

We have some freedom in forming an iterative method from

$$\mathbb{D}\mathbf{x} - \mathbb{L}\mathbf{x} - \mathbb{U}\mathbf{x} = \mathbf{b}. \tag{11.16}$$

We could, for example, take

$$(\mathbb{D} - \mathbb{L})\mathbf{x}_{m+1} = \mathbb{U}\mathbf{x}_m + \mathbf{b}. \tag{11.17}$$

This is known as "Gauss–Seidel," but we have treated the lower triangular matrix preferentially: It would be just as valid to reverse the roles of $\mathbb{L}$ and $\mathbb{U}$ above. A more general approach is obtained by introducing a parameter $\omega$ which tunes the amount of $\mathbb{L}$ and $\mathbb{U}$ we use on the left or right of the iteration. The "successive-over-relaxation" (SOR) approach is defined by the update:

$$(\mathbb{D} - \omega\mathbb{L})\mathbf{x}_{m+1} = ((1 - \omega)\mathbb{D} + \omega\mathbb{U})\mathbf{x}_m + \omega\mathbf{b}, \tag{11.18}$$

with relaxation parameter $\omega > 1$ (for $\omega < 1$, the splitting gives an "under-relaxation"). For $\omega = 1$ (11.18) is precisely "Gauss–Seidel" (11.17). The role of the parameter $\omega$ is clear: We are continuously selecting the weight of our splitting and tuning the various spectra of our update matrices. We are attempting to enforce a spectral radius less than one on the analogue of $\mathbb{S}$ (which would be $(\mathbb{D} - \omega\mathbb{L})^{-1}((1 - \omega)\mathbb{D} + \omega\mathbb{U})$ in this general case). Successive over-relaxation requires both matrix-vector products, and an inverse, for its update. But again, the matrix we need to invert (appearing on the left of (11.18)) is lower triangular, and hence requires the same number of operations as the matrix-vector product appearing on the right. Each step of SOR requires $O(n^2)$ calculations, and then what we want is to "end" while $m \ll n$, in order to beat the timing of a full inversion.

---

**Examples**

Let's think about the discrete analogue of Poisson's problem for electrostatics:

$$\nabla^2 V = -\frac{\rho}{\epsilon_0} \longrightarrow \mathbb{D}\mathbf{f} = \mathbf{s} - \mathbf{b} \tag{11.19}$$

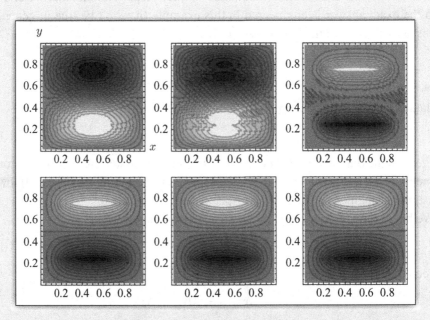

Figure 11.1 Iterations of the Poisson problem using Jacobi's method. Here we have a $40 \times 40$ grid with a positive and negative line of charge symmetrically placed. The top left figure is the solution after 200 iterations of the method, and then the approximations after 400, 600, 800 and 1000 iterations are shown – the lower right image is the solution as determined by explicit inversion.

as in Chapter 4 (the right-hand side has a source component **s**, and a boundary-value component **b** as described in Section 4.2). Here we have to invert the operator matrix $\mathbb{D}$ (a sparse matrix), and we can do this iteratively. Take the parallel plate setup from Section 4.4.2 with a $40 \times 40$ grid. The iterations, using the Jacobi method, with $\mathbf{x}_0 = \mathbf{s} - \mathbf{b}$ (any starting vector will do) are shown in Figure 11.1. There, we see that by 800 iterations, the approximate solution looks close to the "true" inverse – this represents a savings, since the inversion is taking place with a matrix that is $1600 \times 1600$.

As another example, set the source to zero on a square domain, and hold the walls at specified potentials – now the right-hand side of $\mathbb{D}\mathbf{f} = \mathbf{s} - \mathbf{b}$ has only **b** contribution, but it's still an inverse problem. Using Gauss–Seidel, and a $40 \times 40$ grid, the iterative solution to the Laplace problem with:

$$V_\ell = \sin(2\pi y) \quad V_r = \sin(5\pi y) \quad V_b = \sin(3\pi x) \quad V_t = \sin(\pi x) \tag{11.20}$$

(for the left, right, bottom and top boundaries) is shown in Figure 11.2.

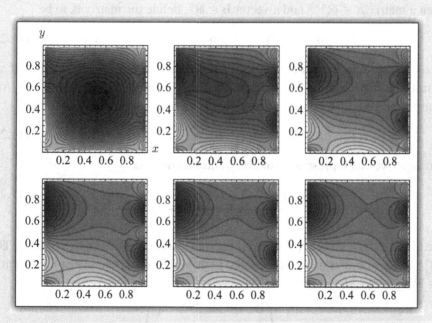

Figure 11.2 Gauss–Seidel used to solve the Laplace problem with boundary potentials: $V_\ell = \sin(2\pi y)$, $V_r = \sin(5\pi y)$, $V_b = \sin(3\pi x)$, and $V_t = \sin(\pi x)$ on a $40 \times 40$ grid. Again, every 200 iterations is shown, and the plot on the bottom right is the explicit solution that comes from the matrix inversion (via Gaussian elimination, or QR factorization, for example).

## 11.3 Krylov subspace

Now we turn to methods that solve problems (meaning inverse and eigenvalue problems) "exactly" (numerically) in smaller spaces than the original problem – these aren't iterative methods in the same sense as SOR is. The advantage lies in turning an $O(n^3)$ calculation (say) into an $O(p^3)$ one, with $p < n$. The particular subspace of interest is the "Krylov" subspace; the solutions found in this special type of smaller space share certain properties with the solution in the full space – we obtain fewer eigenvalues, for example, but the ones we do get are good approximations to some subset of the eigenvalues of a given matrix. A Krylov subspace can be defined easily, but its use inspires its definition. That is to say, we're about to define a Krylov subspace, but it won't look interesting until you see it used to solve problems.

### *11.3.1 Definition*

Given a matrix $\mathbb{A} \in \mathbb{R}^{n \times n}$ and a vector $\mathbf{b} \in \mathbb{R}^n$, define the matrix $\mathbb{K}$ to be

$$\mathbb{K} \doteq \begin{bmatrix} \mathbf{b} & | & \mathbb{A}\mathbf{b} & | & \mathbb{A}^2\mathbf{b} & | & \dots & | & \mathbb{A}^{n-1}\mathbf{b} \end{bmatrix} \tag{11.21}$$

where we are indicating the columns of the matrix $\mathbb{K} \in \mathbb{R}^{n \times n}$. Notice that if we multiply the matrix $\mathbb{K}$ by $\mathbb{A}$, we just shift the columns over one, lose the first column, and pick up a new, final column:

$$\mathbb{A}\mathbb{K} \doteq \begin{bmatrix} \mathbb{A}\mathbf{b} & | & \mathbb{A}^2\mathbf{b} & | & \mathbb{A}^3\mathbf{b} & | & \dots & | & \mathbb{A}^n\mathbf{b} \end{bmatrix}, \tag{11.22}$$

a situation we can represent by matrix multiplication via

$$\mathbb{A}\mathbb{K} = \mathbb{K} \underbrace{\begin{bmatrix} \mathbf{e}_2 & | & \mathbf{e}_3 & | & \mathbf{e}_4 & | & \dots & | & \mathbf{c} \end{bmatrix}}_{\equiv \mathbb{G}} \tag{11.23}$$

for $\mathbf{c} \equiv \mathbb{K}^{-1}\mathbb{A}^n\mathbf{b}$ (assuming $\mathbb{K}$ is invertible) and the normalized basis vectors of $\mathbb{R}^n$, $\{\mathbf{e}_j\}_{j=1}^n$ (where $\mathbf{e}_j$ has a one at location $j$ and zeroes everywhere else). Explicitly, $\mathbb{G} \in \mathbb{R}^{n \times n}$ looks like:

$$\mathbb{G} \doteq \begin{pmatrix} 0 & 0 & 0 & 0 & c_1 \\ 1 & 0 & 0 & 0 & c_2 \\ 0 & 1 & 0 & 0 & c_3 \\ 0 & 0 & \ddots & 0 & \vdots \\ 0 & 0 & \dots & 1 & c_n \end{pmatrix}. \tag{11.24}$$

A matrix of this form is called "upper Hessenberg," meaning that it has entries only on and above the leading subdiagonal, so that all entries of $\mathbb{G}$ with $G_{ij}$ for $i > j + 1$

are zero (think of upper triangular, just shifted down one). Our definitions so far allow us to factor the matrix $\mathbb{A}$ into "upper Hessenberg form":

$$\mathbb{K}^{-1}\mathbb{A}\mathbb{K} = \mathbb{G}. \tag{11.25}$$

We can find a basis for the space spanned by $\mathbb{K}$ by performing a QR decomposition on $\mathbb{K}$ – that is provided by the usual Gram–Schmidt procedure (Algorithm 9.1, or its modified form Algorithm 9.4) using the columns of $\mathbb{K}$: $\mathbb{K} = \mathbb{Q}\mathbb{R}$, and then:

$$Q^T \mathbb{A} Q = \mathbb{R}\mathbb{G}\mathbb{R}^{-1} \equiv \mathbb{T} \tag{11.26}$$

noting that $Q^T = Q^{-1}$ (since $Q$ is orthogonal). Now for an important digression.

---

**$\mathbb{T}$ is upper Hessenberg**

The inverse of an upper triangular matrix is itself upper triangular. Using that, we can show that the product $\mathbb{R}\mathbb{G}\mathbb{R}^{-1}$ for $\mathbb{G}$ defined in (11.24) is upper Hessenberg. Let $\mathbb{S} \equiv \mathbb{R}^{-1}$, then the $i - j$ entry of $\mathbb{T}$ is:

$$T_{ij} = \sum_{k=1}^{n} \sum_{\ell=1}^{n} R_{ik} G_{k\ell} S_{\ell j} \tag{11.27}$$

and we know that:

$$G_{k\ell} = \begin{cases} \delta_{\ell k-1} & \ell < n \\ c_k & \ell = n \end{cases}. \tag{11.28}$$

Using this in the sum, we have:

$$T_{ij} = \sum_{k=1}^{n} \left[ R_{ik} S_{k-1 j} + R_{ik} c_k S_{nj} \right], \tag{11.29}$$

and using the upper triangularity of $R_{ik}$ (all entries zero for $i > k$) and $S_{k-1 j}$ (all entries zero for $k - 1 > j$) and the fact that $S_{nn}$ is the only non-zero entry of the $n$th row (since $\mathbb{S}$ is upper triangular), we have:

$$T_{ij} = \sum_{k=i}^{j+1} R_{ik} S_{k-1 j} + \delta_{jn} S_{nn} \sum_{k=i}^{n} R_{ik} c_k. \tag{11.30}$$

The first term contributes only when $i \leq j + 1$, meaning that all entries with $i > j + 1$ are zero (this is the definition of upper Hessenberg form), and the second term only contributes when $j = n$, so that $T_{in}$ takes on values for $i = 1 \longrightarrow n$ – but this is the final column of the matrix $\mathbb{T}$, and that final column can have non-zero entries for all values of $i$ according to the definition of upper Hessenberg. We conclude, by direct examination, that $\mathbb{T}$ is upper Hessenberg.

We'll specialize to the case in which $\mathbb{A} = \mathbb{A}^T$, then the matrix $\mathbb{T}$ is both upper Hessenberg and lower Hessenberg (meaning it only has non-zero values below the sub-diagonal), and so is tridiagonal and symmetric, a very simple form. Our factorization gives:

$$\mathbb{Q}^T \mathbb{A} \mathbb{Q} \doteq \begin{pmatrix} \alpha_1 & \beta_1 & 0 & \cdots & 0 \\ \beta_1 & \alpha_2 & \beta_2 & \cdots & 0 \\ 0 & \ddots & \ddots & \ddots & 0 \\ 0 & 0 & \beta_{n-2} & \alpha_{n-1} & \beta_{n-1} \\ 0 & \cdots & 0 & \beta_{n-1} & \alpha_n \end{pmatrix}. \tag{11.31}$$

Cute, but what have we actually gained? We still have a matrix $\mathbb{Q} \in \mathbb{R}^{n \times n}$ to construct, and then we've factored $\mathbb{A}$ exactly into tridiagonal form (useful but not cheap). The final stroke lies in the observation that we can truncate $\mathbb{Q}$ – instead of finding all $n$ orthogonal vectors spanning the space of $\mathbb{K}$, we can just find $m < n$ of them. The same argument leads us to:

$$\mathbb{Q}_m^T \mathbb{A} \mathbb{Q}_m = \mathbb{T}_m, \tag{11.32}$$

where this time, $\mathbb{Q}_m \in \mathbb{R}^{n \times m}$ are the orthogonal vectors spanning just the first $m$ columns of $\mathbb{K}$, and $\mathbb{T}_m \in \mathbb{R}^{m \times m}$. This, finally, is the definition of a Krylov subspace – it is the vector space spanned by the vectors:

$$\mathbf{b} \quad \mathbb{A}\mathbf{b} \quad \mathbb{A}^2\mathbf{b} \quad \cdots \quad \mathbb{A}^{m-1}\mathbf{b}, \tag{11.33}$$

a basis for which is $\mathbb{Q}_m$. The vector space is denoted $\mathcal{K}_m(\mathbb{A}, \mathbf{b})$.

### 11.3.2 Constructing the subspace (Lanczos iteration)

We know roughly how the procedure will go – given $\mathbb{A}$ and $\mathbf{b}$, we'll form the first $m$ columns of $\mathbb{K}$, and perform QR on those columns only. The procedure can be simplified and illuminated by exploiting the orthogonality of the columns of $\mathbb{Q}_m$. For the full matrix equation

$$\mathbb{A}\mathbb{Q} = \mathbb{Q}\mathbb{T}, \tag{11.34}$$

we have, taking the $i$th column of $\mathbb{Q}$:

$$\mathbb{A}\mathbf{q}_i = \beta_{i-1}\mathbf{q}_{i-1} + \alpha_i\mathbf{q}_i + \beta_i\mathbf{q}_{i+1}, \tag{11.35}$$

where the right-hand side comes from (11.31). Dot $\mathbf{q}_j$ on both sides to get

$$\mathbf{q}_j^T \mathbb{A}\mathbf{q}_i = \beta_{i-1}\delta_{ji-1} + \alpha_i\delta_{ji} + \beta_i\delta_{ji+1}. \tag{11.36}$$

Suppose we had $\mathbf{q}_i$ and $\mathbf{q}_{i-1}$, $\beta_{i-1}$, then we could determine $\alpha_i$ by taking the inner product:

$$\alpha_i = \mathbf{q}_i^T \mathbb{A} \mathbf{q}_i . \tag{11.37}$$

With $\alpha_i$ known, we can invert (11.35) to solve for both $\mathbf{q}_{i+1}$ and $\beta_i$:

$$\mathbf{v} \equiv \beta_i \mathbf{q}_{i+1} = \mathbb{A}\mathbf{q}_i - \beta_{i-1}\mathbf{q}_{i-1} - \alpha_i \mathbf{q}_i$$

$$\mathbf{q}_{i+1} = \frac{\mathbf{v}}{\sqrt{\mathbf{v} \cdot \mathbf{v}}} \tag{11.38}$$

$$\beta_i = \mathbf{q}_{i+1} \cdot \mathbf{v},$$

where we have used the fact that the vector $\mathbf{q}_{i+1}$ is normalized to extract $\beta_i$. That's enough to proceed to the next step.

The "Lanczos iteration" shown in Algorithm 11.1 carries out this procedure, starting with $\mathbf{q}_0 = 0$, $\beta_0 = 0$ and a vector $\mathbf{b}$ – the first step is to construct $\mathbf{q}_1$ as the normalized version of $\mathbf{b}$, then we have $\mathbf{q}_1$ and $\mathbf{q}_0$, and we can iterate. What we

---

**Algorithm 11.1** Lanczos

set $\mathbf{q}_0 = 0$, $\beta_0 = 0$, $\mathbf{q}_1 = \mathbf{b}/\sqrt{\mathbf{b} \cdot \mathbf{b}}$
**for** $i = 1 \rightarrow m$ **do**
  $\mathbf{w} = \mathbb{A}\mathbf{q}_i$
  $\alpha_i = \mathbf{q}_i \cdot \mathbf{w}$
  $\mathbf{v} = \mathbf{w} - \beta_{i-1}\mathbf{q}_{i-1} - \alpha_i \mathbf{q}_i$
  $\mathbf{q}_{i+1} = \mathbf{v}/\sqrt{\mathbf{v} \cdot \mathbf{v}}$
  $\beta_i = \mathbf{q}_{i+1} \cdot \mathbf{v}$
**end for**

---

have, in the end, is the first $m$ columns of the matrix $\mathbb{Q}$, or precisely what we called $\mathbb{Q}_m$. In addition, we get the matrix $\mathbb{T}_m$ (defined by the $\{\alpha_i\}_{i=1}^m$ and $\{\beta_i\}_{i=1}^{m-1}$, using the truncated form of (11.31)). Note that the true statement here is $\mathbb{Q}_m^T \mathbb{A} \mathbb{Q}_m = \mathbb{T}_m$, and the inverse statement, $\mathbb{A}\mathbb{Q}_m = \mathbb{Q}_m \mathbb{T}_m$ which "should" have been our starting point, is only approximately true.

### *11.3.3 The eigenvalue problem*

How can we use the Krylov subspace to solve the eigenvalue problem? What we have from the Lanczos iteration are the matrices $\mathbb{Q}_m$ and $\mathbb{T}_m$ in:

$$\mathbb{Q}_m^T \mathbb{A} \mathbb{Q}_m = \mathbb{T}_m \tag{11.39}$$

where $\mathbb{T}_m \in \mathbb{R}^{m \times m}$ is a tridiagonal symmetric matrix (since $\mathbb{A}$ is symmetric). The idea is to compute the eigenvalues of the matrix $\mathbb{T}_m$ and hope that these eigenvalues are similar to some set of eigenvalues of $\mathbb{A}$.

To sketch the idea – if we had the full space, $m = n$, then $\mathbb{Q}^T \mathbb{A} \mathbb{Q} = \mathbb{T}$ would indicate that the eigenvalues of $\mathbb{T}$ are the same as the eigenvalues of $\mathbb{A}$ – suppose we have $\mathbb{T} = \mathbb{V}\mathbb{E}\mathbb{V}^T$ as the decomposition of $\mathbb{T}$, where $\mathbb{E}$ is a diagonal matrix with the eigenvalues of $\mathbb{T}$ along its diagonal, and (since $\mathbb{T}$ is symmetric), $\mathbb{V}^T \mathbb{V} = \mathbb{I}$. Then our equation relating $\mathbb{T}$ to $\mathbb{A}$ becomes:

$$\mathbb{A} = \mathbb{Q}\mathbb{V}\mathbb{E}\mathbb{V}^T \mathbb{Q}^T, \tag{11.40}$$

or, defining $\mathbb{W} \equiv \mathbb{Q}\mathbb{V}$,

$$\mathbb{A} = \mathbb{W}\mathbb{E}\mathbb{W}^T \tag{11.41}$$

and we can see that $\mathbb{A}$ itself has eigenvalues that are the diagonal elements of $\mathbb{E}$, and eigenvectors that are the columns of $\mathbb{W} = \mathbb{Q}\mathbb{V}$.

When we take the reduced form, we expect something like:

$$\mathbb{A} \approx \mathbb{Q}_m \mathbb{V}_m \mathbb{E}_m \mathbb{V}_m^T \mathbb{Q}_m^T \tag{11.42}$$

with $\mathbb{Q}_m \in \mathbb{R}^{n \times m}$, $\mathbb{V}_m, \mathbb{E}_m \in \mathbb{R}^{m \times m}$, and then $\mathbb{W}_m \equiv \mathbb{Q}_m \mathbb{V}_m \in \mathbb{R}^{n \times m}$. Remember that aside from its sparsity and symmetry, $\mathbb{T}_m \in \mathbb{R}^{m \times m} = \mathbb{V}_m \mathbb{E}_m \mathbb{V}_m^T$, so that the problem of solving for its eigenvalues is (for $m < n$) cheaper even using methods like QR not optimized for the tridiagonal structure of $\mathbb{T}_m$.

The argument breaks down when we realize that while $\mathbb{Q}_m^T \mathbb{Q}_m = \mathbb{I} \in \mathbb{R}^{m \times m}$, the converse is not true, $\mathbb{Q}_m \mathbb{Q}_m^T \neq \mathbb{I} \in \mathbb{R}^{n \times n}$, it can't since $\mathbb{Q}_m$ doesn't have enough columns to span $\mathbb{R}^n$. So the error, once $m < n$, is in taking the correct statement: $\mathbb{Q}_m^T \mathbb{A} \mathbb{Q}_m = \mathbb{T}_m$ and associating $\mathbb{A} \approx \mathbb{Q}_m \mathbb{T}_m \mathbb{Q}_m^T$ (the truncation is similar to the one we used in Section 10.4.3 for image compression). We expect the eigenvalues of $\mathbb{T}_m$ to bear some resemblance to those of $\mathbb{A}$, becoming precisely those of $\mathbb{A}$ when $m \to n$. The eigenvalues of $\mathbb{T}_m$ are called "Rayleigh–Ritz" values, and are optimal approximations to the eigenvalues of $\mathbb{A}$ (see [14]).

---

**Example**

To give an idea of the convergence using the Lanczos approach, take a matrix $\mathbb{B} \in \mathbb{R}^{10 \times 10}$, made up of random entries $\in [-1, 1]$. We need a symmetric, positive definite matrix, so let $\mathbb{A} = \mathbb{B}^T \mathbb{B}$. Then we can run $m$ from, say, 2 to 10 and see how we both increase the number of eigenvalues we get using Algorithm 11.1, and also improve the accuracy. In Figure 11.3, we can see the progression. Notice that the "extreme" eigenvalues, the largest and smallest, converge earlier than the "middle" ones – this is typical of the method (see [14, 45]). In this example, we just find the eigenvalues of $\mathbb{T}_m$ for each value of $m$ via QR iteration, without using any simplifying (symmetric tridiagonal) properties.

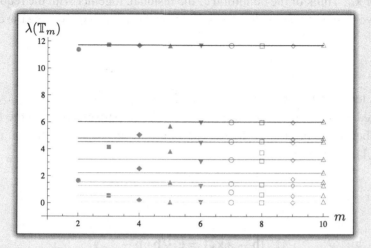

Figure 11.3 The solid lines represent the actual eigenvalues of the matrix $\mathbb{A}$. For each $m$, we see the eigenvalues of the matrix $\mathbb{T}_m$ (plotted as points) – remember that $\mathbb{T}_m \in \mathbb{R}^{m \times m}$, so the number of eigenvalues increases with $m$. At $m = 10$, we are finding all the eigenvalues, matching the "exact" case.

### 11.3.4 Matrix inversion

Now we want to use our Krylov subspace to find approximations to $\mathbf{x}$ solving $\mathbb{A}\mathbf{x} = \mathbf{b}$, given $\mathbb{A} = \mathbb{A}^T \in \mathbb{R}^{n \times n}$ and $\mathbf{b} \in \mathbb{R}^n$. The most general vector we can construct from our basis vectors $\mathbb{Q}_m$ is:

$$\mathbf{x}_m = \sum_{i=1}^m y_i \mathbf{q}_i = \mathbb{Q}_m \mathbf{y}, \tag{11.43}$$

for coefficients $\{y_i\}_{i=1}^m$ forming the vector $\mathbf{y} \in \mathbb{R}^m$, and $\mathbf{q}_i$ the $i$th column of $\mathbb{Q}_m$.

Since we are in a subspace of $\mathbb{R}^n$, we will not necessarily achieve $\mathbf{x}_m = \mathbf{x}$ (that would be a special case) – how are we to evaluate the success of $\mathbf{x}_m$ without knowing $\mathbf{x}$ itself? There are a variety of ways to control the error $\mathbf{r}_m \equiv \mathbf{b} - \mathbb{A}\mathbf{x}_m$, and each leads to its own algorithm for approximating $\mathbf{x}$. The most popular, for symmetric, positive definite matrices[1] is the conjugate gradient algorithm. In this

---

[1] A symmetric, positive definite matrix has all eigenvalues greater than zero.

case, our requirement is that the residual vector $\mathbf{r}_m$ have no component in the Krylov subspace $\mathcal{K}_m(\mathbb{A}, \mathbf{b})$ (yes, *that* $\mathbf{b}$), just a fancy way of saying that $\mathbb{Q}_m^T \mathbf{r}_m = 0$.[2]

Our starting point will be the expression, which we can define symbolically, for the vector $\mathbf{x}_m$. Think of the definition of the residual, together with the requirement that it have no component within the subspace whose basis vectors are the columns of $\mathbb{Q}_m$:

$$\mathbb{Q}_m^T [\mathbf{b} - \mathbb{A}\mathbf{x}_m] = \mathbb{Q}_m^T \mathbf{b} - \mathbb{Q}_m^T \mathbb{A}\mathbf{x}_m. \tag{11.44}$$

The first term on the right is just $\mathbf{e}_1 \sqrt{\mathbf{b} \cdot \mathbf{b}}$, since the first column of $\mathbb{Q}_m$ is the normalized version of $\mathbf{b}$ (and all other columns are perpendicular to the first one – remember our definition of the starting point for the Lanczos iteration in Algorithm 11.1). For the second term, note that

$$\mathbb{Q}_m^T \mathbb{A}\mathbf{x}_m = \mathbb{Q}_m^T \mathbb{A}\mathbb{Q}_m \mathbf{y} = \mathbb{T}_m \mathbf{y} \tag{11.45}$$

so that

$$\mathbb{Q}_m^T [\mathbf{b} - \mathbb{A}\mathbf{x}_m] = \mathbf{e}_1 \sqrt{\mathbf{b} \cdot \mathbf{b}} - \mathbb{T}_m \mathbf{y}. \tag{11.46}$$

Finally, we can see how to choose $\mathbf{x}_m$ so as to make this expression zero – assuming $\mathbb{T}_m^{-1}$ exists (which it must if $\mathbb{A}$ is invertible and symmetric, positive definite), we should set:

$$\mathbf{y} = \mathbb{T}_m^{-1} \mathbf{e}_1 \sqrt{\mathbf{b} \cdot \mathbf{b}}, \qquad \mathbf{x}_m = \mathbb{Q}_m \mathbf{y}, \tag{11.47}$$

uniquely determining $\mathbf{x}_m$.

The goal is to generate $\mathbf{x}_m$ starting from $\mathbf{x}_{m-1}$ (each satisfying (11.47) with appropriate index). We want to expand the dimension of the Krylov subspace, and find, within this new $\mathcal{K}_m(\mathbb{A}, \mathbf{b})$ subspace, the optimal solution given the optimal solution in $\mathcal{K}_{m-1}(\mathbb{A}, \mathbf{b})$. Ideally, we would take the solution $\mathbf{x}_{m-1}$ from (11.47), and then generate the solution $\mathbf{x}_m$ from (11.47) by relating these two vectors in a recursive way. The hard part in setting up the association is that $\mathbb{T}_m^{-1}$ is dense and difficult to write in terms of $\mathbb{T}_{m-1}^{-1}$ (also dense). By contrast, the matrix $\mathbb{Q}_m$ is easy to relate to $\mathbb{Q}_{m-1}$; it's just $\mathbb{Q}_{m-1}$ with an additional, orthogonal column vector tacked on to the end.

So we'll approach the algorithm obliquely, by showing that the iteration (of three separate vectors, as it turns out) defined by the "conjugate gradient" method satisfies, at each stage, the requirement that $\mathbb{Q}_m^T \mathbf{r}_m = 0$, at which point we know that we have the unique solution $\mathbf{x}_m$ in (11.47).

---

[2] We will see this idea again in Chapter 15, where we take an infinite-dimensional space, approximate it with a finite-dimensional subspace, and then require that any error lie in the unapproximated portion of the original vector space.

### *11.3.5 Conjugate gradient*

The algorithm itself is driven by an "update" for $\mathbf{x}_m$:

$$\mathbf{x}_m = \mathbf{x}_{m-1} + \alpha_m \mathbf{p}_{m-1}, \tag{11.48}$$

for a vector $\mathbf{p}_{m-1}$ that we'll define in a moment, and some number $\alpha_m$. As it stands, there is no surprise here, we just iterate by adding a new piece to the old $\mathbf{x}_{m-1}$. The residual, then, should get updated as well:

$$\mathbf{r}_m = \mathbf{b} - \mathbb{A}\mathbf{x}_m = \mathbf{b} - \mathbb{A}(\mathbf{x}_{m-1} + \alpha_m \mathbf{p}_{m-1})$$
$$= \mathbf{r}_{m-1} - \alpha_m \mathbb{A}\mathbf{p}_{m-1}, \tag{11.49}$$

so that whatever the mysterious $\mathbf{p}_{m-1}$ is, its effect on $\mathbf{x}_m$ gets propagated through to the residual. We'll define the update for $\mathbf{p}_m$ via:

$$\mathbf{p}_m = \mathbf{r}_m + \beta_m \mathbf{p}_{m-1}, \tag{11.50}$$

and choose $\beta_m$ so as to achieve certain desirable orthogonality properties.

We'll start with $\mathbf{x}_{m-1} \in \mathcal{K}_{m-1}(\mathbb{A}, \mathbf{b})$, and we want to construct an update $\mathbf{x}_m \in \mathcal{K}_m(\mathbb{A}, \mathbf{b})$ – that means that the vector $\mathbf{p}_{m-1}$ had better be in $\mathcal{K}_m(\mathbb{A}, \mathbf{b})$ if we want to explore the expanded space – in particular, it should have a component in the "new" direction represented by the move from $m-1$ to $m$.

We can see, in an inductive sense (following the sketch in [45]), that the update vectors have the desired property – suppose that $\mathbf{x}_{m-1} \in \mathcal{K}_{m-1}(\mathbb{A}, \mathbf{b})$, $\mathbf{r}_{m-1} \in \mathcal{K}_m(\mathbb{A}, \mathbf{b})$ and $\mathbf{p}_{m-1} \in \mathcal{K}_m(\mathbb{A}, \mathbf{b})$, then $\mathbf{x}_m \in \mathcal{K}_m(\mathbb{A}, \mathbf{b})$ from (11.48), $\mathbf{r}_m \in \mathcal{K}_{m+1}(\mathbb{A}, \mathbf{b})$ since the additional piece in that update (11.49) is $\mathbb{A}\mathbf{p}_{m-1} \in \mathcal{K}_{m+1}(\mathbb{A}, \mathbf{b})[3]$, and then from the update for $\mathbf{p}_m$ (11.50), which has a component in the $\mathbf{r}_m$ direction, we have $\mathbf{p}_m \in \mathcal{K}_{m+1}(\mathbb{A}, \mathbf{b})$.

So our proposed update supports the notion that at each iteration, we are constructing $\mathbf{x}_m \in \mathcal{K}_m(\mathbb{A}, \mathbf{b})$, certainly a natural target for the iteration. Now we will tune the coefficients $\alpha_m$ and $\beta_m$ – motivated by the idea that $\mathbb{Q}_m^T \mathbf{r}_m = 0$, we would like for the residuals to satisfy an orthogonality condition:

$$\mathbf{r}_m \cdot \mathbf{r}_j = 0 \quad j < m. \tag{11.51}$$

This says that each new residual is orthogonal to all previous residuals – that's a good idea, we want to make our errors in new and interesting directions, eventually squeezing out all of the error as $m \longrightarrow n$.[4] In addition, we require:

$$\mathbf{p}_m^T \mathbb{A}\mathbf{p}_j = 0 \quad j < m, \tag{11.52}$$

a type of $\mathbb{A}$-orthogonality.

---

[3] Remember that the Krylov subspace $\mathcal{K}_m(\mathbb{A}, \mathbf{b})$ is formed from the basis of $[\mathbf{b}, \mathbb{A}\mathbf{b}, \ldots, \mathbb{A}^{m-1}\mathbf{b}]$, so any vector in the $m$th subspace, when multiplied by $\mathbb{A}$, is in the $(m+1)$st subspace.

[4] Not strictly speaking possible due to numerical errors, but a noble goal.

Let's work on this pair inductively (following [45]) to see how they can be used to define $\alpha_m$ and $\beta_m$. Assume (11.51) and (11.52) hold for all vectors with $j < m - 1$, then:

$$\mathbf{r}_m \cdot \mathbf{r}_j = \mathbf{r}_{m-1} \cdot \mathbf{r}_j - \alpha_m \mathbf{p}_{m-1}^T \mathbb{A} \mathbf{r}_j \tag{11.53}$$

(using $\mathbb{A}^T = \mathbb{A}$) from the update (11.49). Now notice that $\mathbf{r}_j = \mathbf{p}_j - \beta_j \mathbf{p}_{j-1}$ from (11.50), so that

$$\mathbf{r}_m \cdot \mathbf{r}_j = \mathbf{r}_{m-1} \cdot \mathbf{r}_j - \alpha_m \mathbf{p}_{m-1}^T \mathbb{A} (\mathbf{p}_j - \beta_j \mathbf{p}_{j-1}) . \tag{11.54}$$

As long as $j < m - 1$, the inductive assumptions hold and $\mathbf{r}_m \cdot \mathbf{r}_j = 0$ (the first term is zero by (11.51), the second by (11.52)). When $j = m - 1$, we have:

$$\mathbf{r}_m \cdot \mathbf{r}_{m-1} = \mathbf{r}_{m-1} \cdot \mathbf{r}_{m-1} - \alpha_m \mathbf{p}_{m-1}^T \mathbb{A} \mathbf{p}_{m-1} \tag{11.55}$$

directly from (11.54) and understanding $\mathbf{p}_{m-1}^T \mathbb{A} \mathbf{p}_{m-2} = 0$ by the inductive assumption using (11.52). We can get (11.55) to be zero by setting:

$$\alpha_m \equiv \frac{\mathbf{r}_{m-1} \cdot \mathbf{r}_{m-1}}{\mathbf{p}_{m-1}^T \mathbb{A} \mathbf{p}_{m-1}} \tag{11.56}$$

and we have established that $\mathbf{r}_m \cdot \mathbf{r}_j = 0$ for all $j < m$.

Turning to (11.52) – using the update (11.50), we have:

$$\mathbf{p}_m^T \mathbb{A} \mathbf{p}_j = \mathbf{r}_m^T \mathbb{A} \mathbf{p}_j + \beta_m \mathbf{p}_{m-1}^T \mathbb{A} \mathbf{p}_j . \tag{11.57}$$

This time, from (11.49), we see that

$$\mathbf{r}_{j+1} = \mathbf{r}_j - \alpha_{j+1} \mathbb{A} \mathbf{p}_j \longrightarrow \mathbb{A} \mathbf{p}_j = -\frac{1}{\alpha_{j+1}} (\mathbf{r}_{j+1} - \mathbf{r}_j) \tag{11.58}$$

and we can write (11.57) as

$$\mathbf{p}_m^T \mathbb{A} \mathbf{p}_j = -\mathbf{r}_m \cdot \left[ \frac{1}{\alpha_{j+1}} (\mathbf{r}_{j+1} - \mathbf{r}_j) \right] + \beta_m \mathbf{p}_{m-1}^T \mathbb{A} \mathbf{p}_j . \tag{11.59}$$

For $j < m - 1$, the right-hand side is zero from the inductive assumption – the first term dies because of (11.51) (which we have just extended), and the second is zero from (11.52). When $j = m - 1$, we have

$$\mathbf{p}_m^T \mathbb{A} \mathbf{p}_{m-1} = -\frac{1}{\alpha_m} \mathbf{r}_m \cdot \mathbf{r}_m + \beta_m \mathbf{p}_{m-1}^T \mathbb{A} \mathbf{p}_{m-1} \tag{11.60}$$

(using $\mathbf{r}_m \cdot \mathbf{r}_{m-1} = 0$, which we established first). If we want this to be zero, we must set:

$$\beta_m = \frac{\mathbf{r}_m \cdot \mathbf{r}_m}{\alpha_m \mathbf{p}_{m-1}^T \mathbb{A} \mathbf{p}_{m-1}} = \frac{\mathbf{r}_m \cdot \mathbf{r}_m}{\mathbf{r}_{m-1} \cdot \mathbf{r}_{m-1}} \tag{11.61}$$

using our choice for $\alpha_m$ from (11.56).

The algorithm, finally, is shown in Algorithm 11.2 – there, the method is written to continue while the norm of $\mathbb{A}\mathbf{x}_m - \mathbf{b}$ is greater than some user-specified $\epsilon$. In theory, since we are approximating the solution in the Krylov subspace of dimension $m$, we shouldn't need to iterate past $m = n$, since $\mathcal{K}_n(\mathbb{A}, \mathbf{b})$ is $\mathbb{R}^n$. But, numerically, we accrue error that could end up making $m > n$.

---

**Algorithm 11.2** Conjugate gradient

---

set $\mathbf{x}_0 = 0$, $\mathbf{r}_0 = \mathbf{b}$, $\mathbf{p}_0 = \mathbf{b}$, $m = 0$
**while** $\|\mathbb{A}\mathbf{x}_m - \mathbf{b}\| \geq \epsilon$ **do**
    $m = m + 1$
    $\alpha_m = (\mathbf{r}_{m-1} \cdot \mathbf{r}_{m-1})/(\mathbf{p}_{m-1}^T \mathbb{A} \mathbf{p}_{m-1})$
    $\mathbf{x}_m = \mathbf{x}_{m-1} + \alpha_m \mathbf{p}_{m-1}$
    $\mathbf{r}_m = \mathbf{r}_{m-1} - \alpha_m \mathbb{A} \mathbf{p}_{m-1}$
    $\beta_m = (\mathbf{r}_m \cdot \mathbf{r}_m)/(\mathbf{r}_{m-1} \cdot \mathbf{r}_{m-1})$
    $\mathbf{p}_m = \mathbf{r}_m + \beta_m \mathbf{p}_{m-1}$
**end while**

---

The property (11.51) of the conjugate gradient algorithm provides the connection to the unique solution (11.47). For a Krylov subspace of size $m$, we know that the residual vector $\mathbf{r}_m$ coming from the algorithm has $\mathbf{r}_m \cdot \mathbf{r}_j = 0$ for all $j < m$. Now the vectors $\{\mathbf{r}_j\}_{j=0}^{m-1}$ are in $\mathcal{K}_m(\mathbb{A}, \mathbf{b})$, and the matrix $\mathbb{Q}_m$ has columns that span $\mathcal{K}_m(\mathbb{A}, \mathbf{b})$, so that

$$\mathbf{r}_m \cdot \mathbf{r}_j = 0 \text{ for all } j < m \longrightarrow \mathbb{Q}_m^T \mathbf{r}_m = 0, \tag{11.62}$$

precisely the requirement that led to $\mathbf{x}_m$ in (11.47).

So conjugate gradient is generating the correct $\mathbf{x}_m$. This unique solution minimizes the size of the difference $\mathbf{x} - \mathbf{x}_m$ – specifically, you will show (Problem 11.11) that at each step, the residual magnitude

$$\|\mathbf{x} - \mathbf{x}_m\|_{\mathbb{A}} \equiv \sqrt{(\mathbf{x} - \mathbf{x}_m)^T \mathbb{A}(\mathbf{x} - \mathbf{x}_m)} \tag{11.63}$$

is minimized (with respect to the $\mathbb{A}$-norm).

How about the timing of the method? The largest cost per step is the matrix-vector multiplication, $A\mathbf{p}_{m-1}$, which is a priori $O(n^2)$, then any convergence that occurs with $m > n$ takes at least as much time as a direct solution (using LU or QR factorization). But for sparse matrices, the matrix-vector product is relatively inexpensive, making the method competitive. Besides the sparsity, the structure of $A$ may allow for convergence with $m \ll n$ (the analysis of convergence for this method is interesting, and bounds can be found in [14]).

---

**Example**

We'll construct a sparse, strongly diagonal matrix $A$ as follows – start with a matrix $B \in \mathbb{R}^{500 \times 500}$, then form $A = B^T B$, ensuring that $A$ has real, positive, eigenvalues. Now, following a similar test described in [45], we'll take all entries of $A$ that are greater than 0.2, and set them to zero, this is to introduce the sparsity, and then, finally, we'll add the identity matrix, that helps with invertibility. The log of the residual, $\|A\mathbf{x}_m - \mathbf{b}\|$, for the first 40 iterations of conjugate gradient is shown in Figure 11.4. At $m = 40$, the error is: $\|A\mathbf{x}_{40} - \mathbf{b}\| \sim 10^{-12}$.

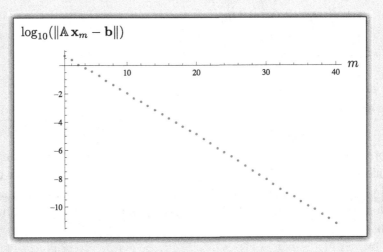

Figure 11.4 The norm of the residual for the first 40 iterations of conjugate gradient. Our test matrix $A \in \mathbb{R}^{500 \times 500}$ is sparse and strongly diagonal.

---

## Further reading

1. Demmel, James W. *Applied Numerical Linear Algebra*. Siam, 1997.
2. Giordano, Nicholas J. *Computational Physics*. Prentice Hall, 1997.
3. Golub, Gene H. & Charles F. Van Loan. *Matrix Computations*. The Johns Hopkins University Press, 1996.
4. Isaacson, Eugene & Herbert Bishop Keller. *Analysis of Numerical Methods*. Dover, 1994.
5. Trefethen, Loyd N. & David Bau III. *Numerical Linear Algebra*. Siam, 1997.

## Problems

### Problem 11.1
What is the potential for a charge $q$ a height $a$ above the center of an infinite grounded conducting sheet (both above and below the sheet)? Use the method of images, or rely on the uniqueness of solutions to the Poisson problem with boundary conditions.

### Problem 11.2
If we have a single particle in a square (two-dimensional) box with grounded walls, how will the particle move (qualitatively) and where will it end up?

### Problem 11.3
For the finite difference approach to hydrogen described in Section 10.4.2, we know that the matrix is symmetric, but it is clearly *not* positive definite (the bound states of hydrogen are included, and these have $\tilde{E}_n = -1/n^2$, negative, in the continuous case). We can shift all the discrete energies upward by adding $\tilde{E}_0 \psi(q)$ to both sides of (10.33) for constant $\tilde{E}_0$ – how must you change the resulting finite difference matrix to ensure that all eigenvalues are positive (i.e. what value of $\tilde{E}_0$ should you take, and how would you incorporate the change in the matrix)?

### Problem 11.4
Consider the continuous eigenvalue problem: $-\nabla^2 f(x, y) = \lambda f(x, y)$ in two dimensions, with boundary conditions: $f(x, 0) = f(x, a) = f(0, y) = f(a, y) = 0$ (i.e. $f = 0$ on a square with side length $a$) (the minus sign and boundary conditions are to make this problem relevant for the infinite square well in two dimensions). Using separation of variables, find the eigenvalues $\lambda$ for this partial derivative operator.

### Problem 11.5
Sparse matrices are relatively easy to implement. In `Mathematica`, you can generate a sparse matrix using the `SparseArray` command. In order to make an empty (sparse) matrix $\mathbb{A}$ in $\mathbb{R}^{n \times n}$, you would set: `A = SparseArray[{}, {n,n},0.0]`. Then `A` can have elements assigned just as if it were a normal `Table`. Modify the `DXmat`, `DYmat` and `MakeSource` functions from the Chapter 4 notebook to use sparse matrices rather than full ones.

### Problem 11.6
Show that the Jacobi method from Section 11.2.1 with $\mathbb{P}$ the diagonal of the matrix $\mathbb{A}$ and $\mathbb{Q} = \mathbb{P} - \mathbb{A}$ has update (11.5) that can be written as:

$$x_k = \frac{1}{A_{kk}} \left[ b_k - \sum_{j=1, j \neq k}^{n} A_{kj} x_j \right] \tag{11.64}$$

for the $k$th element of $\mathbf{x}$, given $\mathbb{A} \in \mathbb{R}^{n \times n}$ and $\mathbf{b} \in \mathbb{R}^n$.

**Problem 11.7**

Take the orthogonal matrices:

$$\mathbb{A} \doteq \begin{pmatrix} \cos\theta & \sin\theta \\ -\sin\theta & \cos\theta \end{pmatrix}. \tag{11.65}$$

Find the eigenvalues of the matrix $\mathbb{S}$ appearing in $\mathbf{x}_{m+1} = \mathbb{S}\mathbf{x}_m + \omega\, [\mathbb{D} - \omega\mathbb{L}]^{-1}\mathbf{b}$ for SOR (11.18) (you do *not* have to do this by hand). Given $\theta = \pi/3$, find the values of $\omega$ for which SOR will converge (a plot of $\omega \in [0, 2]$ will tell you quickly).

**Problem 11.8**

For Lanczos iteration used to find eigenvalues, convergence occurs for large and small eigenvalues first – it can be useful to shift the eigenvalues of a matrix so as to target a particular one (or one near a known or estimated value) using the Lanczos decomposition. Given a matrix $\mathbb{A} \in \mathbb{R}^{n\times n}$ with eigenvalues $\{\lambda_j\}_{j=1}^n$ and eigenvectors $\{\mathbf{v}_j\}_{j=1}^n$, show that the matrix $\mathbb{B} \equiv (\mathbb{A} - \mu\mathbb{I})^{-1}$ has the same eigenvectors as $\mathbb{A}$, and eigenvalues given by $\{(\lambda_j - \mu)^{-1}\}_{j=1}^n$. This means that $\mathbb{B}$ has a spectrum that can be tuned, by the choice of $\mu$, to have its largest eigenvalue that is associated with any of the eigenvalues of $\mathbb{A}$ (don't worry about the potential invertibility crisis that could occur when subtracting $\mu\mathbb{I}$ from $\mathbb{A}$).

**Problem 11.9**

Inverse iteration refers to using a variant of the power method to isolate the eigenvector associated with an eigenvalue near a constant $\mu$ (rather than the largest absolute eigenvalue that is the natural target of the power method). Given a symmetric, invertible matrix $\mathbb{A} \in \mathbb{R}^{n\times n}$ and a constant $\mu$, and some initial vector $\mathbf{x}$ we're going to multiply by $(\mathbb{A} - \mu\mathbb{I})^{-1}$ over and over again, and this will pick out the largest eigenvalue of the matrix $\mathbb{B} \equiv (\mathbb{A} - \mu\mathbb{I})^{-1}$, namely (from the previous problem) $(\lambda_\mu - \mu)^{-1}$, where $\lambda_\mu$ is the eigenvalue of $\mathbb{A}$ closest to $\mu$. Operationally, starting from $\mathbf{x}_0$, define the update $\mathbf{x}_i$ via:

$$(\mathbb{A} - \mu\mathbb{I})\mathbf{y} = \mathbf{x}_{i-1}$$
$$\mathbf{x}_i = \frac{\mathbf{y}}{\sqrt{\mathbf{y}\cdot\mathbf{y}}}. \tag{11.66}$$

Implement inverse iteration and use it to find the eigenvalue closest to $\mu = \pi \approx 3.14159$ in a matrix $\mathbb{A}$ that you generate by symmetrizing a random matrix $\mathbb{C} \in \mathbb{R}^{100\times100}$ with entries drawn from a flat distribution on $[-1, 1]$ (generate $\mathbb{C}$, then take $\mathbb{A} = \mathbb{C}^T\mathbb{C}$). Use `LinearSolve` to perform the inverse in (11.66), ending when $\|\mathbf{x}_i - \mathbf{x}_{i-1}\| < 10^{-5}$, say, and check your eigenvalue using `Eigenvalues`.

**Problem 11.10**

Prove that for a symmetric, positive definite matrix, $\mathbb{A} \in \mathbb{R}^{n\times n}$, the object:

$$\|\mathbf{x}\|_\mathbb{A} \equiv \sqrt{\mathbf{x}^T\mathbb{A}\mathbf{x}} \tag{11.67}$$

is a norm. Hint: start by showing that $\|\mathbf{x}\|_A \geq 0$ for all $\mathbf{x}$ and equal to zero only for $\mathbf{x} = 0$. That, together with $A = A^T$ and the definition of matrix-vector multiplications, establishes $\langle \mathbf{x}, \mathbf{y} \rangle \equiv \mathbf{x}^T A \mathbf{y}$ as an inner product. Then the Cauchy–Schwarz inequality, $|\langle \mathbf{x}, \mathbf{y} \rangle|^2 \leq \|\mathbf{x}\|_A^2 \|\mathbf{y}\|_A^2$ holds, and you can use this to get the triangle inequality.

## Problem 11.11

Show that the conjugate gradient iterate $\mathbf{x}_m$ minimizes the residual $\mathbf{r} \equiv \mathbf{x} - \mathbf{x}_m$ (where $\mathbf{x}$ is the solution to $A\mathbf{x} = \mathbf{b}$) with respect to the $A$ norm, i.e.

$$\|\mathbf{x} - \mathbf{x}_m\|_A \tag{11.68}$$

is a minimum for $\mathbf{x}_m$ coming from Algorithm 11.2. Do this by taking $\mathbf{y} = \mathbf{x}_m + \boldsymbol{\epsilon}$ for arbitrary $\boldsymbol{\epsilon} \in \mathcal{K}_m(A, \mathbf{b})$ and showing that (11.68) is smaller (the important element from CG is the property (11.51)).

## Lab problems

## Problem 11.12

Implement the SOR method using the update defined in (11.18) – your function can call `LinearSolve` to perform the matrix inversion (as opposed to using the `LDBacksolve` routine you developed for Chapter 9). For inputs, take the matrix $A$, the vector $\mathbf{b}$, the SOR parameter $\omega$ and the number of steps to take. Use a vector with random entries $\in [-1, 1]$ for the starting point, $\mathbf{x}_0$.

(a) We'll solve the Poisson problem from Chapter 4, make an $N_x = N_y = 29$ Laplacian finite difference matrix (use the functions from the Chapter 4 notebook) with $\Delta x = \Delta y = \frac{1}{N_x+1}$, and for the source, take a point charge (with $\frac{q}{4\pi\epsilon_0} = 1$) at grid location $i = j = 15$. Using grounded boundaries, make the matrix, call it $A$, and the right-hand side vector (a combination of the source and boundary conditions as always), $\mathbf{b}$. Use your SOR routine instead of matrix inversion to find the approximate solution to $A\mathbf{x} = \mathbf{b}$ with $\omega = 1.9$ and 300 steps of iteration. What is $\max_i |(A\mathbf{x}_{300} - \mathbf{b})_i|$ for your final approximation to the inverse?

(b) Using the same matrix, $A$ and $\mathbf{b}$ as in part (a), make a plot of the maximum absolute value of the residual $A\mathbf{x}_M - \mathbf{b}$ for $M = 10$ to 300 in steps of 10. Make plots for $\omega = 0.5$, 1 and 1.5. In your SOR solver, prior to generating the initial vector $\mathbf{x}_0$, use `SeedRandom[1]` to ensure that each of your tests uses the same (random) starting point.

## Problem 11.13

Make the two-dimensional Laplacian matrix $\mathbb{D}$ from Section 4.3 with $N_x = N_y = 50$ – we know that the eigenvalues of the negative of this matrix should approximate the spectrum you found in Problem 11.4. Using the `Lanczos` routine from the chapter notebook, generate $\mathbb{T}_{500}$ (the $m = 500$ subspace approximation to $\mathbb{T}$) from $-\mathbb{D}$ and find the eigenvalues of $\mathbb{T}_{500}$ using the built-in `Eigenvalues` function. Do the values match relevant values from the actual eigenvalues of $-\nabla^2$?

**Problem 11.14**

Using the finite difference setup from Section 10.4.2, generate the finite difference matrix associated with hydrogen – use $q_\infty = 40$ and $N_s = 1000$ (the matrix there is symmetric, but *not* positive definite – use your result from Problem 11.3 to modify the potential to ensure that the eigenvalues are all positive). The Lanczos method will factor the matrix into the appropriate tridiagonal form given $m = 300$ (the size of the Krylov subspace) – find this factorization using Lanczos from the chapter notebook, and then use Eigenvalues (the built-in routine) to find the eigenvalues of the resulting matrix. What must you do to the eigenvalues to compare with the bound-state spectrum of hydrogen? As you will see, only some of the resulting eigenvalues correspond to bound states – the Lanczos routine generates a matrix whose eigenvalues approximate a subset of the original matrix's, but which ones it approximates is not obvious (beyond the rule of thumb that convergence occurs first for the extreme values).

**Problem 11.15**

We'll repeat the previous problem, but this time use the result from Problem 11.8 to target specific eigenvalues without having too large a Krylov subspace. Construct the same matrix as above (with the offset you used to make all eigenvalues positive, $\tilde{E}_0$), call it $\mathbb{A}$. Using Inverse construct $\mathbb{B} \equiv (\mathbb{A} - \mu\mathbb{I})^{-1}$ for $\mu$ near the ground state (now at $\tilde{E}_0 - 1$ for $\mathbb{A}$) – take, for example, $\mu = \tilde{E}_0 - 1 + \frac{1}{100}$. Now form $\mathbb{T}_4$ for a Krylov subspace of size 4, this will be a four-by-four matrix – use the built-in routine Eigenvalues to compute the eigenvalues of $\mathbb{T}_4$ and find the associated four eigenvalues of $\mathbb{A}$ from these. You should capture the ground state eigenvalue quite well. Try it again with $\mu = \tilde{E}_0 - \frac{1}{9} + \frac{1}{100}$ to try to hone in on the third eigenvalue in the hydrogen spectrum.

**Problem 11.16**

For the same matrix $\mathbb{A}$ and right-hand side $\mathbf{b}$ as in Problem 11.12, use conjugate gradient (you can use CG from the chapter notebook) to find $\mathbf{x}$ – take 10 iterations, what is the maximum element (in absolute value) of the residual $\mathbb{A}\mathbf{x}_{10} - \mathbf{b}$? Try it for 50 and 100 steps (note the savings here, your matrix is in $\mathbb{R}^{841 \times 841}$, and would require time $O(841^3)$ using QR, for example). You can check the time using the Timing command – how long does it take to solve the problem using the built-in function LinearSolve? How long does it take using conjugate gradient with 70 steps? How's the accuracy at 70 steps?

**Problem 11.17**

Use your sparse matrix functions from Problem 11.5 to make the discrete Laplacian matrix for a grid with $N_x = N_y = 100$, $\Delta x = \Delta y = \frac{1}{N_x+1}$ and boundaries that are grounded. As your source, take a line of charge (with $\frac{\rho_g}{\epsilon_0} = 1$) extending vertically from the top of your grid to the bottom, and with horizontal location: $n = 20$. Using 1000 steps of conjugate gradient to perform the inversion, find the approximate solution to Poisson's problem inside the square domain and make a contour plot of it. Find the maximum (in absolute value) element of the residual vector $\mathbb{D}\mathbf{f} - (\mathbf{s} - \mathbf{b})$ from the matrix inversion.

**Problem 11.18**

Get the matrix "ch11.Circle.mat" from the book's website – this is a matrix that has ones in it, except along the boundary of a circle (minus a small hole), so it can be used to define a conducting boundary as described in Section 4.3. Use the routines `HashMat`, `cMakeD2`, and `cMakeSource` from the Chapter 4 notebook to set up a sparse matrix representing the Laplacian that grounds the boundary defined in "ch11.Circle.mat". In `cMakeD2` and `cMakeSource`, define `retmat` and `svec` to be sparse matrices. For the source, set a point charge (of strength $\frac{\rho_g}{\epsilon_0} = 1$) at grid location $n = 100$ (horizontal index), $m = 150$ (vertical index), this will put it near the hole in the grounded ring. Use one thousand steps of conjugate gradient to find the potential and make a contour plot of it.

**Problem 11.19**

Use your sparse setup for the Laplacian difference matrix and your conjugate gradient solver to repeat Problem 5.19 (replacing `LinearSolve` with the conjugate gradient routine), a two-dimensional solution for the time-dependent Schrödinger equation. Use the same physical parameters, but this time we'll make grid twice as fine. Take a $50 \times 50$ grid with $\Delta q = 1/51$, $\Delta r = 1/51$ and $\Delta s = 5 \times 10^{-6}$. Experiment with the number of steps needed in the conjugate gradient solution to the implicit Euler matrix inversion – make sure you have both accuracy and speed (a hundred iterations should suffice, but check for yourself). Run for 2000 steps, and make a movie of the probability density (contour plot) using every hundredth step. Check norm preservation for your solution.

# 12

# Minimization

Many problems that we have seen already require that we "minimize" a certain function, or find the "minimum" of a function of many variables. Formally, the problem is: given $F(x)$, find a value $x_0$ such that $F(x) > F(x_0)$ for all $x$, for a global minimum, or $F(x) > F(x_0)$ where $x \in [a, b]$ (some constrained range) for a local minimum. We will review these familiar problems, and introduce some new ones. Our first job will then be to write a function that can find the minimum of a function of one variable, a sort of minimization analogue to the bisection method from Section 3.2.1 for finding roots of $F(x)$. Then we will study methods that can be used to minimize a function $u(\mathbf{x})$ where $\mathbf{x} \in \mathbb{R}^n$.

## 12.1 Physical motivation

There are many places in physics where finding the minimum of a function can be useful. We have already done linear minimization in solving the least squares problem for data fitting from Section 10.2 – there we turned the minimization of the Pythagorean length of the residual vector into a matrix inverse problem with a unique solution. There are nonlinear extensions of this process that rely on iterative techniques directly (as opposed to the matrix inversion that solves the least squares problem, which can be performed iteratively or not).

Recall the harmonic approximation from Chapter 8: Given the potential energy $U(x)$, some complicated function, we know that in the vicinity of a local minimum, the motion of a particle will be harmonic with effective spring constant set by the second derivative of the potential evaluated at the minimum. But we need to *find* the minima if we are to explore motion in their vicinity, so again, a minimization issue.

We used the "steepest descent" method already in Chapter 3 to minimize a norm in the context of quantum mechanical scattering, and we will see this method (and its extensions) again. As for new applications, we'll start with another quantum mechanical minimization problem: the variational method, then define and isolate minimization in the nonlinear least squares problem, and finally, consider a classical mechanics minimization problem.

### *12.1.1 Variational method in quantum mechanics*

Given a probability density $\rho(x)$, and a function $E(x)$ that gives the energy (of something, a particle, say) as a function of position, the average energy is defined as:

$$\langle E \rangle = \int_{-\infty}^{\infty} \rho(x)E(x)dx. \tag{12.1}$$

In quantum mechanics, the "Hamiltonian operator", H, represents the energy function $E(x)$, and is just the left-hand side of the time-independent Schrödinger equation from (10.20) – for any function $\psi(x)$, we define the action of H on $\psi(x)$ by:

$$H\psi(x) \equiv -\frac{\hbar^2}{2m}\psi''(x) + V(x)\psi(x), \tag{12.2}$$

given some potential $V(x)$. We think of $\psi(x)^*\psi(x)$ as a probability density, so that the expectation value for energy, for a particle in the state $\psi(x)$ is, as above:

$$\langle E \rangle = \int_{-\infty}^{\infty} \psi(x)^* H\psi(x)dx = \int_{-\infty}^{\infty} \psi(x)^* \left( -\frac{\hbar^2}{2m}\psi''(x) + V(x)\psi(x) \right) dx. \tag{12.3}$$

If we had the solutions to Schrödinger's equation, a set of $\{\psi_n(x)\}_{n=1}^{\infty}$ with $H\psi_n(x) = E_n\psi_n(x)$, then the state with the lowest energy is $\psi_1(x)$ and its "ground state" energy is $E_1$ (assuming no degeneracy, so that $E_1 < E_2 < \ldots$). This is the minimum value for average energy – an arbitrary function $\psi(x)$ will then have $\langle E \rangle \geq E_1$. To see this, note that the set of functions $\{\psi_n(x)\}_{n=1}^{\infty}$ solving Schrödinger's equation is complete, meaning that any function $\psi(x)$ can be written as

$$\psi(x) = \sum_{j=1}^{\infty} \alpha_j \psi_j(x) \tag{12.4}$$

(think of the solutions to the infinite square well: $\{\psi_n(x) = \sqrt{2/a}\sin(n\pi x/a)\}_{n=1}^{\infty}$). Then acting on our arbitrary $\psi(x)$ with the operator $H$ gives:

$$H\psi(x) = \sum_{j=1}^{\infty} \alpha_j H\psi_j(x) = \sum_{j=1}^{\infty} \alpha_j E_j \psi_j(x). \tag{12.5}$$

Multiplying this expression by $\psi(x)^*$ and integrating gives

$$\langle E \rangle = \sum_{j=1}^{\infty} \alpha_j^* \alpha_j E_j, \tag{12.6}$$

using the orthogonality of the functions $\{\psi_n(x)\}_{n=1}^{\infty}$: $\int_{-\infty}^{\infty} \psi_j(x)^* \psi_k(x) dx = \delta_{jk}$. If $\psi(x)$ is normalized, then $\sum_{j=1}^{\infty} \alpha_j^* \alpha_j = 1$, and we can put a bound on the expectation value:

$$\langle E \rangle = \sum_{j=1}^{\infty} \alpha_j^* \alpha_j E_j \geq E_1 \sum_{j=1}^{\infty} \alpha_j^* \alpha_j = E_1. \tag{12.7}$$

So, for any $\psi(x)$, we have $\langle E \rangle \geq E_1$, the ground state energy. This observation is the basis of the "variational method" in quantum mechanics, used to put bounds on the ground state energy for systems where the ground state is not known. The idea is to pick a wave function with an arbitrary parameter (or parameters) $\beta$, $\psi(\beta, x)$, compute the expectation value of energy:

$$\langle E(\beta) \rangle = \int_{-\infty}^{\infty} \psi^*(\beta, x) H \psi(\beta, x) dx, \tag{12.8}$$

and then find the minimum value of $\langle E(\beta) \rangle$, a minimization problem in the parameter $\beta$. That minimum value will still be greater than the ground state energy, in general, but by introducing better (or more) guesses, we can sometimes get a very tight bound on the ground state energy. In many cases, the minimum can be found directly by solving $\frac{d \langle E(\beta) \rangle}{d\beta} = 0$ – for everything else, there is a numerical technique.

### 12.1.2 Data fitting

Remember our linear least squares problem from Section 10.2 – given a set of data points $\{d_j\}_{j=1}^{n}$, take a set of functions $\{f_k(t)\}_{k=1}^{m}$ and find the linear combination of these such that:

$$r_j = d_j - \sum_{k=1}^{m} \alpha_k f_k(t_j) \quad j = 1 \longrightarrow n \tag{12.9}$$

is minimized (in norm). Formally, we want to choose the set of values $\{\alpha_k\}_{k=1}^{m}$ such that the vector $\mathbf{r}(\alpha)$ has the smallest possible magnitude. We solved that case exactly by turning the problem into an inverse problem (the normal equations (10.7)).

There was a certain inefficiency built into that process. Suppose the data comes from the discretization of $d(t) = \sin(2\pi 4t)$, an oscillating signal. How should we choose our functions $f_k(t)$? One obvious choice would be $f_k(t) = \sin(2\pi kt)$, which would give us the answer exactly provided $m \geq 4$. But we might equally well take $f_k(t) = \sin(kt)$ – that would miss the exact solution. I guess what we would do is bracket the values of $k$ in this second set of functions that worked best, and try to refine our choice of functions – we would note, for example, that values of $k$ near 25 worked well, and maybe take a new set of functions $f_k(t) = \sin((25 + 0.01k)t)$, perform another minimization via the normal equations, and continue this refinement process until satisfied.

Another way to approach this type of data fitting problem is to try to *find* a value $x$ in $f(x, t) = \sin(xt)$ that matches the data – so rather than a linear combination of functions, we'd like to generate a single function by finding the best parameter $x$.[1] More generally, we will attempt to find a set of parameters $\mathbf{x} \in \mathbb{R}^m$ in a function $F(\mathbf{x}, t)$, such that the residual:

$$r_j = d_j - F(\mathbf{x}, t_j) \tag{12.10}$$

has the smallest possible norm. So we want to find $\mathbf{x}$ minimizing

$$u(\mathbf{x}) \equiv \sum_{j=1}^{n} [d_j - F(\mathbf{x}, t_j)]^2. \tag{12.11}$$

We can still take $F(\mathbf{x}, t)$ as a linear combination of other functions by writing, for example:

$$F(\mathbf{x}, t) = \sum_{j=2,4,\dots}^{m/2} x_j \sin(x_{j-1} t). \tag{12.12}$$

Here, the even components of $\mathbf{x}$ represent the coefficients in a sum, while the odd components are probing the frequencies.

### 12.1.3 Action extremization

In the action formulation of classical dynamics, we define $S[x(t)]$ to be:

$$S[x(t)] = \int_0^T \left( \frac{1}{2} m \dot{x}^2 - U(x) \right) dt \tag{12.13}$$

given potential energy $U(x)$, and a particle of mass $m$ moving under its influence. This number, called the action, takes a function $x(t)$ as its argument. The dynamical trajectory (the one satisfying Newton's second law, $m\ddot{x} = -\frac{dU}{dx}$) is the one that *minimizes* the action in the sense that, given a perturbation $\delta x(t)$ to the trajectory $x(t)$,

$$S[x(t) + \delta x(t)] > S[x(t)] \tag{12.14}$$

---

[1] Never mind that this example should really be Fourier transformed prior to any fitting, the point is, we have a parameter in our fit, and we want to use the data to find a good value for that parameter.

(actually, the requirement is that $x(t)$ be an extremum, min or max, but we'll consider problems in which the relevant extremum is a minimum). We can think of dynamics as a high-dimensional minimization problem.

In fact, the form of this problem, an integral over a function of a single variable, is similar to the variational method from Section 12.1.1. The target quantity is different – in this case what we really want is the minimizing trajectory $x(t)$, whereas for quantum mechanics, we want a bound on the ground state energy (i.e. a number), but the idea is similar: In both problems, the function to minimize takes a function as its argument and returns a number.

## 12.2 Minimization in one dimension

Suppose we have a function $F(x) \in \mathbb{R}$ with $x \in \mathbb{R}$. We know that there is a minimum in between $x_\ell$ and $x_r$ (with $x_r > x_\ell$), and we'd like to find the numerically exact location of this minimum, i.e. we want $x$ such that $F(x + \epsilon) \geq F(x)$ and $F(x - \epsilon) \geq F(x)$ for small $\epsilon$. This problem could come, for example, from an interest in the location of minima for some complicated potential. One way to proceed is to find the roots of $F'(x)$ using bisection, say. But $F'(x)$ has roots at both minima and maxima, so additional checking is necessary. Instead we will develop a simple method that hones in on the minimum in much the same way that bisection found roots (see Section 3.2.1).

Remember that for bisection, we had to have a pair of values, $\{x_\ell, x_r\}$, with $F(x_\ell)F(x_r) < 0$, so that a root was in between the two points. Then we chose the middle of the interval, determined in which half the root lived, and proceeded recursively to refine the bracketing until $x_r - x_\ell$ was small (alternatively, $|F(x_\ell)| < \epsilon$ for some small $\epsilon$). The method was motivated by: 1. successive bracketing while shrinking the interval, and 2. equal intervals for each step (that is why we cut the original interval in half). Our minimum finding routine, called the "golden section search," will have the same basic structure (the presentation here follows [40]). In this case, where we want a minimum rather than a zero, we will take triples of numbers: $\{x_\ell, x_m, x_r\}$ with $x_\ell < x_m < x_r$ (so the subscripts stand for "left," "middle," and "right"). We need three values instead of two because we want more information than in the root-finding case – a minimum of $F(x)$ is defined as having points on either side that have larger value, so we need to check both sides – a root has one side positive, one negative, so requires only two points.

At each stage, we will enforce the relationships:

$$F(x_\ell) \geq F(x_m) \text{ and } F(x_r) \geq F(x_m), \tag{12.15}$$

and we will refine $x_\ell$, $x_m$ and $x_r$ so that the interval $[x_\ell, x_r]$ shrinks. Define the lengths $w \equiv x_r - x_\ell$, $\ell \equiv x_m - x_\ell$, as shown in Figure 12.1. We'll agree to the

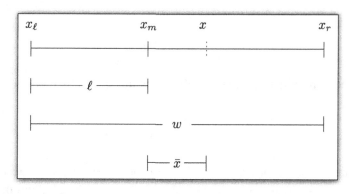

Figure 12.1 A triplet $\{x_\ell, x_m, x_r\}$ and proposed test point $x$, with the lengths $\ell$, $w$, and $\bar{x}$ defined.

convention that the test point $x$ is always taken in the larger of the two intervals, $x_r - x_m$ and $x_m - x_\ell$ so that in Figure 12.1, we've put $x$ in $[x_m, x_r]$ and defined the positive length $\bar{x} \equiv x - x_m$.

Assume that we have a valid initial bracketing, so that $F(x_\ell) > F(x_m)$ and $F(x_r) > F(x_m)$. Pick the test point $x$ in the larger interval. All we need to do is check the value of $F(x)$ – if $F(x) < F(x_m)$, then the new bracket set will be $\{x_m, x, x_r\}$, since now the minimum is between $x_m$ and $x_r$ ($F(x_r) > F(x_m)$ by assumption, so that $F(x_r) > F(x)$ in this case). If $F(x) > F(x_m)$, then the new set will be $\{x_\ell, x_m, x\}$, the minimum is now between $x_\ell$ and $x$ ($F(x_\ell) > F(x_m)$ by assumption). We repeat this procedure until the length of the bracket: $x_r - x_\ell$ is within some tolerance.

To choose the point $x$, we'll work by analogy with bisection. There, the test point was $x = \frac{1}{2}(x_\ell + x_r)$, midway between the left and right points, so either interval, left or right, had the same size. We can do the same thing here; our two options are the bracket $\{x_m, x, x_r\}$ and $\{x_\ell, x_m, x\}$. The lengths are then $x_r - x_m = w - \ell$ and $x - x_\ell = \ell + \bar{x}$. For these two lengths to be the same, we must have:

$$w - \ell = \ell + \bar{x} \longrightarrow \bar{x} = w - 2\ell. \tag{12.16}$$

In addition to letting both intervals have the same length, we'll preserve the ratio $\ell/w$ in the update. We need an additional constraint to fix $\bar{x}$, and requiring that the ratio of the smaller interval to the entire interval length (i.e. $\ell/w$) be constant is as good as any. For the partitioning shown in Figure 12.1, the constraint is:

$$\frac{\ell}{w} = \frac{\bar{x}}{w - \ell} \tag{12.17}$$

where the right-hand side represents the ratio of the new small interval to the new total interval (for either bracketing, since they have the same length). This fixes the ratio $\ell/w$ once we use $\bar{x} = w - 2\ell$:

$$\frac{\ell}{w} = \frac{w - 2\ell}{w - \ell} \longrightarrow \frac{\ell}{w} = \frac{2}{3 + \sqrt{5}}, \qquad (12.18)$$

taking the geometrically relevant ($w > \ell$) root of this quadratic relation. That gives the length

$$\bar{x} = w - 2\ell = \frac{1}{2}\left(-1 + \sqrt{5}\right)\ell, \qquad (12.19)$$

so that we are taking $w$ and splitting it into this new $\bar{x} \approx 0.61803\ell$ piece and "the rest," $w - \ell - \bar{x}$ – these two lengths are then in the golden ratio, and that ratio is enforced at each stage.

Numerology aside, we now know how to pick the test point $x$:

$$x = x_m + \frac{1}{2}\left(-1 + \sqrt{5}\right)\ell, \qquad (12.20)$$

and generate the new bracket by testing. The only snag is that we have to check which interval is larger. We generated $x$ assuming $x_r - x_m \geq x_m - x_\ell$ as in Figure 12.1. If $x_m - x_\ell$ is the larger interval, we run the same argument with $x = x_m - \bar{x}$. That is the source of the mess in Implementation 12.1 which is otherwise a straightforward recursive algorithm.

---

**Implementation 12.1** Golden ratio minimization

```
GSMin[f_, Xbrack_, eps_] := Module[{retval, x, xl, xm, xr},
  xl = Xbrack[[1]];
  xm = Xbrack[[2]];
  xr = Xbrack[[3]];
  If[xm - xl < eps,
   Return[xm];
  ];
  If[(xr - xm) > (xm - xl),
   x = N[xm + (1/2) (-1 + Sqrt[5]) (xm - xl)];
   If[f[x] > f[xm],
    retval = GSMin[f, {xl, xm, x}, eps];
    ,
    retval = GSMin[f, {xm, x, xr}, eps];
   ];
   ,
   x = N[xm - (1/2) (3 - Sqrt[5]) (xm - xl)];
   If[f[x] < f[xm],
    retval = GSMin[f, {xl, x, xm}, eps];
    ,
    retval = GSMin[f, {x, xm, xr}, eps];
   ];
  ];
  Return[retval];
 ]
```

There is one final issue – we need an initial set $\{x_\ell, x_m, x_r\}$ that itself satisfies the correct ratio for $\ell/w$, and the condition $F(x_\ell) > F(x_m)$, $F(x_r) > F(x_m)$. When we worked on bisection, we performed this initial bracketing "by eye," plotting the function of interest and isolating, however coarsely, the root. In the minimization case, we will be more careful, and eliminate the need for "looking," while enforcing the appropriate ratio for $\ell/w$. The idea is to start at some $x_\ell$, and take a user-specified step $\Delta x$ to define the first $x_m$, then $x_r$ is fixed by:

$$\frac{w}{\ell} = \frac{1}{2}\left(3 + \sqrt{5}\right) \longrightarrow x_r = \frac{1}{2}\left(3 + \sqrt{5}\right)(x_m - x_\ell) + x_\ell. \qquad (12.21)$$

We continue this process, taking additional steps to fix $x_m$ until the relation $F(x_\ell) > F(x_m)$ and $F(x_r) > F(x_m)$ is satisfied. The process is shown in Implementation 12.2.

---

**Implementation 12.2** Initial golden ratio bracket

```
PreGSMin[f_, xl_, step_] := Module[{xr, xm, done},
  done = False;
  xm = xl + step;
  While[done == False,
   xr = N[(1/2) (3 + Sqrt[5])] (xm - xl) + xl;
   If[f[xl] > f[xm] && f[xr] > f[xm],
    done = True;

    ,
    xm = xm + step;
   ];
  ];
  Return[{xl, xm, xr}];
 ]
```

---

### *Variational ground state of hydrogen*

The search for minima can be performed on purely numerical functions, as well as continuous $F(x)$. In the continuous case, there is not much to say; you can try out the routine in Implementation 12.1 on simple cases.

Let's return to the variational method, and use one-dimensional minimization to put a bound on the ground state of hydrogen. In nondimensionalized form, we had the one-dimensional Schrödinger equation (from Section 10.4.2):

$$-\frac{d^2\psi(q)}{dq^2} - \frac{2}{q}\psi(q) = \tilde{E}\psi(q) \qquad (12.22)$$

so that

$$\tilde{H}\psi(q) = \left[-\frac{d^2}{dq^2} - \frac{2}{q}\right]\psi(q). \qquad (12.23)$$

We'll take our trial wave function, with parameter $\beta$, to be $\psi(q, \beta) = 2\beta^{3/2}qe^{-\beta q}$, a natural choice even if we didn't know the answer – we want a function that is finite at $q = 0$, goes to zero as $q \longrightarrow \infty$, doesn't oscillate (ground state wave functions typically do not oscillate), and has one "hump" – so $qe^{-\beta q}$ is motivated. Our parameter is $\beta$, and the coefficient out front of our ansatz comes from the normalization condition, $\int_0^\infty |\psi(\beta, q)|^2 dq = 1$. In this case, we could run our continuous form through

$$\langle \tilde{E}(\beta) \rangle = \int_0^\infty \psi(q)^* \tilde{H} \psi(q) dq, \tag{12.24}$$

and take the derivative of $\langle E(\beta) \rangle$ to find the minimizing value for $\beta$. Let's instead work purely numerically; we'll define a spatial grid $q_j = j \Delta q$ for $N$ steps, and use a numerical approximation to the integral. The function we want to integrate is:

$$\psi(\beta, q)^* \tilde{H} \psi(\beta, q) = -4\beta^3 e^{-2\beta q} q \left( 2 - 2\beta + \beta^2 q \right), \tag{12.25}$$

and we'll use Simpson's rule from Section 6.2 to integrate numerically. The result is predictable, if we perform the integration from $q = 0 \longrightarrow 20$ (where the right-hand endpoint is a reasonable approximation to infinity here) with $N = 1000$ steps, and proceed to minimize the result with $\epsilon = 0.0001$ in Implementation 12.1, we get $\beta \approx 0.999976$, and $\langle \tilde{E}(\beta) \rangle \approx -1.0$ for this minimizing value. That energy is, of course, precisely $-1/n^2$ for $n = 1$, and $\beta = 1$ corresponds to the ground state wave function of hydrogen.

## 12.3 Minimizing $u(\mathbf{x})$

We'll start by thinking about the minimization of a function $u(\mathbf{x}) \in \mathbb{R}$ where $\mathbf{x} \in \mathbb{R}^m$, say. We want $\bar{\mathbf{x}}$ such that $u(\bar{\mathbf{x}}) \leq u(\mathbf{x})$ for any $\mathbf{x}$ (at least within some domain – we may be looking for local minima). The specific function we want to minimize in the least squares setting is of this form. We'll eventually focus our attention on that case, but first we'll develop the methods of minimization for more generic $u(\mathbf{x})$.

### 12.3.1 Steepest descent

We have already seen one method for finding minima – steepest descent, from Section 3.2.4, is an iterative way to go downhill. We note that $\frac{\partial u}{\partial \mathbf{x}}$ at any point is the direction of greatest increase, so if we are at **x** and want to make $u(\mathbf{x} + \Delta\mathbf{x})$ smaller, we take $\Delta\mathbf{x} \parallel -\frac{\partial u}{\partial \mathbf{x}}$.[2]

The steepest descent method takes some initial point, $\mathbf{x}_0$, and moves downhill until $\frac{\partial u}{\partial \mathbf{x}}$ is zero. We saw the two-dimensional version of this, tailored for a function with zero as a minimum, in Algorithm 3.1. The general version is shown in Algorithm 12.1 – we give a parameter $\eta$, the step size,[3] and end when the change in **x** from one step to the next is less than $\epsilon$ (some provided tolerance) in magnitude, indicating that $\frac{\partial u}{\partial \mathbf{x}}$ is small.

---

**Algorithm 12.1** Steepest descent

> $\mathbf{y} \leftarrow \mathbf{x}_0$
> $\mathbf{z} \leftarrow 0$
> **while** $\|\mathbf{y} - \mathbf{z}\| \geq \epsilon$ **do**
>    $\mathbf{z} \leftarrow \mathbf{y}$
>    $\mathbf{y} \leftarrow \mathbf{y} - \eta \frac{\partial u}{\partial \mathbf{x}}|_{\mathbf{x}=\mathbf{y}}$
> **end while**

---

The inner loop in Algorithm 12.1 looks familiar by now: we are iteratively updating a vector of values, **y** becomes $\mathbf{y} - \eta \frac{\partial u}{\partial \mathbf{x}}|_{\mathbf{x}=\mathbf{y}}$, where the derivative is evaluated at the current **y**. Suppose we have the special form:

$$u(\mathbf{x}) = \frac{1}{2}\mathbf{x}^T \mathbb{A}\mathbf{x} - \mathbf{x} \cdot \mathbf{b}, \tag{12.26}$$

for a matrix $\mathbb{A} \in \mathbb{R}^{n \times n}$ and a vector $\mathbf{b} \in \mathbb{R}^n$. In this quadratic case, we can calculate the derivative exactly:

$$\frac{\partial u}{\partial \mathbf{x}} = \mathbb{A}\mathbf{x} - \mathbf{b} \tag{12.27}$$

and this will be zero when $\mathbf{x} = \mathbb{A}^{-1}\mathbf{b}$. We already have an update-routine that solves this problem, precisely the conjugate gradient method, Algorithm 11.2 – there, we have a loop that involves updating $\mathbf{x}_m$, and the choice of step size is known. Indeed, conjugate gradient can be viewed as a minimizing algorithm when applied to this quadratic $u(\mathbf{x})$ – part of its power comes in its choice of update direction, $\mathbf{p}_{m-1}$, not just the negative gradient.

---

[2] Remember that we are using the expression $\frac{\partial u}{\partial \mathbf{x}}$ as shorthand – what we really mean is the vector $\mathbf{G} \in \mathbb{R}^m$ with entries $G_j \equiv \frac{\partial u}{\partial x_j}$ for $j = 1 \longrightarrow m$.

[3] We can normalize the descent direction, making $\eta$ slightly more relevant. Let $\mathbf{G} \equiv \frac{\partial u}{\partial \mathbf{x}}$, then replace the update in Algorithm 12.1 with $\mathbf{y} \leftarrow \mathbf{y} - \eta \hat{\mathbf{G}}(\mathbf{y})$ where $\hat{\mathbf{G}}(\mathbf{y})$ is a unit vector.

### *12.3.2 Newton's method*

Newton's method, familiar from Section 3.2.2, can be applied to the gradient of $u$ to find minima. Suppose we are at $\mathbf{x}$, and we want to find $\Delta\mathbf{x}$ such that $u(\mathbf{x} + \Delta\mathbf{x})$ is a minimum – that means that

$$\left.\frac{\partial u}{\partial \mathbf{x}}\right|_{\mathbf{x}+\Delta\mathbf{x}} = 0. \qquad (12.28)$$

Let $\mathbf{G}(\mathbf{x}) \equiv \frac{\partial u}{\partial \mathbf{x}}$, then the requirement is $\mathbf{G}(\mathbf{x} + \Delta\mathbf{x}) = 0$. We can Taylor expand this vector function:

$$\mathbf{G}(\mathbf{x} + \Delta\mathbf{x}) \approx \mathbf{G}(\mathbf{x}) + \frac{\partial \mathbf{G}}{\partial \mathbf{x}}\Delta\mathbf{x} + O(\Delta\mathbf{x}^2) = 0 \longrightarrow \mathbf{G}(\mathbf{x}) \approx -\frac{\partial \mathbf{G}}{\partial \mathbf{x}}\Delta\mathbf{x}, \quad (12.29)$$

or, back to $u$:

$$\frac{\partial u}{\partial \mathbf{x}} \approx -\frac{\partial^2 u}{\partial \mathbf{x}^2}\Delta\mathbf{x} \qquad (12.30)$$

where $\frac{\partial^2 u}{\partial \mathbf{x}^2}$ stands for the Hessian matrix $\mathbb{H}$, with entries $H_{ij} = \frac{\partial^2 u}{\partial x_i \partial x_j}$ (evaluated at $\mathbf{x}$).

The approximation (12.30) implies an iteration of the form:

$$\mathbf{x}_{j+1} = \mathbf{x}_j + \left(-\mathbb{H}^{-1}\mathbf{G}\right), \qquad (12.31)$$

(where, as usual, the Hessian, and derivative on the right are evaluated at the current $\mathbf{x}_j$) solving (12.30) for $\Delta\mathbf{x}$. This update is, in a way, "smarter" than steepest descent, in that it uses the second derivative information (in the form of the Hessian) to inform the step size. Note that we still generally allow a fudge factor, to control the size of the step, since we obtained $\Delta\mathbf{x}$ by assuming that we could step directly to the minimum, which may not be a good approximation.

Putting it all together, if we have a function $u(\mathbf{x})$, and a function $\mathbf{G}(\mathbf{x}) \equiv \frac{\partial u}{\partial \mathbf{x}}$ (to calculate the gradient), *and* access to $\mathbb{H}(\mathbf{x})$, the second derivative matrix for $u$, then Newton's method is defined by Algorithm 12.2. There, we use $\eta$, again, to control the size of the step, and we exit when there has been little significant change in a step, which will be true when $\mathbf{y} \sim \mathbf{z}$ (i.e. $\mathbf{G} \sim 0$).

---

**Algorithm 12.2** Newton's method

$\quad \mathbf{y} \leftarrow \mathbf{x}_0$
$\quad \mathbf{z} \leftarrow 0$
$\quad$ **while** $\|\mathbf{y} - \mathbf{z}\| \geq \epsilon$ **do**
$\quad\quad \mathbf{z} \leftarrow \mathbf{y}$
$\quad\quad \mathbf{y} \leftarrow \mathbf{y} - \eta\mathbb{H}(\mathbf{y})^{-1}\mathbf{G}(\mathbf{y})$
$\quad$ **end while**

---

## 12.4 Nonlinear least squares

Let's now take the specific least squares case – given a set of data $\{d_j\}_{j=1}^n$, which we'll represent as a vector $\mathbf{d} \in \mathbb{R}^n$, we want to find a function $F(\mathbf{x}, t)$ such that

$$u(\mathbf{x}) = \sum_{j=1}^n \left(d_j - F(\mathbf{x}, t_j)\right)^2 \tag{12.32}$$

is minimized. Here, we are thinking of $t_j$ as the discrete "time" (so keeping with our example setup, $d_j = d(t_j)$, the signal evaluated at time $t_j$), and it is not a parameter that we control, so will not play a role in the minimization. We'll take $\mathbf{x} \in \mathbb{R}^m$, so that we have $m$ independent parameters.

This is our familiar "least squares" problem, but now we have not assumed that $F(\mathbf{x}, t)$ is linear in its parameters $\mathbf{x}$. Let $\mathbf{F}(\mathbf{x})$ be the vector in $\mathbb{R}^n$ whose entries are $F_j = F(\mathbf{x}, t_j)$, and $\mathbf{d}$ be the data vector with entries $d_j$. We can write (12.32) in vector form

$$u(\mathbf{x}) = r^2(\mathbf{x}) \quad \mathbf{r} \equiv \mathbf{d} - \mathbf{F}(\mathbf{x}), \tag{12.33}$$

so that it is clear, when we *minimize* $u(\mathbf{x})$ w.r.t. $\mathbf{x}$, we are minimizing the residual, $\mathbf{d} - \mathbf{F}(\mathbf{x})$.

Let's develop the form of Newton's method appropriate to this specific $u(\mathbf{x})$, called the "Gauss–Newton algorithm" (although that more traditionally refers to its application to vector functions, not our scalar $u$). For $\mathbf{G}(\mathbf{x}) \equiv \frac{\partial u}{\partial \mathbf{x}}$ (so that $\mathbf{G} \in \mathbb{R}^m$), if we go from $\mathbf{x}$ to $\mathbf{x} + \Delta\mathbf{x}$, a minimum, we have $\mathbf{G}(\mathbf{x} + \Delta\mathbf{x}) = 0$, giving us (12.30). Writing: $u = (\mathbf{d} - \mathbf{F})^T (\mathbf{d} - \mathbf{F})$, we can take the derivatives:

$$\mathbf{G} \equiv \frac{\partial u}{\partial \mathbf{x}} = -2\frac{\partial \mathbf{F}^T}{\partial \mathbf{x}}\mathbf{d} + 2\frac{\partial \mathbf{F}^T}{\partial \mathbf{x}}\mathbf{F} \tag{12.34}$$

where now, we define the matrix $\mathbb{J} \equiv \frac{\partial \mathbf{F}}{\partial \mathbf{x}}$ with entries $J_{ij} = \frac{\partial F_i}{\partial x_j}$ (so that $\mathbb{J} \in \mathbb{R}^{n \times m}$) – remember that $F_i = F(\mathbf{x}, t_i)$, so we can think of $\mathbb{J}$ as the gradient of $F(\mathbf{x}, t)$ evaluated at all times $t_i$. The derivative can be written simply as:

$$\mathbf{G} \equiv \frac{\partial u}{\partial \mathbf{x}} = -2\mathbb{J}^T\mathbf{d} + 2\mathbb{J}^T\mathbf{F} = -2\mathbb{J}^T(\mathbf{d} - \mathbf{F}) \tag{12.35}$$

and the second derivative is:

$$\mathbb{H} \equiv \frac{\partial^2 u}{\partial \mathbf{x}^2} = -2\frac{\partial \mathbb{J}^T}{\partial \mathbf{x}}\mathbf{d} + 2\mathbb{J}^T\mathbb{J} + 2\frac{\partial \mathbb{J}^T}{\partial \mathbf{x}}\mathbf{F}. \tag{12.36}$$

Putting these together, the two sides of (12.30) become:

$$\mathbb{J}^T[\mathbf{d} - \mathbf{F}] \approx \left[-\frac{\partial \mathbb{J}^T}{\partial \mathbf{x}}\mathbf{d} + \mathbb{J}^T\mathbb{J} + \frac{\partial \mathbb{J}^T}{\partial \mathbf{x}}\mathbf{F}\right]\Delta\mathbf{x}. \tag{12.37}$$

We now have a matrix-vector equation that could, in theory, be solved for $\Delta\mathbf{x}$ to define an iteration similar to (12.31). That, in the end, is *all* (12.37) does: provides an estimate for $\Delta\mathbf{x}$ that is informed by the fact that $u(\mathbf{x} + \Delta\mathbf{x})$ must have vanishing gradient at a minimum. But there is nothing sacred about (12.37) – in particular, it doesn't look like much fun to compute the triply-indexed quantity: $\frac{\partial\mathbb{J}}{\partial\mathbf{x}}$, which expands to:

$$\frac{\partial\mathbb{J}}{\partial\mathbf{x}} \sim \frac{\partial^2 F(\mathbf{x}, t_i)}{\partial x_j \partial x_k}, \tag{12.38}$$

the second derivatives of the function $F(\mathbf{x}, t)$ evaluated at various times. For this, if for no other reason, we will rid ourselves of these second derivative terms in (12.37), leaving:

$$\mathbb{J}^T [\mathbf{d} - \mathbf{F}] = \mathbb{J}^T \mathbb{J} \Delta\mathbf{x}. \tag{12.39}$$

This is a manageable equation, and one we have seen before – it represents the normal equations for a matrix $\mathbb{J}$ – this is heartening even as it is not surprising – we expect our fancy "nonlinear" method to recover the linear case from Section 10.2 (throwing out the second derivatives in (12.37) re-enforces this association). We know how to solve these equations for $\Delta\mathbf{x}$ using the QR decomposition, or, since the matrix $\mathbb{J}^T \mathbb{J}$ is symmetric and positive definite, we could use conjugate gradient.

Before we perform additional surgery, let's take a moment to review. All we have done, in the end, is taken $u(\mathbf{x})$, assumed that we are near enough to a minimum to use a linear approximation to $F(\mathbf{x}, t)$, i.e. a quadratic approximation to $u$ – then, from the definition of minimum, we obtain a linear equation for $\Delta\mathbf{x}$. We can solve this equation at each step, and iterate that procedure until we stop moving. Note that the matrix we need to invert in (12.39) to get $\Delta\mathbf{x}$ is $\mathbb{J}^T \mathbb{J} \in \mathbb{R}^{m \times m}$ – its size (and hence the cost of the inverse problem) is set by the number of parameters $m$ in the function $F(\mathbf{x}, t)$, not the size of the data.

Let's turn now to the same problem approached from the steepest descent point of view. The update described in Algorithm 12.1 is informed primarily by an observation about the nature of derivatives. In the present case, steepest descent would have an update, in terms of $\mathbf{F}$, given by:

$$\mathbf{x}_{j+1} = \mathbf{x}_j + \eta\mathbb{J}^T(\mathbf{d} - \mathbf{F}) \tag{12.40}$$

where we have used the first derivative from (12.35), and dropped the two in favor of our arbitrary step size $\eta$. In this case,

$$\eta\mathbb{J}^T [\mathbf{d} - \mathbf{F}] = \Delta\mathbf{x}, \tag{12.41}$$

which should be compared with the Newton-inspired (12.39) – the matrix inverse is gone, and the parameter $\eta$ appears on the left, but otherwise, a very similar type of equation.

In the Levenberg–Marquardt approach to least squares minimization, the starting point is an equation for determining $\Delta \mathbf{x}$ that interpolates smoothly between (12.39) and (12.41). Take

$$\mathbb{J}^T [\mathbf{d} - \mathbf{F}] = \left[\mathbb{J}^T \mathbb{J} + \lambda \mathbb{I}\right] \Delta \mathbf{x} \tag{12.42}$$

as the defining equation for $\Delta \mathbf{x}$. When $\lambda = 0$, we recover (12.39), and as $\lambda$ goes to infinity, we get back (12.41) with $\eta \equiv 1/\lambda$. The final observation is that $\lambda$ must have the same dimension as $\mathbb{J}^T \mathbb{J}$, which is not clearly possible for a single number – the values of the parameters $\mathbf{x}$ may not even have the same dimension. In order to restore balance to the equation, replace the identity matrix with the diagonal components of $\mathbb{J}^T \mathbb{J}$. This defines the final Levenberg–Marquardt update:

$$\boxed{\mathbb{J}^T [\mathbf{d} - \mathbf{F}] = \left[\mathbb{J}^T \mathbb{J} + \lambda \mathrm{Diag}(\mathbb{J}^T \mathbb{J})\right] \Delta \mathbf{x}} . \tag{12.43}$$

The addition of the diagonal elements of $\mathbb{J}^T \mathbb{J}$ doesn't spoil the steepest descent interpretation of the large $\lambda$ limit, rather it allows the descent to proceed in directions with varying step size.

We're ready to define the algorithm. Given the machinations involved in motivating (12.43) (dropping second derivatives of $\mathbf{F}$, adding an interpolating piece), we are not assured that $\Delta \mathbf{x}$ solving (12.43) actually moves us in a decreasing direction. Fortunately, we have the parameter $\lambda$, which can be used to *guarantee* that $u(\mathbf{x} + \Delta \mathbf{x}) < u(\mathbf{x})$, since the method becomes steepest descent as $\lambda$ gets large. So we'll build the choice of $\lambda$ into the algorithm.

Let's work through Levenberg–Marquardt, Algorithm 12.3 – we start at some point $\mathbf{x}_0$, and we'll run the method until the norm of the parameters changes less than $\epsilon$ (in absolute value) from one step to the next. We must provide the initial choice of $\lambda$, called $\lambda_0$, and a factor $\lambda_\times$ by which to increase or decrease the current value of $\lambda$. In addition to these inputs, we need functions that will return $u(\mathbf{x})$, $\mathbf{F}(\mathbf{x})$, and $\mathbb{J}(\mathbf{x})$. At each step, we construct the matrices and vectors necessary to solve (12.43) for $\Delta \mathbf{x}$ using the current value of $\lambda$. Once we have the solution, we check to see if the resulting parameter set yields a value for $u(\mathbf{y} + \Delta \mathbf{x})$ that is greater than or less than $u(\mathbf{y})$ – if the new value is larger than the old, we increase $\lambda$ by the factor $\lambda_\times$, making the method more like steepest descent and try again. If the new value is less than the old value, indicating success in decreasing the value of the function $u$, then we divide $\lambda$ by $\lambda_\times$ in preparation for the next step, and set $\mathbf{y}$ equal to this new, improved, value.

**Algorithm 12.3** Levenberg–Marquardt

$\mathbf{y} \leftarrow \mathbf{x}_0$
$\mathbf{z} \leftarrow 0$
$\lambda \leftarrow \lambda_0$
**while** $\|\mathbf{y} - \mathbf{z}\| \geq \epsilon$ **do**
  $\mathbb{B} \leftarrow \mathbb{J}^T(\mathbf{y})\mathbb{J}(\mathbf{y})$
  $\mathbf{w} \leftarrow \mathbf{y}$
  **while** $u(\mathbf{y}) \leq u(\mathbf{w})$ **do**
    $\mathbb{A} \leftarrow \mathbb{B} + \lambda \text{Diag}(\mathbb{B})$
    $\Delta \mathbf{x} \leftarrow \mathbb{A}^{-1}\mathbb{J}^T(\mathbf{y})(\mathbf{d} - \mathbf{F}(\mathbf{y}))$
    $\mathbf{w} \leftarrow \mathbf{y} + \Delta \mathbf{x}$
    **if** $u(\mathbf{y}) \leq u(\mathbf{w})$ **then**
      $\lambda \leftarrow \lambda \lambda_\times$
    **else**
      $\lambda \leftarrow \lambda / \lambda_\times$
    **end if**
  **end while**
  $\mathbf{z} \leftarrow \mathbf{y}$
  $\mathbf{y} \leftarrow \mathbf{w}$
**end while**

---

**Example**

Suppose we have an oscillatory, decaying signal (a damped harmonic oscillator of some sort) with:

$$d(t) = 8e^{-\sqrt{2}t}\sin(2\pi t). \tag{12.44}$$

We'll run the Levenberg–Marquardt algorithm with discretized data $\{d_j = d(t_j)\}_{j=0}^{n-1}$ for $n = 10$ with $\Delta t = 1/(n-1)$ (i.e. $\mathbf{d} \in \mathbb{R}^{10}$). As our function to fit, we'll take the most natural damped harmonic oscillator choice:

$$F(\mathbf{x}) = x_1 e^{x_2 t}\sin(x_3 t) \tag{12.45}$$

so that our vector $\mathbf{x} \in \mathbb{R}^3$. Running an implementation of Algorithm 12.3 with $\lambda_0 = 0.001$, $\lambda_\times = 10$ and starting at $\mathbf{x}_0 \doteq \begin{pmatrix} 1 & 1 & 1 \end{pmatrix}^T$, we recover:

$$x_1 = 8.00000202309151$$

$$x_2 = -1.4142169820981867 \tag{12.46}$$

$$x_3 = 6.283184686947074$$

using $\epsilon = 0.01$.

## 12.5 Line minimization

We can put together the steepest descent, or Newton's method with one-dimensional minimization. In both Algorithm 12.1 and Algorithm 12.2, we have an arbitrary step $\eta$ that we have taken to be fixed. But the update in, for example, steepest descent can be viewed as a one-dimensional function of $\eta$ – i.e. we could minimize the function:

$$F(\eta) = u\left(\mathbf{y} - \eta\frac{\partial u}{\partial \mathbf{x}}|_{\mathbf{x}=\mathbf{y}}\right). \tag{12.47}$$

This procedure, of turning the step size $\eta$ into a one-dimensional minimization is known as "line minimization." It is easy to introduce such a scheme, and Implementation 12.3 provides the details.[4] Since we know that the direction $\frac{\partial u}{\partial \mathbf{x}}$ points downhill, we need only consider $\eta > 0$, and so we start our search for an initial bracketing at zero.

---

**Implementation 12.3** Line minimization with steepest descent

```
Descent[u_, du_, x_, minstep_, eps_] := Module[{y, z, F, etamin, grad, bracket},
  y = x;
  z = Table[0.0, {i, 1, Length[y]}];
  While[Norm[y - z] ≥ eps,
    grad = du[y];
    F[eta_] := u[y - eta grad];
    bracket = PreGSMin[F, 0.0, minstep];
    etamin = GSMin[F, bracket, eps];
    z = y;
    y = y - etamin grad;
  ];
  Return[y];
]
```

---

The same idea works with any method that provides a search direction, so that in general, we could minimize:

$$F(\eta) = u(\mathbf{y} - \eta\Delta\mathbf{y}) \tag{12.48}$$

given $\Delta\mathbf{y}$ – for steepest descent, we have $\Delta\mathbf{y} = \mathbf{G}(\mathbf{y}) \equiv \frac{\partial u}{\partial \mathbf{x}}|_{\mathbf{x}=\mathbf{y}}$, for Newton's method, we have $\Delta\mathbf{y} = \mathbb{H}^{-1}(\mathbf{y})\mathbf{G}(\mathbf{y})$. We can see the problem with this latter case; the calculation of the inverse of the Hessian could be time-consuming, making it a bottleneck for the minimization (that has always been true, incidentally). There are a variety of methods that start by approximating $\mathbb{H}^{-1}(\mathbf{y})$ and/or generating iterative updates for this matrix that follow the minimization procedure itself (those are the Powell class of methods – see [40] for more on these). In the end, it doesn't much

---

[4] The only parameter in Implementation 12.3 that isn't directly involved in the steepest descent is `minstep`, and this is used to set the step size in the initial golden section bracketing.

matter *how* you get $\Delta\mathbf{y}$ as long as it is motivated vaguely by a minimizing direction (as with steepest descent).

## Action minimization

Let's take the action minimization case from Section 12.1.3 as an example of line minimization for many parameters. We'll start by defining the numerical action, an approximation to (12.13), given a vector $\mathbf{x}$ with $x_j = x(t_j)$ describing the (one-dimensional) trajectory of a particle, moving in the presence of $U(x)$. For a temporal grid with $t_j = j\Delta t$, we'll use a finite difference approximation for $\dot{x}$ appearing in $S$, and a simple box sum to approximate the integral,

$$S \approx S_N \equiv \sum_{j=1}^{N-1} \left[ \frac{1}{2}m \left( \frac{x_{j+1} - x_{j-1}}{2\Delta t} \right)^2 - U(x_j) \right] \Delta t, \qquad (12.49)$$

where $N\Delta t = T$. This numerical $S_N(\mathbf{x})$ depends on $\mathbf{x}$, and we want to minimize with respect to this trajectory (i.e. the values approximating $x(t_j)$ are what we will be changing) given some initial $x_0$ and final $x_N$, the boundary values, which will remain unchanged.

In order to use Implementation 12.3, we need both the function to minimize, $S_N$, and its derivative with respect to all the components of $\mathbf{x}$: $\frac{\partial S_N}{\partial \mathbf{x}}$. From its definition, we can compute the derivative of $S_N$ with respect to the $k$th element of $\mathbf{x}$:

$$\frac{\partial S_N}{\partial x_k} = \sum_{j=1}^{N-1} \left[ m \frac{x_{j+1} - x_{j-1}}{(2\Delta t)^2} (\delta_{kj+1} - \delta_{kj-1}) - \frac{\partial U}{\partial x_k} \right] \Delta t$$

$$= \left[ \frac{m}{(2\Delta t)^2} (2x_k - x_{k+2} - x_{k-2}) + F(x_k) \right] \Delta t. \qquad (12.50)$$

We can use this expression for $\frac{\partial S_N}{\partial x_k}$ as the entries of the gradient vector $\frac{\partial S_N}{\partial \mathbf{x}}$, and now we just specify a potential energy function $U(x)$, the associated force $F = -\frac{dU}{dx}$, and we're ready to run Implementation 12.3.

Take $U = (1/2)2x^2$ so that $F = -2x$, a linear force. We'll set $x_0 = 0$, $x_T = 1$ for $T = 1$, and $N = 100$. The correct trajectory, given this force and boundary conditions, is:

$$x(t) = \frac{\sin(\sqrt{2}t)}{\sin(\sqrt{2})}. \qquad (12.51)$$

This exact solution, together with the result of line minimization for this problem are shown in Figure 12.2 – the stopping point had $\epsilon = 10^{-4}$.

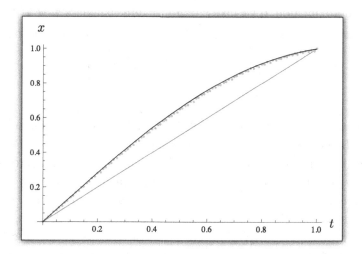

Figure 12.2 Action minimization for a harmonic oscillator potential – the exact solution (12.51) is shown in black, with the numerical minimizing solution shown as gray dots. The initial trajectory used in the iteration is the straight gray line, for reference.

## 12.6 Monte Carlo minimization

Many minimization routines proceed from a rational choice of update step, but in the end, if there are elements of the update that are time-consuming, we might use a well-defined starting point with the unwieldy elements dropped (think of the move from (12.37) to (12.39)). Our final technique makes the ultimate in uninformed update choice – random. There are a class of problems for which this works, and works better than a more motivated choice, particularly when the parameter space is very large.

In the case of variational bounds in quantum mechanics (or the direct action minimization from the previous section), the guarantee is that the true ground state energy is below any guess we might make. That provides some freedom, even releasing us from rational choices altogether. Let's work out the variational integral setup for a simple harmonic oscillator and see how making random changes to the wave function can be used to establish a fairly tight upper bound on the ground state energy. The expectation value (12.3) for this problem, where $V = \frac{1}{2}m\omega^2 x^2$, is

$$\langle E \rangle = \frac{\int_{-\infty}^{\infty}\left[-\frac{\hbar^2}{2m}\psi^*\frac{d^2\psi}{dx^2} + \frac{1}{2}m\omega^2 x^2\psi^*\psi\right]dx}{\int_{-\infty}^{\infty}\psi^*\psi\,dx}. \qquad (12.52)$$

The integral in the denominator is one if the wave function is normalized, but we leave this factor in place so that we need not worry about normalizing our numerical approximation. If we take $x \equiv aq$ for dimensionless $q$, $\psi \equiv 1/\sqrt{a}\bar{\psi}$ for dimensionless $\bar{\psi}$,[5] and set $a \equiv \sqrt{\frac{\hbar}{\sqrt{2m\omega}}}$ then the expectation value becomes

$$\langle E \rangle = \frac{\hbar\omega}{\sqrt{2}} \frac{\int_{-\infty}^{\infty} \left[ -\bar{\psi}^* \frac{d^2\bar{\psi}}{dq^2} + \frac{1}{2}q^2\bar{\psi}^*\psi \right] dq}{\int_{-\infty}^{\infty} \bar{\psi}^*\bar{\psi}\, dq}. \tag{12.53}$$

Remember the idea, *any* $\bar{\psi}$ will have $\langle E \rangle$ greater than the ground state energy of the system. In the case of a harmonic oscillator, that lowest energy is $\frac{1}{2}\hbar\omega$.

What we'll do here is define $\bar{\psi}$ for a grid of values $q_j = j\Delta q$, approximate the second derivative using finite difference, and the integral using a box sum. We'll take $N_\infty \Delta q$ to be a numerical value for infinity (meaning, among other things, that the wave function should be zero near these grid values), and denote $\bar{\psi}_j \approx \bar{\psi}(q_j)$ as always. If we assume that $\bar{\psi}$ is real (it is, in the case of the true harmonic oscillator ground state), we can drop the complex conjugates, and our approximate expression for $\langle E \rangle$ is

$$\bar{E} \equiv \frac{1}{\sqrt{2}} \frac{\sum_{j=-N_\infty}^{N_\infty} \left[ -\frac{\bar{\psi}_{j+1} - 2\bar{\psi}_j + \bar{\psi}_{j-1}}{\Delta q^2}\bar{\psi}_j + \frac{1}{2}q_j^2\bar{\psi}_j^2 \right]\Delta q}{\sum_{j=-N_\infty}^{N_\infty} \bar{\psi}_j^2 \Delta q}. \tag{12.54}$$

If we start with any set of values $\{\bar{\psi}_j\}_{j=-N_\infty}^{N_\infty}$, then we know that $\bar{E} \geq \frac{1}{2}$ (we have omitted the overall factor of $\hbar\omega$ in going from (12.53) to the approximate expression above, and compare with the dimensionless ground state energy, $\frac{1}{2}$). Any method for updating the values of $\bar{\psi}_j$ will also produce $\bar{E} \geq \frac{1}{2}$. Now the Monte Carlo element: Update the values of $\bar{\psi}_j$ randomly, and if the value of $\bar{E}$ for the updated wave function is less than the value of $\bar{E}$ for the original, take the new values and repeat the process.

One way to do this is to modify each element $\bar{\psi}_j$ by a small random amount, compute $\bar{E}$, and accept or reject based on whether the energy expectation value has decreased or increased. Let $\bar{\boldsymbol{\psi}}$ be the vector whose entries are the $\{\bar{\psi}_j\}_{j=-N_\infty}^{N_\infty}$. We take $\bar{\psi}_{\pm N_\infty} = 0$, and in order to enforce these boundaries, our random update will not proceed over the entire grid – we'll go from $-N_\infty + N_0$ to $N_\infty - N_0$, modifying values randomly – but those first and last $N_0$ entries we'll set to zero and leave there. We have the choice of $N_0$, the number of points that make up the boundaries, as a parameter in the method.

Assuming you have a function to compute $\bar{E}(\bar{\boldsymbol{\psi}})$ given the values $\bar{\boldsymbol{\psi}}$, a single pass through all grid values starting with a wave function (in vector form) $\bar{\boldsymbol{\psi}}$ is shown in Algorithm 12.4. There, $G(0, \alpha)$ means that we take a random number drawn from a Gaussian distribution with mean zero, standard deviation $\alpha$.

---

[5] The density $\psi^*\psi$ has dimension of inverse length.

We go through every entry of the input $\bar{\psi}$ (except the first and last $N_0$), modify by a random amount, then see if the result has smaller energy expectation value – if it does not, we retain the original value, else, we keep the new, modified value.

At the end of a single pass, the vector $\hat{\psi}$ is an approximation to a wave function with energy expectation value less than the original $\bar{\psi}$. You can continue the process until the energy expectation value "doesn't change much." We set up the problem with an eye towards the harmonic oscillator, but of course, any function $\bar{E}$ will do; the idea is always the same.

---

**Algorithm 12.4** Variational Monte Carlo

---

**for** $j = -N_\infty + N_0 \rightarrow N_\infty - N_0$ **do**
   $\hat{\psi}_j \leftarrow \bar{\psi}_j + G(0, \alpha)$
   **if** $\bar{E}(\hat{\psi}) \geq \bar{E}(\bar{\psi})$ **then**
      $\hat{\psi}_j \leftarrow \bar{\psi}_j$
   **end if**
**end for**

---

Monte Carlo minimization represents a relatively slow way to proceed, but it can be beneficial when working on many-body problems in quantum mechanics, where the wave function contains many coordinate degrees of freedom.

## Further reading

1. Giordano, Nicholas J. *Computational Physics*. Prentice Hall, 1997.
2. Gould, Harvey, Jan Tobochnik, & Wolfgang Christian. *Computer Simulation Methods: Applications to Physical Systems*. Pearson, 2007.
3. Pang, Tao. *An Introduction to Computational Physics*. Cambridge University Press, 2006.
4. Press, William H., Saul A. Teukolsky, William T. Vetterling, & Brian P. Flannery. *Numerical Recipes in C*. Cambridge University Press, 1996.

## Problems

**Problem 12.1**
Find the minima and maxima of the function:

$$F(x) = (x - 1)(x - 2)(x - 3) \tag{12.55}$$

for $x \in [1, 3]$.

**Problem 12.2**

The expectation value (12.3) for a generic potential $V(x)$ in one dimension is:

$$\langle E \rangle = \frac{\int_{-\infty}^{\infty} \left[ -\frac{\hbar^2}{2m} \psi^* \frac{d^2\psi}{dx^2} + V(x)\psi^*\psi \right] dx}{\int_{-\infty}^{\infty} \psi^*\psi \, dx}. \tag{12.56}$$

Let $x = aq$, $\psi = \bar{\psi}/\sqrt{a}$ for a length $a$, and define dimensionless $\tilde{E} \equiv \frac{2ma^2}{\hbar^2} E$, $\tilde{V} \equiv \frac{2ma^2}{\hbar^2} V$. Express $\langle \tilde{E} \rangle$ in terms of these dimensionless values.

**Problem 12.3**

We assume $\psi^*\psi$ has dimension of inverse length, but if $\psi$ solves Schrödinger's equation, then $A\psi$ does as well – where, then, does the dimensional requirement for $\psi$ come from?

**Problem 12.4**

We can find the bound on $\langle \tilde{E}(\beta) \rangle$ for hydrogen using $\psi(\beta, q) = 2\beta^{3/2}qe^{-\beta q}$ directly. Compute the integral of (12.25), and then take the derivative of the resulting expression (w.r.t. $\beta$). Set that to zero and solve for the minimizing $\beta$. Compare your value of $\langle \tilde{E}(\beta) \rangle$ for this minimum with the correct expression.

**Problem 12.5**

For the infinite square well, extending from $q = 0 \longrightarrow 1$ (in the usual parametrization of, for example, Section 10.4.2), the expression $H\psi(q) = -\psi''(q)$, so that (assuming $\psi(q)$ is real)

$$\langle E \rangle = -\frac{\int_0^1 \psi''(q)\psi(q)dq}{\int_0^1 \psi(q)^2 dq}. \tag{12.57}$$

**(a)** Given that one of the defining properties of the infinite square well is that $\psi(0) = \psi(1) = 0$, show that the expectation value above can be written as:

$$\langle E \rangle = \frac{\int_0^1 \psi'(q)^2 dq}{\int_0^1 \psi(q)^2 dq}. \tag{12.58}$$

**(b)** Set $\psi(q) = \alpha + \beta q + \gamma q^2$ – force this $\psi(q)$ to have the correct boundary behavior, and normalize it. This should fix all constants. What is the resulting expression for $\langle E \rangle$? How does it compare with the true ground state energy?

**Problem 12.6**

For a charged particle moving in a magnetic field, the force is $\mathbf{F} = q\mathbf{v} \times \mathbf{B}$.

**(a)** Write the equations of motion for a particle of mass $m$ and charge $q$ traveling in the $x - y$ plane under the influence of a magnetic field: $\mathbf{B} = -B_0 \frac{r}{a}\hat{z}$ (where $r$ is the distance from the origin, and $a$ is a length). Set $x = au$, $y = av$, and $t = \alpha s$

with $u$, $v$, and $s$ dimensionless, and rewrite the equations of motion, getting rid of all extraneous constants (namely $m$, $q$, and $B_0$) by appropriately defining $\alpha$.

**(b)** If the particle starts at $u = \frac{1}{2}$, $v = 0$ when $s = 0$, find the initial velocity (i.e. $u'(0)$ and $v'(0)$) that will cause circular motion.

## Problem 12.7

In the lab problems, you will use a maximizer, in addition to a minimizer. The development of a maximum-finder parallels that of the minimization routine in Section 12.2. Assume that you have a bracketing, $\{x_\ell, x_m, x_r\}$ such that $F(x_\ell) \leq F(x_m)$ and $F(x_r) \leq F(x_m)$. Suppose the setup shown in Figure 12.1 holds – with the test point $x$ from (12.20)) – what bracketings should you return for $F(x) > F(x_m)$ and $F(x) < F(x_m)$?

## Problem 12.8

For a complicated or non-analytic function $u(\mathbf{x})$, the derivatives needed by, for example, steepest descent can be calculated numerically. Write a function that generates a numerical approximation to the gradient of $u(\mathbf{x})$ evaluated at a vector location $\mathbf{x}_0$ by approximating the derivative via:

$$\frac{\partial u}{\partial x_i} \approx \frac{u(\mathbf{x}_0 + \Delta x \mathbf{e}_i) - u(\mathbf{x}_0 - \Delta x \mathbf{e}_i)}{2\Delta x} \tag{12.59}$$

(where $\mathbf{e}_i$ has zeroes at all locations except for $i$, where it has a 1). Your function should take in the function $u$, the vector $\mathbf{x}_0$, and a user-specified $\Delta x$ giving the step-size for the derivative approximation – it should return a vector approximating $\frac{\partial u}{\partial \mathbf{x}}|_{\mathbf{x}=\mathbf{x}_0}$.

## Problem 12.9

Using a finite difference approach as in Chapter 10, with $N = 99$, find the (approximate) ground state energy for an infinite square well (extending for spatial $q = 0$ to 1) with a finite square bump in the middle, i.e. with

$$\tilde{V} = \begin{cases} \infty & q < 0 \\ 0 & 0 < q < .4 \\ 50 & 0.4 \leq q \leq .6 \\ 0 & 0.6 < q < 1 \\ \infty & q > 1 \end{cases} \tag{12.60}$$

Plot the associated eigenvector.

## Lab problems

## Problem 12.10

Referring to the function (12.55):

**(a)** Find the location of the minimum for $x \in [1, 3]$ using the `PreGSMin` and `GSMin` functions from the chapter notebook (set $\epsilon = 10^{-5}$).

**(b)** Modify those functions to find maxima, and find the location of the maximum for $x \in [1, 3]$. In both cases, check against the actual answer you found in Problem 12.1.

### Problem 12.11

For the induced matrix norm described in Section 10.4.3, we have a "maximization" problem – given $A \in \mathbb{R}^{n \times n}$, we must find the unit vector $x$ that maximizes $\|Ax\|$ (the Pythagorean norm, $\|y\| = \sqrt{y \cdot y}$). Use a "steepest ascent" approach – implement an appropriately re-written form of Algorithm 12.1 that goes uphill, and use that with an appropriate $u$ and your numerical gradient calculator from Problem 12.8 to write a function that returns $\|A\| = \max_{x:x=1} \|Ax\|$. Check your function on the matrix

$$A \doteq \begin{pmatrix} 1 & 2 & 3 & 4 \\ 5 & 6 & 7 & 8 \\ 9 & 10 & 11 & 12 \\ 13 & 14 & 15 & 16 \end{pmatrix}, \tag{12.61}$$

and compare with the built-in `Norm` command in `Mathematica`.

### Problem 12.12

Suppose we have the magnetic field described in Problem 12.6, we would like to shoot a particle so that it hits the origin (our "target").

**(a)** Using your RK4 routine from Chapter 2, write a function that solves the (dimensionless) equations of motion from part (a) of Problem 12.6 with initial positions $u(0) = \frac{1}{2}$, $v(0) = 0$ – test your solver by using your circular motion initial velocity from part (b) of Problem 12.6. Run with $\Delta s = 0.01$ and 2000 steps. Be sure you recover a circular trajectory.

**(b)** Our particle cannon fires with constant speed 1, and our goal is to find the angle $\theta$ at which we should fire the particle so that it goes as close to the origin as possible. Write a function $F(\theta)$ that provides the minimum distance to the origin given an initial angle (this function will call your ODE solver with initial conditions $u(0) = \frac{1}{2}$, $v(0) = 0$, $u'(0) = \cos\theta$, $v'(0) = \sin\theta$, and you should continue to use $\Delta s = 0.01$ and 2000 steps). Using the golden section minimizer from the chapter notebook, find the angle $\theta \in [0, \frac{1}{2}\pi]$ that ensures our particle passes through the origin (to within $\epsilon = 10^{-5}$) – plot the particle trajectory for the minimizing angle. Try plotting the trajectory at twice this angle to see what "missing" the origin looks like.

### Problem 12.13

Using the data "ch10.dat" from Problem 10.11 together with the reasonable starting point (for that data)

$$F(x) = x_1 e^{x_2 t} \left[ \cos(x_3 t) + x_4 \sin(x_5 t) \right], \tag{12.62}$$

find the coefficients **x** that match the input using the Levenberg–Marquardt routine from the chapter notebook. This is a more efficient way than the method presented in Problem 10.11. You will have to cut the data in half (the tail of the exponential here is difficult to fit since it is small) – be sure to use the correct total time and time step $\Delta t$, refer to the example at the end of Section 12.4 to get the setup (you should use $\lambda_0 = 0.1$ and $\lambda_\times = 10$ with $\mathbf{x}_0 = \{1, 2, 3, 4, 5\}$ and $\epsilon = 10^{-6}$). Again find $b$ and $\omega$ from your solution.

## Problem 12.14

Import the data "ch12.dat", this is a sinusoidal signal with noise. Use Levenberg–Marquardt to fit the data with the function:

$$F(\mathbf{x}) = x_1 \sin(2\pi x_2 t + x_3), \tag{12.63}$$

so that we include a phase shift, in addition to frequency and amplitude. Start at $x_1 = 1$, $x_2 = 1$, $x_3 = 1$ with $\lambda_0 = 0.1$ and $\lambda_\times = 10$, $\epsilon = 10^{-4}$. Plot your continuous function on top of the data. Try starting at $x_1 = 1$, $x_2 = 2$, $x_3 = 1$; you should get a different set of parameters. Plot the fit curve on top of the data again. Which of the "minima" are correct (how do you know), and why is there a difference?

## Problem 12.15

For the function:

$$u(x, y) = J_0(x) \sin(x + y) \tag{12.64}$$

where $J_0$ is the zeroth Bessel function of the first kind (accessed using the built-in command BesselJ), use steepest descent with line minimization (the function Descent from the chapter notebook) to find minima starting at $x = 0$, $y = 0$ and $x = 4$, $y = -4$ (do you end up at the same spot?). In both cases, use minstep= 0.01 and eps= $10^{-5}$. For the gradient, you may use the relation $\frac{d J_0(x)}{dx} = -J_1(x)$.

## Problem 12.16

There is a class of vector norms that take the form (for a vector $\mathbf{x} \in \mathbb{R}^n$):

$$\|\mathbf{x}\|_p = \left[ \sum_{j=1}^{n} |x_j|^p \right]^{1/p}. \tag{12.65}$$

When we solve least squares problems, linear or not, we use $p = 2$, but this is not necessary. Given $\mathbb{A} \in \mathbb{R}^{n \times m}$, $\mathbf{b} \in \mathbb{R}^n$, suppose we define the residual $\mathbf{r} = \mathbb{A}\mathbf{x} - \mathbf{b}$ as usual, but now we take the $p = 1$ norm as the object to minimize:

$$u(\mathbf{x}) = \sum_{j=1}^{n} |r_j|. \tag{12.66}$$

**(a)** Show that for this function $u$, the derivative with respect to the $i$th element of $\mathbf{x}$ is

$$\frac{\partial u}{\partial x_i} = \sum_{j=1}^{n} \left[ 2\theta(y_j) - 1 \right] A_{ji}, \qquad y_j \equiv \sum_{k=1}^{n} A_{jk} x_k - b_j \qquad (12.67)$$

where $\theta(x)$ is the step function, $\theta(x) = 0$ for $x < 0$ and $1$ for $x > 0$.

**(b)** Let's re-do the data-fitting example from Section 10.2 for this new $p = 1$ norm minimization. Starting from $\mathbf{x}$ solving the least squares problem there (available in the chapter notebook), use steepest descent with line minimization and the derivative in (12.67) to find a vector $\bar{\mathbf{x}}$ that has $u(\bar{\mathbf{x}}) < u(\mathbf{x})$ (i.e. one that is more of a minimum of $u$ than the original, least squares starting point) – use $10^{-9}$ for minstep, and run until the tolerance eps$= 10^{-7}$ is achieved. Plot your solution – it should look similar to the least squares case, which is not particularly surprising. Nevertheless, the two solutions are quite different.

**Problem 12.17**

Modify the function EQM from the chapter notebook to handle generic potentials (use your dimensionless form from Problem 12.2). For the potential in Problem 12.9, run one thousand steps of UpdateQM (also in the chapter notebook) with Nzero$= 2$, and mag$= 10$ (this is the standard deviation for the Gaussian, $\alpha$ in Algorithm 12.4) starting with an initial psibar that is generated randomly via:

```
psibar = Table[{q, RandomReal[{0, 1}]}, {q, dq, 1 - dq, dq}];
```

with dq $= 1/100$, and the first two and last two points of the resulting table set to zero. Plot the resulting (after one thousand steps) psibar, what is $\langle \tilde{E} \rangle$? Run for an additional thousand steps at mag $= 1$, and again find $\langle \tilde{E} \rangle$ and plot the wave function. Run for another thousand steps, this time with mag $= 0.1$. Compare your final state and energy with the ground state energy and wave function you computed using finite difference in Problem 12.9.

# 13

# Chaos

This chapter is slightly different than our previous discussions. Our subject is not a numerical method, exactly, but a property of nonlinear differential equations that was brought to people's attention by attempts at numerical solution. Roughly speaking, a chaotic system is characterized by aperiodic behavior that is determined, in a sensitive way, by initial conditions. Wait, you say, all solutions to ODEs depend on initial data, and in a very sensitive way – for example, if we have a harmonic oscillator, then we know the solution to the equation of motion, given initial position $x_0$ and velocity $v_0$, is:

$$x(t) = x_0 \cos(\omega t) + \frac{v_0}{\omega} \sin(\omega t) \tag{13.1}$$

where $\omega$ is the frequency of oscillation, and is based on the physical parameters provided in the problem. See the $x_0$ and $v_0$ dependence in $x(t)$? Looks pretty sensitive to initial conditions. It is, but not in quite the way we have in mind – after all, if you make a small perturbation to $x_0$ or $v_0$, certain fundamental properties of the system remain unchanged – the frequency, for example. In addition, the error remains bounded as a function of time. If we take $x(t) = x_0 \cos(\omega t)$ and $y(t) = (x_0 + \epsilon) \cos(\omega t)$, then the difference between $y(t)$ and $x(t)$, as a function of time, is:

$$y(t) - x(t) = \epsilon \cos(\omega t), \tag{13.2}$$

and this difference is at most $\epsilon$ for *any* value of $t$.

Think instead of a function like $x(t) = x_0 e^{\gamma t}$ – exponential growth. Again let $y(t) = (x_0 + \epsilon)e^{\gamma t}$ for small $\epsilon$. Then the difference between $y(t)$ and $x(t)$ grows exponentially as well:

$$y(t) - x(t) = \epsilon e^{\gamma t} \tag{13.3}$$

323

and this means that even for small $\epsilon$, there exists a time at which the difference is arbitrarily large.

Who cares? We'll just agree to not make any errors in our initial conditions, then all of our problems are solved. Well, we have been working on numerical solutions, and while a computer is good at approximating real numbers, it is still *approximating* them. The precision of a machine is on the order of $10^{-13}$, although the actual number doesn't matter – the point is, given an $x_0$ chosen from the Platonic form of real numbers, our representation on the computer will *always* contain some error. That error, $\epsilon$, cannot be made arbitrarily small; it has a fixed, finite size. This is our motivating interest in the problem of initial values – although it extends to errors made in the parametric values involved in various equations, $\gamma$ in the case of exponential growth, for example.

We'll start by looking at "maps," rules for generating sequences of numbers – these rules take the form of recursions: $u_{n+1} = G(u_n)$ for some function $G(u)$. We'll define some basic ideas on the map side. Most of our numerical methods for solving ODEs generate discrete maps that depend on both the method being used, and the physical problem of interest – so these maps are useful both by themselves, and as approximations to solutions of continuous problems.

The various solutions of nonlinear maps (and, when they exist, the associated ODE) have interesting geometric properties – fractals are an example. We will look briefly at fractals and the manner in which we can characterize these discrete shapes.

### 13.1 Nonlinear maps

The simplest possible map, in one dimension, is defined by the recursion: $u_{n+1} = au_n + b$ – we can solve this one explicitly:

$$u_n = a^n u_0 + b \sum_{j=0}^{n-1} a^j = a^n u_0 + \frac{a^n - 1}{a - 1} b. \tag{13.4}$$

There are a few things we can say about such a solution – it exists, and is predictable for all values of $u_0$. For $a \in [0, 1)$, the solution limits to:

$$u_\infty = \frac{b}{1 - a}, \tag{13.5}$$

and for $a > 1$, the solution runs off to infinity. In the case that $b = 0$, we can also identify a "fixed point" of the map, a value that is unchanged by the map's action – that is, of course, $u_0 = 0$. For $u_0 = 0$, $u_n = 0$ for all $n$. If we could characterize all maps by such complete solutions, life would be easier.

The next natural map to study would be a quadratic one:

$$u_{n+1} = u_n^2 \longrightarrow u_n = u_0^{2^n}. \tag{13.6}$$

We can ask: What are possible "end" values as $n \longrightarrow \infty$? There are three in this case – if $u_0 = 0$, we get $u_\infty = 0$. If $u_0 = 1$, then $u_\infty = 1$, and if $u_0 > 1$, we get $u_\infty = \infty$. The ending values are determined by the fixed points 0 and 1. There is clearly a difference between 0 and 1 as ending values. If $u_0 \in [0, 1)$, then $u_\infty = 0$, while $u_\infty = 1$ *only if* $u_0 = 1$. The set of all points from which a final state can be reached is called the "basin of attraction" of that final state. Here, $[0, 1)$ is the basin of attraction for $u_\infty = 0$, and $(1, \infty]$ is the basin of attraction for $u_\infty = \infty$.

We can consider fixed points and the limit solutions even in the absence of explicit closed forms. For the cubic map

$$u_{n+1} = 3u_n - u_n^3, \tag{13.7}$$

there is a fixed point at $u_n = 0$, so that if we start with $u_0 = 0$, we will remain at zero. What if we start near zero? If we think of $u_0 = \epsilon$, for small $\epsilon$, then the linear term dominates, and $u_{n+1} \approx 3u_n$, the solution will move *away* from zero. So unlike the quadratic case, values near zero are not in the basin of attraction for zero. A fixed point is defined to be "stable" if values close to the fixed point are in the basin of attraction. From this point of view, 0 is not a stable fixed point of the map. In fact finding the values that lead to $u_\infty = 0$ as a limiting solution is not easy. It is clear that $u_0 = 0$ will give $u_\infty = 0$, but so will $u_0 = \pm\sqrt{3}$ since then $u_1 = 0$ (and then all other values will be zero as well).

More complicated nonlinear functions exhibit more varied behavior. Take, for example, the sine map:

$$u_{n+1} = a \sin(u_n). \tag{13.8}$$

We know that 0 is a fixed point of this map, and there are others. If we take $a = 2.2$, and $u_0 = 0.1$, then we see that after a few jumps, the solution settles down to a number (a plot of the first fifty iterates is shown on the left in Figure 13.1) – if we make $a = 2.3$, for the same starting $u_0 = 0.1$, then the solution oscillates between two different numbers, a "period two" solution (shown on the right in Figure 13.1). So a small change in the parameter $a$ is causing very different behavior.

In higher dimensions, we have the same basic questions and can explore the answers in a similar manner. Take the Hénon map for two variables, defining sequences $x_n$ and $p_n$ (see [25]):

$$\begin{pmatrix} x_{n+1} \\ p_{n+1} \end{pmatrix} = \begin{pmatrix} a - bp_n - x_n^2 \\ x_n \end{pmatrix}. \tag{13.9}$$

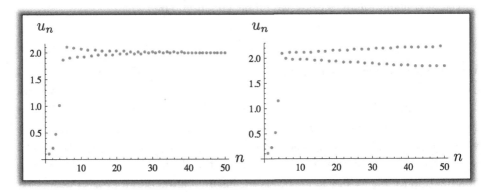

Figure 13.1 The sine map (13.8) with $a = 2.2$ (left) and $a = 2.3$ (right) – in both cases, $u_0 = 0.1$. For the left solution, we end up with a single number as a fixed point of the map, while on the right is a solution that oscillates between two different values – this is an example of a pair of "period two" solutions.

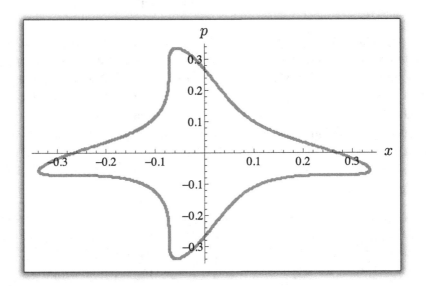

Figure 13.2 One thousand steps of the Hénon map for $x_0 = 0.1$, $p_0 = 0.1$, with $a = 0$, $b = 1$.

This pair of equations is written in a form reminiscent of the vectorization of a second-order ODE, where we have position and momentum, say. If we take $x_0 = p_0 = 0.1$, $a = 0$ and $b = 1$ (following [44]), then a two-dimensional plot (a phase space plot of $x$ vs. $p$) can be generated from the data, and is shown in Figure 13.2. Notice its crazy shape. The shape persists, so that continued iteration traverses the blob over and over. This recurrence of both $x$ and $p$ values means that there is some scale built into the map, determining "when" (in $n$) the values return to the starting point.

## 13.2 Periodicity and doubling

We can generalize the discussion of the specific one-dimensional examples from the previous section. Given a function of a single variable, $G(u)$, define an update map via:

$$u_{n+1} = G(u_n). \tag{13.10}$$

We use the function $G$ to generate a sequence of values $\{u_j\}_{j=1}^{\infty}$ given $u_0$.

Fixed points of the map can be found by solving:

$$G(u^*) = u^*, \tag{13.11}$$

so that if $u_n = u^*$, then $u_{n+1} = u^*$, and in fact, all successive values of $u_k$ ($k \geq n$) are precisely $u^*$. This type of fixed point has period one, meaning that it returns at each step. Period two solutions satisfy:

$$G(G(u^*)) = u^*, \tag{13.12}$$

and here, if $u_n = u^*$, then $u_{n+2} = u^*$.[1] If we define $k$ iterations of the map to be: $G^{(k)}(u)$, then the period $k$ fixed point is the solution to

$$G^{(k)}(u^*) = u^*. \tag{13.14}$$

A fixed point $u^*$ is "stable" if nearby values, $u^* + \epsilon$ (for $\epsilon$ small), are driven towards $u^*$ by the map, and unstable if nearby values are driven away. We can determine the stability of a fixed point by analyzing the function $G(u)$ directly. Take a fixed point $u^*$, and suppose we generate a sequence $\bar{u}_n = u^* + \epsilon_n$ ($n = 0, 1, \ldots$) where $\epsilon_n$ is small (starting with an initial perturbation $\epsilon_0$). If we run $\bar{u}_n$ through $G$, we get:

$$\bar{u}_{n+1} = G(\bar{u}_n) = G(u^* + \epsilon_n) \approx G(u^*) + \epsilon_n G'(u^*) = u^* + \epsilon_n G'(u^*) \tag{13.15}$$

where we Taylor expand, and then use the definition of fixed point, $G(u^*) = u^*$ to get the right-hand side. Now $\epsilon_{n+1} = \bar{u}_{n+1} - u^*$, so (13.15) represents the update, for $\epsilon_n$,

$$\epsilon_{n+1} = \epsilon_n G'(u^*). \tag{13.16}$$

---

[1] The $u^*$ in (13.12) is a period two solution for $G(u)$, but it is a fixed point for the map:

$$u_{n+1} = \tilde{G}(u_n) \tag{13.13}$$

with $\tilde{G}(u) = G(G(u))$, so we'll refer to periodic solutions as fixed points with appropriate period in general.

The solution to this recursion is given by:

$$\epsilon_n = \epsilon_0 \left[ G'(u^*) \right]^n \tag{13.17}$$

and $\epsilon_n \longrightarrow 0$ if $|G'(u^*)| < 1$. This gives us a stability condition: The fixed point $u^*$ is stable if $|G'(u^*)| < 1$.

### 13.2.1 The logistic map

The stable periodic solutions will be the limiting solutions for a sequence $\{u_j\}_{j=0}^\infty$ – if a value $u_n$ is near one of these fixed points, it will be sucked towards it, and then remain there (with appropriate period). The form of $G(u)$ determines where the fixed point is, with what period it occurs, and its stability. Some of the most interesting maps are those with a tunable parameter that can alter these properties (so that a fixed point with period one exists for some values of the parameter, but fixed points with period two emerge for other values). As an example, we'll take a particular quadratic map, the "logistic map," defined by:

$$G(u) = au(1 - u), \tag{13.18}$$

with parameter $a$. Our sequence $\{u_j\}_{j=0}^\infty$ is then defined by

$$u_{n+1} = au_n(1 - u_n) \tag{13.19}$$

given a starting value $u_0$.

---

**Motivation for the logistic map**

Many of the update maps we can generate come from the discretization of some first-order ODE, but some of them come directly from physical cases of interest. The logistic map is meant to provide a simple model of population growth. Let $N(t)$ be the population of some animal at time $t$. We'd like to model the population growth of this animal with two simple assumptions: 1. more animals implies more offspring, and 2. there is a maximum number of animals that the environment can support.

The first assumption would be satisfied by taking $\frac{dN(t)}{dt} \propto N(t)$, so that the larger $N(t)$ is, the more growth you have. If we call the proportionality constant $\alpha$, so that $\frac{dN}{dt} = \alpha N$, then the solution is:

$$N(t) = N_0 e^{\alpha t}, \tag{13.20}$$

---

i.e. exponential growth from some starting population. But this solution violates our second assumption, that there is a finite supportable population. Take the new model:

$$\frac{dN(t)}{dt} = \alpha \underbrace{\left(1 - \frac{N(t)}{N_{\text{max}}}\right)}_{\equiv \beta} N(t), \qquad (13.21)$$

which is the simplest (read "quadratic") modification that gives a growth "constant," $\beta$, of zero when $N(t) = N_{\text{max}}$, and enforces exponential *decay* when $N(t) > N_{\text{max}}$ (where $\beta$ is negative).

If we define $u \equiv N(t)/N_{\text{max}}$, then (13.21) becomes

$$\frac{du}{dt} = \alpha u(1 - u), \qquad (13.22)$$

just a continuous version of $u_{n+1} = au_n(1 - u_n)$.

We can find a fixed point of the map by solving $G(u^*) = u^*$, for the logistic map:

$$au^*(1 - u^*) = u^* \longrightarrow u^* = 1 - \frac{1}{a}. \qquad (13.23)$$

We'll require that $u \in [0, 1]$ to limit our focus, so the parameter $a$ must have $a \geq 1$. Let's examine the stability of this fixed point – are values of $u_n$ that are close to $u^*$ attracted towards $u^*$ or repelled? The derivative of $G(u)$ is

$$G'(u) = a - 2au \longrightarrow G'(u^*) = 2 - a \qquad (13.24)$$

and the stability requirement is $|G'(u^*)| < 1$, or here, $|2 - a| < 1$ giving a range $1 < a < 3$. For this range of values for the parameter $a$, $u^* = 1 - 1/a$ is a stable fixed point, and we expect solutions starting at any $u_0$ to limit to this value.

Graphically, we can see the behavior of the fixed points as a function of $a$ by looking at a plot of $u_{n+1}$ vs. $u_n$ – in Figure 13.3, we see the intersection of the line $u_{n+1} = u_n$ (representing a period one fixed point) with $G(u_n)$ for $a = 1, 2$, and 3. In all three cases, there is a stable fixed point at $u_n = 0$, but for $a = 1$, that is the *only* intersection. At $a = 2$, there is an intersection at $u_n = 1/2$, corresponding to $1 - 1/a$, and that is stable, meaning graphically that the slope (derivative) of $G(u)$ there is less than one. For $a = 3$, there is still an intersection, but the slope is steeper, with $|G'(u)| = 1$.

We can also find period two solutions for the logistic map, $G(G(u^*)) = u^*$ becomes:

$$a^2(1 - u^*)u^*(1 - a(1 - u^*)u^*) = u^* \qquad (13.25)$$

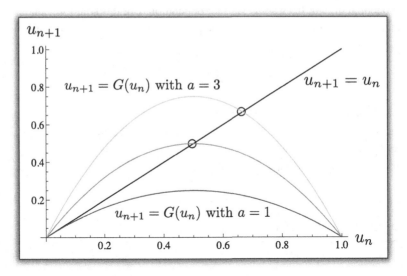

Figure 13.3 The period one fixed points for the logistic map are the points at which $u_{n+1} = G(u_n)$ and $u_{n+1} = u_n$ intersect (circled above, except for $u_n = 0$). For the logistic map (13.18), there are no fixed points (other than zero) for $a = 1$, and there is one each for $a = 2$ and $a = 3$.

so that

$$u^* = \frac{1}{2a}\left(1 + a \pm \sqrt{-3 - 2a + a^2}\right). \tag{13.26}$$

The plot analagous to Figure 13.3 is shown in Figure 13.4 for $a = 4$. There are three intersections of $u_{n+2} = G(G(u_n))$ with $u_{n+2} = u_n$. One of them is just the period one fixed point at $1 - 1/a$ – that of course has the property that $u_{n+2} = u_n$, since $u_{n+1} = u_n$ for it. The period one fixed point is unstable, since $a > 3$ (this is clear from the slope at the fixed point in Figure 13.4). There is also a pair representing the two solutions to (13.26), and we can have sequences that flip back and forth between these two, with $u_{n+2} = u_n$ for $n$ even taking on one of the values, and the odd $n$ taking the other one.

What is interesting is that the period two solutions (13.26) can only exist for $a > 3$ (otherwise, the square root will return a complex number). In fact, additional periodic solutions occur for $a > 3$. In Figure 13.5, we see the values of the parameter $a$, and the corresponding periodic points (i.e. the limiting values of the sequence) – for $a < 3$, there is only a single fixed point, but at values slightly greater than three, we have two period two solutions. The point $a = 3$ is called a "pitchfork bifurcation," and another one occurs near 3.5. In fact, the bifurcations continue, with period four, eight, etc. solutions showing up. At a particular value, $a \approx 3.5699456$,

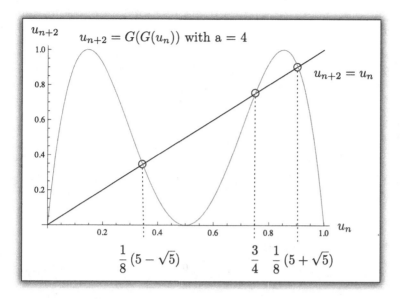

Figure 13.4 The period two fixed points for the logistic map are the points at which $u_{n+2} = G(G(u_n))$ and $u_{n+2} = u_n$ intersect.

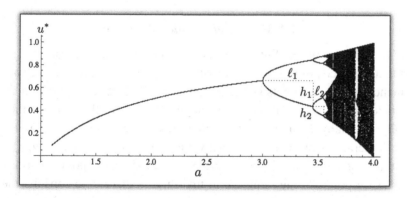

Figure 13.5 Period doubling in the logistic map – the "final" values of $u_n$ are shown for various values of $a$.

the sequence $\{u_n\}_{n=1}^{\infty}$ is not periodic at all; the values that come out of the logistic map are, apparently, random. These aperiodic solutions are referred to as "chaotic."

A picture of some of the actual sequences, generated for period one, two and four solutions is shown in Figure 13.6. We can also see a parameter choice that is in the chaotic region – in this case, the sequence of $u_n$ looks more or less random, and does not repeat.

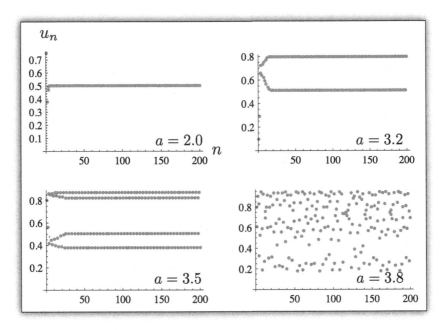

Figure 13.6 The logistic map with period one, two and four solutions – on the bottom right is a chaotic solution with $a = 3.8$.

### 13.2.2 Making bifurcation diagrams

In order to make a plot like Figure 13.5, called a bifurcation diagram, we rely on the fact that the fixed point solutions (of whatever period) are the limiting values of the sequence $\{u_n\}_{n=0}^{\infty}$ as $n$ gets large. Since there are stable fixed points, it doesn't really matter where we start, $u_0$ can be anything, and we'll end up at the fixed points. Operationally, then, we start at some location (choose a value for $u_0$), and generate $N$ iterations of the map, so that we have the finite sequence $\{u_n\}_{n=0}^{N}$. We assume that $N$ is large enough that we have the periodic fixed point solutions, and take the last $m$ points of the sequence to find these – that is, we consider the first $N - m$ values to be "transients," during which the sequence settles down. Now we just have to identify, in the set $\{u_n\}_{n=(N-m)+1}^{N}$, the number of unique values for $u_n$ we have – if we have two period two fixed points, then $u_n$ will oscillate between the two values in (13.26), for example. If there are four period four solutions, we'll get four unique values appearing over and over.

There are obvious limitations – we can obtain, at best, solutions with period $k = m$, and it is not clear what these represent – it could be that we have $m$ different values that start over at $m + 1$, or we could be in a chaotic regime for the parameter $a$, in which case the values would never repeat, but we wouldn't know that since we only have $m$ values. In order to make a crude bifurcation diagram, all we need to supply is the sequence $\{u_n\}_{n=0}^N$, and the number of values $m$ to take from the end, then we scan through the list and identify "unique" values for the $u_n$ (unique up to tolerance $\epsilon$, yet another parameter). In Implementation 13.1, we can see this process – the function $\texttt{FindPeriods}$ uses the $\texttt{Mathematica}$ command $\texttt{Union}$ to return the unique elements in a set. If we are working with the logistic map, then we generate the sequence $\{u_n\}_{n=0}^N$ for a variety of values of $a$, and plot, for each $a$, the periodic values that we find to generate Figure 13.5.

---

**Implementation 13.1** Finding periodic solutions

---

```
FindPeriods[inu_, m_, eps_] := Module[{Ns, udata, retvals},
   Ns = Length[inu];
   udata = Table[inu[[j]], {j, Ns - m + 1, Ns}];
   retvals = Union[udata, SameTest -> (Abs[#1 - #2] <= eps &)];
   Return[retvals];
   ]
```

---

## 13.3 Characterization of chaos

It is clear, visually, and somewhat clear numerically, that the four cases in Figure 13.6 represent two very different types of behavior. In the periodic solutions, we just cycle through a set of numbers. In the chaotic solution, there is no repetition, which we can think of as a solution with infinite period. There are a couple of analytical tools that can be used to quantify these observations.

### *13.3.1 Period doubling*

The ratio of successive locations of bifurcation points (in $a$ space) for the logistic map converges to a fixed number called the Feigenbaum number (for a good description, see [43]), denoted $\delta$.[2] Take the parameter $a_n$ to be the location of the $n$th bifurcation (so that $a_1 = 3$, the first point at which we obtain period doubling, $a_2 \approx 3.45$, etc.), then the distance between successive bifurcations is $\ell_n = a_{n+1} - a_n$ and the definition of $\delta$ is:

$$\delta \equiv \lim_{n \to \infty} \frac{a_n - a_{n-1}}{a_{n+1} - a_n} = \lim_{n \to \infty} \frac{\ell_{n-1}}{\ell_n} \approx 4.66920161. \tag{13.27}$$

In generating the coarse bifurcation plot in Figure 13.5, the measured values for the first three bifurcations are:

$$a_1 \approx 2.996 \qquad a_2 \approx 3.448 \qquad a_3 \approx 3.545 \tag{13.28}$$

---

[2] In fact, this number is a constant, like $\pi$ or $e$ – it shows up in the period doubling of any "unimodular" map. Unimodular is just a fancy way to say that $G(u)$ vanishes at $u = 0, 1$ (or whatever the relevant endpoints are), and has a single maximum on the interval $u \in [0, 1]$.

so the first approximation we would make to $\delta$ is:

$$\delta \approx \frac{a_2 - a_1}{a_3 - a_2} = \frac{\ell_1}{\ell_2} \approx 4.6598. \tag{13.29}$$

In addition to the horizontal spacings between bifurcations, we can also characterize the ratio of vertical spacings – referring again to Figure 13.5, the ratio of the heights converges to a number $\alpha$ – for $h_n = u^*_{n+1} - u^*_n$, we have

$$\alpha \equiv \lim_{n \to \infty} \frac{u^*_n - u^*_{n-1}}{u^*_{n+1} - u^*_n} = \lim_{n \to \infty} \frac{h_{n-1}}{h_n} \approx 2.50290787. \tag{13.30}$$

Again, using the approximation from the production of the plot above, we get:

$$h_1 = 0.44 - 0.666 \approx -0.226 \quad h_2 = 0.363 - 0.44 \approx -0.077 \tag{13.31}$$

so our first approximation to $\alpha$ is:

$$\alpha \approx \frac{h_1}{h_2} \approx 2.94 \tag{13.32}$$

These approximations, to $\delta$ and $\alpha$ are coarse – they come from our bifurcation diagram Figure 13.5, which we created with a fixed step size in $a$. That limits the resolution of the Feigenbaum numbers by limiting our ability to accurately pinpoint the value of $a$ at a doubling. You'll use your maximization routine to make better estimates of these fundamental constants in Problem 13.17.

### 13.3.2 Initial conditions

In a non-chaotic region, starting at $u_0$ or at $u_0 + \epsilon$ for $\epsilon$ small will lead to similar "endpoints" – that is, we expect $u_\infty$ to be the same for starting points that are close together. In the case of the single fixed point found for $a < 3$, if we start at $u_0 = u^* = 1 - 1/a$, then we remain there. If we start nearby, at $u_0 = u^* + \epsilon$, then we are driven towards the fixed point (since it is stable). In fact, you can start almost anywhere, and eventually the sequence will enter the vicinity of $u^*$, at which point you'll end up at $u^*$. In Figure 13.7, we see the first ten iterations of the logistic map for initial values ranging from $u_0 = 0.1 \longrightarrow 0.9$ (the parameter $a = 2.5$, so the fixed point is at $u^* = 0.6$). In each case, the solution settles down quickly to the fixed point. Compare that behavior to Figure 13.8, where we take the same initial values, but $a = 3.8$, where we are in a chaotic regime. There is no obvious relation between the first ten iterates of the map in this case; starting from different values of $u_0$ leads to different sequences.

We can quantify the difference in endpoint given a difference in starting point – denote the $n$th application of $G(u)$ to $u_0$ via: $G^{(n)}(u_0)$, then starting from $u_0$ versus

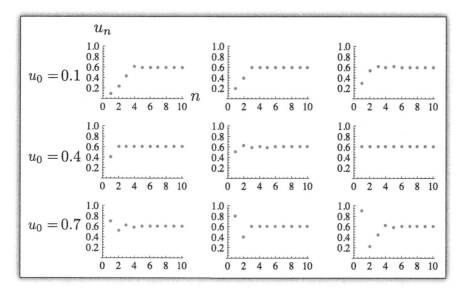

Figure 13.7 For $a = 2.5$, the fixed point is at $u^* = 0.6$ – the plots show the route to the fixed point for initial values $u_0$ ranging from 0.1 to 0.9 in steps of 0.1 (in each row, $u_0$ increases horizontally).

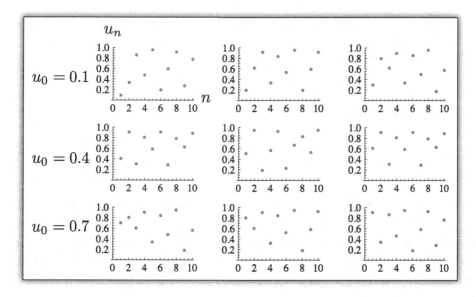

Figure 13.8 In the chaotic regime, with $a = 3.8$. The first ten points are not clearly related, even though we are taking steps of 0.1 in $u_0$ from 0.1 to 0.9.

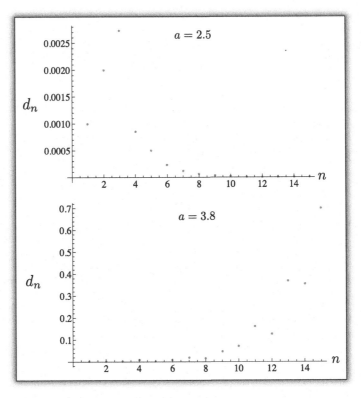

Figure 13.9 Distance at $n$th step, $d_n$, from (13.33) for the logistic map – we use $u_0 = 0.1$, $u_0 + \epsilon = 0.101$ and pick $a = 2.5$ and $a = 3.8$ as representative "non-chaotic" and "chaotic" regimes.

$u_0 + \epsilon$ leads to a difference:

$$d_n \equiv |G^{(n)}(u_0 + \epsilon) - G^{(n)}(u_0)| \tag{13.33}$$

at the $n$th step. In Figure 13.9, we can see the behavior of $d_n$ in the chaotic ($a = 3.8$) and fixed ($a = 2.5$) regimes. In both cases, we start at $u_0 = 0.1$ and use $\epsilon = 0.001$. In the non-chaotic case, there is (roughly) exponential decay to zero as both solutions approach the fixed point. In the chaotic case, the difference grows exponentially (keeping in mind that $u_n \in [0, 1]$, so there is a maximum distance of $\approx 1$ that can be achieved).

To capture the behavior of $d_n$ as $n \longrightarrow \infty$, we assume that $d_n = \epsilon e^{n\lambda}$ and try to put a bound on $\lambda$. This parameter is called the "Lyapunov exponent" and tells us how two initially nearby solutions diverge as a function of $n$ by characterizing the exponential growth or decay of $d_n$. Solving the assumed form of $d_n$ for $\lambda$ gives:

$$\lambda = \frac{1}{n} \log \left[ \frac{|G^{(n)}(u_0 + \epsilon) - G^{(n)}(u_0)|}{\epsilon} \right] \approx \frac{1}{n} \log \left( \left| \frac{d}{du} [G^{(n)}(u)] \right| \right) \Bigg|_{u=u_0}. \quad (13.34)$$

To evaluate the derivative of the iterated map, consider the case $n = 2$:

$$\frac{d}{du} G^{(2)}(u)|_{u=u_0} = \frac{d}{du} (G(G(u_0))) |_{u=u_0} = G'(G(u_0))G'(u_0) \quad (13.35)$$

using the chain rule. But $G(u_0) = u_1$, so

$$\frac{d}{du} G^{(2)}(u)|_{u=u_0} = G'(u_1)G'(u_0), \quad (13.36)$$

or in general:

$$\frac{d}{du} G^{(n)}(u)|_{u=u_0} = \prod_{j=0}^{n-1} G'(u_j). \quad (13.37)$$

Putting this relation in (13.34) gives

$$\boxed{\lambda = \frac{1}{n} \sum_{j=0}^{n-1} \log(|G'(u_j)|)} \quad (13.38)$$

using $\log(a) + \log(b) = \log(ab)$ and re-introducing the absolute values around $G'(u_j)$. This expression is defined for any finite number of steps, but when the limit as $n \longrightarrow \infty$ exists, we take the limit of (13.38) to define the Lyapunov exponent (finite truncations then represent approximations).

Taking the starting point to be $u_0 = 0.1$ (this won't matter much), we calculate the Lyapunov exponent defined above for the logistic map using $a = 2.5$, and get $\lambda \approx -0.692944$ (less than zero, corresponding to exponential decay). We can plot the function $\lambda n + A$ for constant $A$ and compare with the computed $\log(d_n)$; that is shown in Figure 13.10, where the slope of the data and line are quite close after transients have died off (i.e. starting at $n \approx 5$). If we calculate the Lyapunov exponent for $a = 3.8$, we get $\lambda \approx 0.430847$ – this is larger than zero and indicates that the chaotic solution has divergent $d_n$.

The Lyapunov exponent gives us an $n$-scale for the exponential divergence of initially nearby points under the influence of a map. It can be used as a test for chaos ($\lambda > 0$), and provides a time scale, in numerical ODE work, of validity for a solution (weather prediction, for example, cannot be performed reliably for more than a few days). Notice that our definition relies on an initial point $u_0$ – we can define the *average* Lyapunov exponent to be the average over "all" initial values. In higher dimension, the value of $\lambda$ becomes more difficult to compute, although it is defined in a manner similar to the one-dimensional case.

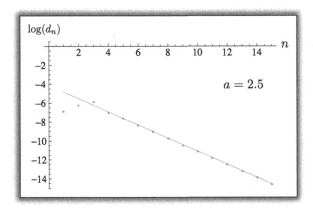

Figure 13.10 The points are $\{\log(d_n)\}_{n=1}^{15}$ for $a = 2.5$, $u_0 = 0.1$ and $u_0 + \epsilon = 0.101$ – the solid curve is $\lambda n + A$ where $\lambda = -0.692944$ is the Lyapunov exponent calculated for the logistic map with $a = 2.5$. The constant $A$ has been chosen so that the line lies on top of (most of) the data for easy slope comparison.

## 13.4 Ordinary differential equations

There is a connection between the discrete maps we have been thinking about and ordinary differential equations. Aside from the fact that we can discretize an ODE, and then define a map based on both the original ODE form and the numerical method, we can analyze differential equations as a sort of continuous form of $u_{n+1} = G(u_n)$ by letting the index $n$ become a continuous parameter (like time).

### 13.4.1 Linear problems

Starting with a simple linear system, suppose we have:

$$\frac{d}{dt} \underbrace{\begin{pmatrix} x(t) \\ y(t) \end{pmatrix}}_{\equiv \mathbf{X}(t)} = \underbrace{\begin{pmatrix} a & b \\ b & c \end{pmatrix}}_{\equiv \mathbb{Q}} \begin{pmatrix} x(t) \\ y(t) \end{pmatrix}. \tag{13.39}$$

We know the exact solution to this pair from Chapter 8,

$$\mathbf{X}(t) = A e^{\alpha t} \mathbf{v}_1 + B e^{\beta t} \mathbf{v}_2 \tag{13.40}$$

where $\mathbf{v}_1$ and $\mathbf{v}_2$ are the eigenvectors of $\mathbb{Q}$ having eigenvalues $\alpha$ and $\beta$, respectively. Since $\mathbb{Q}$ is symmetric and (take it to be) real, the eigenvalues are real quantities. Now there is a fixed point at $\mathbf{X}(t) = 0$; if we start there, we stay there in time. But it is also clear that the origin or infinity are natural ending points of any starting value $\mathbf{X}(0)$. Suppose both $\alpha$ and $\beta$ are less than zero – then $\mathbf{X}(\infty) = 0$ for any $A$ and $B$, and the point 0 is a stable fixed point – points that start nearby tend to fall in. If $\alpha > 0$ and $\beta > 0$, then the origin is an unstable fixed point (if you start there, you stay there, but starting nearby still drives $\mathbf{X}(\infty) \longrightarrow \infty$). If one eigenvalue is positive, the other negative, the origin is a saddle point – we approach it along the eigenvector associated with the negative eigenvalue, and recede from it along the eigenvector associated with the positive eigenvalue.

The point here is that we can identify fixed points of ordinary differential equations, and characterize the behavior of solutions given only the eigenvectors and eigenvalues associated with the matrix appearing in the linearized problem. For nonlinear problems, we can identify fixed points (sometimes), and then perform a local analysis (linearizing the problem near the fixed points) to inform our understanding of the full nonlinear solutions. The approach is similar to the case for linear maps.

In two dimensions, we can have more than just fixed points, though – we can have fixed curves – that is, it may be the case that there is a solution in two dimensions that is stable and traces out a one-dimensional curve in $x - y$ space (an example is the curve shown in Figure 13.2 for the Hénon map).

### 13.4.2 The Lorenz equations

Lorenz used the following, coupled nonlinear set of ODEs as part of a simplified weather prediction model:

$$\frac{dx}{dt} = \sigma(y - x)$$

$$\frac{dy}{dt} = x(r - z) - y \qquad (13.41)$$

$$\frac{dz}{dt} = xy - bz$$

where $\sigma$, $r$, and $b$ are parameters set by physical considerations – the spatial designation given to the variables is for visualization purposes only (i.e. the variables here are not the $x$, $y$, and $z$ coordinates of anything). The parameters $\sigma = 10$ and $b = 8/3$ are fixed, and $r$ is allowed to vary.

Let's start by thinking about the fixed points of this ODE – we are looking for points that remain the same for all time – so that $x = x_0$, $y = y_0$ and $z = z_0$ – then the time derivatives are zero, and we can solve the above set algebraically for $\frac{dx}{dt} = \frac{dy}{dt} = \frac{dz}{dt} = 0$. It is clear that $x = y = z = 0$ is a fixed point, but there are two others:

$$\begin{aligned} x = -\sqrt{b(r - 1)} \quad & y = -\sqrt{b(r - 1)} \quad z = r - 1 \\ x = \sqrt{b(r - 1)} \quad & y = \sqrt{b(r - 1)} \quad z = r - 1. \end{aligned} \qquad (13.42)$$

We can analyze the behavior of these fixed points by looking at the linearization
of the Lorenz equation about them. Around any point $(x_0, y_0, z_0)$, small deviations
obey the linearized equation:

$$\frac{d}{dt} \begin{pmatrix} x \\ y \\ z \end{pmatrix} = \underbrace{\begin{pmatrix} -\sigma & \sigma & 0 \\ r - z_0 & -1 & -x_0 \\ y_0 & x_0 & -b \end{pmatrix}}_{\equiv \mathbb{H}} \begin{pmatrix} x \\ y \\ z \end{pmatrix}. \tag{13.43}$$

Now, if the eigenvalues of $\mathbb{H}$ have *real* part that is negative, the fixed point is
attractive (think of the three-dimensional version of (13.40) – if the eigenvalues
have negative real part, we get exponential decay towards the equilibrium configuration: the fixed point) – for our parameter choice, the top fixed point in (13.42)
is attractive for values of $r$ below $r \approx 24.7$. For $r = 28$, the Lorenz equations are
chaotic.

   We can see the sensitive dependence on initial conditions associated with the
chaotic Lorenz attractor (the set of points along which solutions will cycle over
time) by starting at two nearby points. Take $x(0) = 2 + \epsilon$, $y(0) = z(0) = 5$, we'll
start with the non-chaotic value, $r = 20$, and compare the difference in the two
trajectories as time evolves – that is, we'll take the points generated for each initial
condition ($\epsilon = 0$ and $\epsilon \neq 0$), and find the distance between two solution vectors
for all times. The solution for the $\epsilon = 0$ case is shown for $t = 0 \longrightarrow 100$ on the
bottom in Figure 13.11 – the distance between $\mathbf{r}(t)$ (with $\epsilon = 0$) and $\mathbf{r}_\epsilon(t)$ with
$\epsilon = 1.0 \times 10^{-7}$ is shown on the top – notice that there is a "transient" region,
where the differences grow, but then decay, so that by $t \approx 40$, the two solutions are
"the same" (both attracted to the stable fixed point).

   If we take $r = 28$, in the chaotic regime, and generate the same plot, we get
very different behavior – the corresponding trajectory and residual magnitude are
shown in Figure 13.12. Notice that in this chaotic case, the distance between
the perturbed and unperturbed trajectories oscillates, and is bounded – if we
look at the trajectories themselves, we see that the solution spends most of its
time in one of two lobes – unlike the $r = 20$ case, the solution continually hops
back and forth (since the fixed points at the centers of the lobes are not attractive) – so there is a maximum distance apart that two initially nearby solutions
can be. That is the reason for the bounded difference magnitude in the top plot
of Figure 13.12.

   The chaotic case is interesting in that solutions will continue to cycle between the
two lobes effectively forever, without being attracted to a fixed point, and this causes
a fill-in of the (roughly) planar lobe regions. Unlike a normal "attractor",[3] which

---

[3] Think of the circle in phase space associated with the harmonic oscillator.

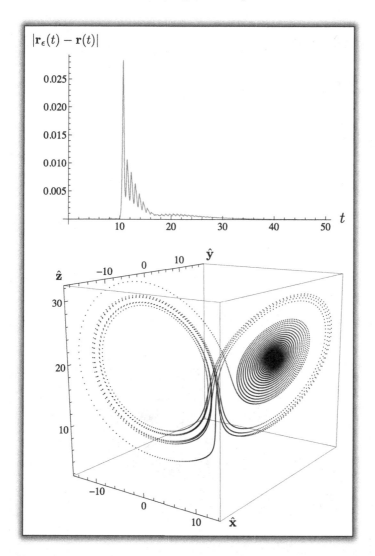

Figure 13.11 Example solution (bottom) and sensitivity to initial conditions (top) for $r = 20$.

is typically a one-dimensional curve, these plots are not clearly one dimensional, nor are they clearly two dimensional. And so, we can ask, what is the dimension of this (potentially) "fractal" object. An example for the $r = 28$ case, with the lobes projected into the $x - z$ plane, is shown in Figure 13.13 – we show the first 2000, 4000, 6000 . . . time steps – the trajectory starts off one dimensional (meaning that we are tracing out a curve), and ends looking more two dimensional (the shape is filled in, it appears to have a continuous "area").

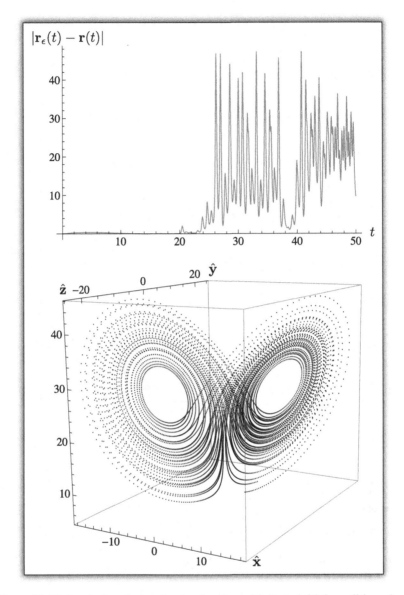

Figure 13.12 Example solution (bottom) and sensitivity to initial conditions (top) for $r = 28$.

## 13.5 Fractals and dimension

What is dimension? From a classical mechanics point of view, dimension is the number of real numbers required to specify a point – in one dimension, we need an origin, and a positive or negative real number to identify any point. In two

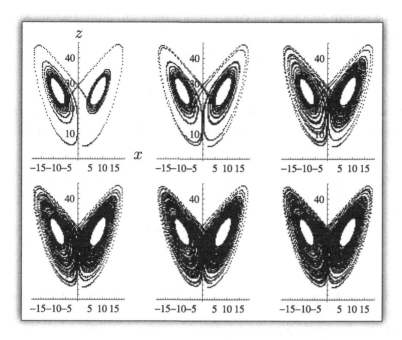

Figure 13.13 Lorenz attractor in the chaotic regime – here we show the values in the $x$–$z$ plane for every 2000 steps of the solution (starting at 2000 and going to 12 000).

dimensions, the Cartesian $x$ and $y$ will uniquely provide a label for all points, etc. So on the one hand, it is easy to determine the dimensionality of a curve embedded in some higher "dimension" – we just count the number of real numbers needed to label a point in the higher-dimensional space, and subtract the number of unique constraints defining the curve. Motion, in three dimensions, follows a one-dimensional path according to Newton's second law, and finding that path is the program of Newtonian dynamics. But there are "pathological" cases that are not the normal dynamics paths – we could, for example, have a bounded curve in two dimensions that has infinite length. There are, similarly, geometrical objects in three dimensions with infinite area that occupy a finite volume (meaning that we can enscribe such shapes in a sphere of finite and definite radius). What dimension should we associate with such objects?

There is another way to define dimension, and we can sketch it briefly en route to its formal definition. If you take a line (a one-dimensional object) of length $L$, you can cover it with line segments of length $d\ell$ – the number of segments you need, end-to-end, is $N = L/d\ell$. A square of area $L^2$ in two dimensions can be covered by small squares of infinitesimal area $d\ell^2$ – there, you need $N = L^2/d\ell^2$

boxes. The pattern continues, to cover an $L^d$ volume in $d$ dimensions with small volumes of size $d\ell^d$, you need:

$$N = \frac{L^d}{d\ell^d} \tag{13.44}$$

box-like objects. This formula can be used to define the "box-counting dimension" (or "capacity dimension") of an object (see [3, 5, 43]). Solving (13.44) for $d$ gives

$$d = \frac{\log N}{\log L + \log\left(\frac{1}{d\ell}\right)}, \tag{13.45}$$

and if we take the limit $d\ell \longrightarrow 0$, we have:

$$\boxed{d_c = \lim_{d\ell \to 0} \frac{\log N}{\log\left(\frac{1}{d\ell}\right)}.} \tag{13.46}$$

This definition matches our physical experience for $D = 1, 2, 3$ etc. but is not confined to the integers.

As an example of an object with fractal dimension, we'll consider the famous Mandelbrot set. This is a set inspired by a quadratic map – for complex $c$, we generate:

$$z_{n+1} = z_n^2 + c, \tag{13.47}$$

with $z_0 = c$, then the Mandelbrot set is

$$\mathbb{M} = \{c : |z_\infty| < \infty\}, \tag{13.48}$$

that is: "the set of all $c$ such that the absolute value of $z_\infty$ exists (is not infinity)." The pretty pictures come from plotting the set in the complex plane. Black points are in the set, any other color is not.[4]

It should be familiar, from nights in high school, that fractals are "self-similar," and the non-integer dimension of fractals is, to a certain extent, the hallmark of this self-similarity. Just for funsies, we pick a segment of Figure 13.14 and zoom in – notice that the picture, shamelessly displayed in Figure 13.15, looks like a rotated version of the original.

---

[4] The grayscale in Figure 13.14 and Figure 13.15 has been introduced by taking the log of the absolute value of the thirtieth iterate, $z_{30}$ – that then defines a grayscale value for the point. There are boundaries (stripes) in those figures that are, I hope clearly, not part of the Mandelbrot set, but are very dark, owing to the grayscale mapping.

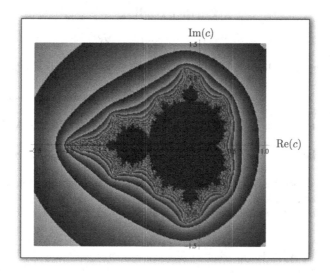

Figure 13.14  The Mandelbrot set.

Figure 13.15  A zoom-in from Figure 13.14.

## Further reading

1. Arfken, George B. & Hans J. Weber. *Mathematical Methods for Physicists*. Academic Press, 2001.
2. Baker, Gregory L. & Jerry P. Gollub. *Chaotic Dynamics: An Introduction*. Cambridge University Press, 1990.
3. Bender, Carl M. & Steven A. Orszag. *Advanced Mathematical Methods for Scientists and Engineers*. McGraw-Hill Book Company, 1978.
4. Goldstein, Herbert, Charles Poole, & John Safko. *Classical Mechanics*. Addison-Wesley, 2002.
5. Gutzwiller, Martin C. *Chaos in Classical and Quantum Mechanics*. Springer-Verlag, 1990.
6. Infeld, Eryk & George Rowlands. *Nonlinear Waves, Solitons and Chaos*. Cambridge University Press, 2000.
7. Pang, Tao. *An Introduction to Computational Physics*. Cambridge University Press, 2006.
8. Strogatz, Steven H. *Nonlinear Dynamics and Chaos*. Addison-Wesley, 1994.
9. Stuart, A. M. & A. R. Humphries. *Dynamical Systems and Numerical Analysis*. Cambridge University Press, 1998.

## Problems

### Problem 13.1

The cubic map, $u_{n+1} = 3u_n - u_n^3$ has fixed points at $\pm\sqrt{2}$ – are these stable or unstable (i.e. does $u_0 = \pm\sqrt{2} + \epsilon$ for small epsilon tend to go towards or away from $\pm\sqrt{2}$)?

### Problem 13.2

Find two additional points in the basin of attraction of zero for the cubic map (13.7) – define the relevant polynomial, and then use your root finding routine (bisection) to find a numerical approximation – work in $u_0 > 0$, and be sure not to reproduce the two we already know about, $u_0 = 0$ and $u_0 = \pm\sqrt{3}$.

### Problem 13.3

In Section 13.2.1, we associated the logistic map, $u_{n+1} = au_n(1 - u_n)$ with the ODE: $\frac{du}{dt} = \alpha u(1 - u)$, but there are a few steps missing from that association. Taking $u_n \equiv u(t_n)$ for continuous $u(t)$, Taylor expand $u_{n+1} \equiv u(t_{n+1}) = u(t_n + \Delta t)$, replace the derivative there using (13.22), and let $u = \bar{u}q$, $t = \bar{t}s$ for dimensionless $q$ and $s$ – what must $\bar{u}$ and $\bar{t}$ be in order to get the map: $q_{n+1} = aq_n(1 - q_n)$?

### Problem 13.4

For the chaotic systems we have studied, there is a notion of aperiodicity that is usually taken as part of the definition of a chaotic system (together with a positive Lyapunov exponent). From that point of view, is the solution to $\dot{x} = x$ "chaotic"?

## Problem 13.5

The "standard map" shows up in a variety of physical applications (see, for example [21]), but it comes most naturally from the equations of motion for a kicked rotor (rigid rod):

$$\frac{dp}{dt} = k \sin\theta \sum_n \delta(t - nT)$$

$$\frac{d\theta}{dt} = \frac{p}{I},$$

(13.49)

where $I$ is the moment of inertia of the rod and $\theta$ is the angle the rod makes w.r.t. vertical. A vertical force is applied instantaneously at times $t = nT$ for integer $n$ – the first equation renders this force in torque form appropriate for the rigid rod. The second equation then tells us the evolution of the angle $\theta$ in terms of the angular momentum $p$. Consider $p$ after the $j$th application of the vertical force – it has some constant value, call it $p_j$ that does not change until the $(j + 1)$st application. Meanwhile, $\theta$ evolves linearly during this time. Just before the $j + 1$ kick, $\theta = \theta_j + p_j T / I$ where $\theta_j$ is the value of $\theta_j$ just after the $j$th kick. So just after $t = (j + 1)T$, we have $\theta_{j+1} = \theta_j + p_j T / I$. By integrating $\frac{dp}{dt}$ across the application of the force, find the value of $p_{j+1}$ after the $(j + 1)$st kick (at which point it is constant) in terms of $p_j$ and $\theta_{j+1}$ – this, together with the $\theta_j$ update define the standard map.

## Problem 13.6

Prove (13.37) by induction – that is, assume the equation holds for $n$, then show that the equation holds for $n + 1$ – we already established the formula for $n = 2$, so the checking of a concrete case is done.

## Problem 13.7

What is the Lyapunov exponent for the linear map: $u_{n+1} = a u_n + b$ for $a$, $b$ real (and positive)? For what range of values of $a$ is the exponent negative? Does this make sense in terms of exponential growth/decay of nearby trajectories?

## Problem 13.8

Find the eigenvalues of the matrix in (13.43) given the parameters $\sigma = 10$, $b = 8/3$, using the second fixed point in (13.42), and give the last value of $r$ for which that fixed point is stable (use the `Eigenvalues` routine, and plot the real part of the resulting functions to get a coarse picture, then use bisection).

## Problem 13.9

The Cantor "middle thirds" set is constructed by taking a line from $0 \longrightarrow 1$, and extracting the middle third, leaving two lines of length $1/3$, then taking out the middle third of each of these, etc. The set is defined by the limit as the number of iterations of this process goes to infinity. A picture of the construction is shown in Figure 13.16.

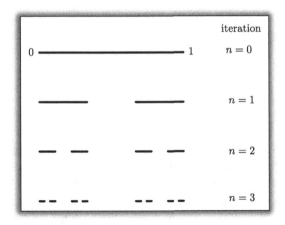

Figure 13.16 Constructing the Cantor set – at each iteration level $n$, we extract the middle third of all line segments.

At the $n$th level, how many line segments are there? How long is each segment? Using these, compute the box-counting dimension of the Cantor set.

**Problem 13.10**
Express $i^i$ in terms of $e$ and $\pi$.

## Lab problems

**Problem 13.11**
Test your stability prediction for the cubic map, from Problem 13.1: Write a function that returns the hundredth iterate of the map given $u_0$ and plot the result for $u_0 = \sqrt{2} + \epsilon$ with $\epsilon = 10^{-10}$ to $10^{-8}$ in steps of $\Delta\epsilon = 10^{-11}$.

**Problem 13.12**
A damped, driven pendulum can be described by the equation of motion (see, for example [39]):

$$\ddot{\theta} = -\frac{g}{L}\sin\theta - \alpha\dot{\theta} + \beta\cos(\omega t), \tag{13.50}$$

where we start with (6.17) and add a damping term $\sim -\dot{\theta}$ and time-dependent forcing function by hand with constants $\alpha$ and $\beta$ setting the strength of these terms. If we set $t = \sqrt{L/g}\,s$ for dimensionless $s$, then

$$\frac{d^2\theta}{ds^2} = -\sin\theta - \underbrace{\alpha\sqrt{\frac{L}{g}}}_{\equiv A}\frac{d\theta}{ds} + \underbrace{\beta\frac{L}{g}}_{\equiv B}\cos\left(\omega\sqrt{\frac{L}{g}}\,s\right), \tag{13.51}$$

and define the dimensionless driving frequency $\Omega \equiv \omega\sqrt{\frac{L}{g}}$.

(a) For $A = 1/(0.5)$, $B = 1.5$ and $\Omega = 2/3$, take $\Delta t = \frac{\Omega}{64}$, and run your RK4 ODE solver for $2^{16}$ steps starting from $\theta(0) = \frac{1}{2}\pi$ and $\dot{\theta} = 0$. Taking just the second half of the numerical solution (to allow transients to decay), plot the angle $\theta$ as a function of dimensionless time $s$. Compute the power spectrum and plot the log of the power spectrum, focusing on the area around the driving frequency $\Omega$.

(b) Re-do part (a) with $A = 1/3$, in the chaotic region for this problem [4].

## Problem 13.13

Write a function that iterates the "standard map" (your solution to Problem 13.5) for $T/I = 1$ (be careful to keep $\theta_j \in [0, 2\pi)$) – given $k$, and an initial $\theta_0$ and $p_0$ return a list of pairs: $\{\theta_j, p_j\}_{j=0}^N$ for user-specified $N$. Using your function:

(a) Plot the trajectory ($p$ vs. $\theta$) for $\theta_0 = 1$, $p_0 = 2\pi$ and $k = 0$ with $N = 100$. Is this what you expected?

(b) Plot trajectories for $\theta_0 = 0.1 \longrightarrow 2\pi$ in steps of 0.1, all with $p_0 = 2\pi$ and $k = 0.1$, each with $N = 100$. Make a single plot of all the points from these trajectories.

(c) Make the same plot for $k = 1$, 1.5, 2, and 2.5 (all at $N = 100$).

(d) Plot the trajectory that starts at $\theta_0 = 0.1$, $p_0 = 2\pi$ with $k = 20$ (use $N = 10\,000$).

## Problem 13.14

As another example of sensitive parameter dependence, here is a driven, nonlinear, oscillatory system that can exhibit "training" behavior (from [43]):

$$\dot{\theta}(t) = \omega + A \sin(\Theta(t) - \theta(t)), \tag{13.52}$$

where $\Theta(t) = \Omega t$ is an angle with frequency $\Omega$, and $\theta$ is an angle that has, for $A = 0$, natural frequency $\omega$. For certain values of $A$, we can force $\theta \longrightarrow \Theta + \phi$, i.e. we can make $\theta(t)$ have the same frequency as the driving $\Omega$, but with a phase offset $\phi$. For other values of $A$, no "training" occurs. Defining $\psi \equiv \Theta - \theta$, and introducing $\tau = At$, $\mu = (\Omega - \omega)/A$, we can write the equation as: $\psi'(\tau) = \mu - \sin(\psi)$. Solve this equation numerically (using RK, for example) for $\mu = 0$, $\mu = 0.9$ and $\mu = 1.1$ and $\psi(0) = 0.1$. What behaviors do you see in these three cases?

## Problem 13.15

Construct a bifurcation diagram for the sine map:

$$u_{n+1} = a \sin(\pi u_n) \tag{13.53}$$

for $a = 0.7$ to 0.86 in steps of 0.0001 (use the last ten percent of the data to identify periodic solutions). Find approximate values for the first three bifurcation points, $a_1$, $a_2$, and $a_3$ as in Section 13.3, and construct the approximation to the Feigenbaum number $\delta$ analogous to (13.29).

**Problem 13.16**

We found the Lyapunov exponent for the logistic map using a single value of the parameter $a$ – compute the Lyapunov exponent for values of $a$ lying between 2.9 and 4.0 using steps of size 0.01 – take 20 000 steps of the map and construct a plot of $\lambda(a)$. At what value, roughly, does the exponent become positive? What happens to $\lambda$ at the bifurcation points? What does it mean for the exponent to have negative "spikes" while otherwise positive (as occurs in the region between 3.8 and 4.0)? Hint: line up your Lyapunov plot with Figure 13.5.

**Problem 13.17**

Write a function that returns the Lyapunov exponent for the sine map, using a random starting location $u_0 \in [0, 1]$ and 10 000 steps, given a parameter $a$. Make a rough plot of $\lambda(a)$ for $a \in [0.6, 1]$. Using your golden section maximizer from Chapter 12, find the first five bifurcation points (you will need to make additional plots of $\lambda(a)$ at different resolutions to figure out good choices for the initial bracketing), use $\epsilon = 10^{-4}$ in the maximizer. For these first five bifurcations, you can compute the first three approximations to $\delta$ from (13.27). You should be able to achieve better accuracy than the bifurcation approach, since there the "grid" in $a$ that you use determines the resolution of the bifurcation point.

**Problem 13.18**

With your RK4 solver, solve the Lorenz equations (13.41) with $\sigma = 10$, $b = 8/3$:

(a) using $r = 20$, and starting from $x = -10$, $y = 10$, $z = 15$. Note that as in Figure 13.11, you will end up honing in on the bottom fixed point in (13.42) – what must you do to approach the top one?

(b) using $r = 20$ and starting from $x = 0$, $y = 0$, $z = 1$, and again from $x = 0$, $y = 10^{-9}$, $z = 1$ – try this case one last time using $r = 28$.

In all cases, you can use $\Delta t = 0.01$ and run for 4000 steps to get a rough picture.

**Problem 13.19**

A "Multibrot" set is defined to be the set of all $c$ such that the map:

$$z_{n+1} = z_n^q + c \qquad z_0 = c \tag{13.54}$$

has absolute value that converges to a finite number, i.e. $|z_\infty| < \infty$ (for $q = 2$, we recover the Mandelbrot set). Construct the points $c$ in the complex plane that are in the Multibrot set with $q = 3$ – use a grid from $x = -1$ to 1 in 400 steps, and $y = -3$ to 3 in 400 steps to find points $c = x + iy$ that are in the set (you only need to run the map out to about ten iterations for a rough idea of the initial point's potential inclusion), and plot those points. The set is defined even for complex powers; $q = 2 + i$, for example, provides an interesting pattern.

# 14

# Neural networks

Coupled, nonlinear sets of ODEs, of the sort that describe most physical processes, do not always have well-behaved numerical solutions (as we saw in the last chapter). In addition, there is certain physical behavior that cannot be (or at any rate has not been) rendered into well-defined, deterministic, ODEs. Whatever the source, there are some problems where we have a large amount of data, but no good rule for generating output from input. Suppose we have an electromagnetic signal that is supposed to propagate through space to an antenna on the ground – as the signal enters the Earth's atmosphere . . . things happen. The signal is scattered by the atmosphere, for example, and the properties of the air that govern that scattering change as the light makes its way to an antenna on the Earth. Now there is nothing physically obscure about this process, and yet one can imagine that the particular model of the atmosphere plays a large role in taking us from an observed signal to properties of its source. If we had associated pairs of known source signals together with ground-based measurements of them, we could side-step the model-dependent portion of the problem, and simply estimate the signal given the measurement by comparing with the previously observed data.

Problems in which we have an incomplete, or unsophisticated physical model, but a lot of associated input–output pairs (from observation of known sources, in this case) can benefit from a neural network model. Neural networks provide a framework for nonlinear fitting where the fitting parameters are determined by "training data," that is, data where both the input and output are known. By fixing the parameters in the neural network model, we have a machine that takes input data and provides output data, where the underlying "model" is determined entirely by the correct examples we have provided.

As another example, take the price of a stock on a given day. We would like to predict the price of the stock a few days later. What physical model should we use?

351

There exist stochastic differential equations meant to model the stock market, but these can be overly simplistic, or based on information that is not robust (as it turns out, for example, you cannot use a fixed-rate mortgage default model to predict the occurrence of defaults for variable rate mortgages). One way to avoid biasing our predictions is to use the data (and only the data) itself. This is a classic case for neural networks; there is a vast amount of stock price data over a variety of time scales, plenty of data to train a neural network.

## 14.1 A neural network model

We will avoid the historical description of this numerical method (covered in [2, 26, 28]). That history clarifies the name for this nonlinear fitting procedure, and informs some of the choices we make, but is otherwise not particularly relevant to an understanding of the construction of a neural network.

### 14.1.1 A linear model

We must "provide correct input and output training data pairs" – what does that mean? And what are the internal parameters of a neural network that allow it to provide model-free predictions? We have seen over and over again that most physical data can be put into discrete vector form. The discretized solution to Newton's second law, for example, might take the form: $x_j \approx x(t_j)$ for $t_j = j\Delta t$, and we have $n$ data points, say. Or, we might have a set of decomposition coefficients from a Fourier transform, $\{P_j\}_{j=0}^{n-1}$, again, representable as a vector of quantities. So by "input" we mean a vector $\mathbf{x} \in \mathbb{R}^n$, and by output we mean a vector $\mathbf{z} \in \mathbb{R}^m$. Now, the most general, linear map between vectors in $\mathbb{R}^n$ and vectors in $\mathbb{R}^m$ can be represented by a matrix – so our linear neural network consists of a matrix $\mathbb{A} \in \mathbb{R}^{m \times n}$, with a priori unknown entries $A_{ij}$. The action of our linear model on an input vector $\mathbf{x}$ is:

$$\mathbf{z} = \mathbb{A}\mathbf{x}, \qquad (14.1)$$

so that $\mathbf{z}$ is an output vector, and $\mathbb{A}$ transforms an input vector into an output vector in an $\mathbb{A}$-dependent way.

How should we set the values of the entries of $\mathbb{A}$? Well, if we were given enough input and associated output pairs, we could construct the linear map that takes every input vector to an output vector.

**Constructing $\mathbb{A}$**

Suppose we are given the following associated input and output vector pairings:

$$\mathbf{x}^1 \doteq \begin{pmatrix} 1 \\ 0 \end{pmatrix} \qquad \mathbf{y}^1 \doteq \begin{pmatrix} 1 \\ 0 \end{pmatrix}$$

$$\mathbf{x}^2 \doteq \begin{pmatrix} 0 \\ 1 \end{pmatrix} \qquad \mathbf{y}^2 \doteq \begin{pmatrix} 1 \\ 1 \end{pmatrix} \tag{14.2}$$

$$\mathbf{x}^3 \doteq \begin{pmatrix} 1 \\ 1 \end{pmatrix} \qquad \mathbf{y}^3 \doteq \begin{pmatrix} 4 \\ 1 \end{pmatrix}.$$

We want to find a matrix $\mathbb{A}$ such that:

$$\mathbf{y}^1 = \mathbb{A}\mathbf{x}^1 \qquad \mathbf{y}^2 = \mathbb{A}\mathbf{x}^2 \qquad \mathbf{y}^3 = \mathbb{A}\mathbf{x}^3 \tag{14.3}$$

where the matrix $\mathbb{A}$ has four unknown entries:

$$\mathbb{A} \doteq \begin{pmatrix} a & b \\ c & d \end{pmatrix}. \tag{14.4}$$

We'll use the first two pairs, $\{\mathbf{x}^1, \mathbf{y}^1\}$ and $\{\mathbf{x}^2, \mathbf{y}^2\}$ to set $\{a, b, c, d\}$ as follows – demand that $\mathbf{y}^1 = \mathbb{A}\mathbf{x}^1$ and $\mathbf{y}^2 = \mathbb{A}\mathbf{x}^2$:

$$\mathbb{A}\mathbf{x}^1 \doteq \begin{pmatrix} a \\ c \end{pmatrix} = \begin{pmatrix} 1 \\ 0 \end{pmatrix}$$

$$\mathbb{A}\mathbf{x}^2 \doteq \begin{pmatrix} b \\ d \end{pmatrix} = \begin{pmatrix} 1 \\ 1 \end{pmatrix}, \tag{14.5}$$

from which we learn that

$$\mathbb{A} \doteq \begin{pmatrix} 1 & 1 \\ 0 & 1 \end{pmatrix} \tag{14.6}$$

and the values of the matrix $\mathbb{A}$ are fixed. But now, we have the problem that:

$$\mathbb{A}\mathbf{x}^3 \doteq \begin{pmatrix} 2 \\ 1 \end{pmatrix} \neq \begin{pmatrix} 4 \\ 1 \end{pmatrix} = \mathbf{y}^3. \tag{14.7}$$

We have turned our target idea into a concrete mathematical one – "training" the model refers to finding the entries of $\mathbb{A}$ that define the mapping from input to output. But as the above example shows, it is not clear how useful a linear map of this sort will be in modeling, for example, *non*-linear behaviour. Linear equations tend to have unique solutions; $x = 1$, for example, has only one solution. Nonlinear equations tend to have multiple solutions ($x^2 = 1$ has two solutions). The first step in generating a more robust model is to make the mapping from $\mathbf{x}$ to $\mathbf{z}$ nonlinear.

That will make the determination of the matrix entries more difficult, but will allow us to capture more interesting input–output relations.

### 14.1.2 A nonlinear model

Suppose we take our linear model with output $\mathbf{z} = \mathbb{A}\mathbf{x}$, and run it through a nonlinear function. Define the action of a function of a single variable, $g(p)$, on a vector by:

$$g(\mathbf{x}) \doteq \begin{pmatrix} g(x_1) \\ g(x_2) \\ \vdots \\ g(x_n) \end{pmatrix}, \tag{14.8}$$

then we could take a new output definition:

$$\mathbf{z} = g(\mathbb{A}\mathbf{x}) \tag{14.9}$$

for our neural network. Notice that if $g(p) = p$, the identity, we recover the linear map from the previous section. Depending on our choice of nonlinear function $g(p)$, we can satisfy all three of the input–output requirements of (14.2), for example. The introduction of the nonlinear function $g(p)$ gives us greater freedom.

---

**Constructing $\mathbb{A}$ redux**

Suppose we take (and this is highly tailored, of course):

$$g(p) = p^2, \tag{14.10}$$

then solving the nonlinear (quadratic) equations associated with $\mathbf{y}^1 = g(\mathbb{A}\mathbf{x}^1)$, $\mathbf{y}^2 = g(\mathbb{A}\mathbf{x}^2)$ and $\mathbf{y}^3 = g(\mathbb{A}\mathbf{x}^3)$ from (14.2) gives us four valid matrices (a lot better than the none we had before). They are:

$$\mathbb{A}_1 = \begin{pmatrix} -1 & -1 \\ 0 & -1 \end{pmatrix} \quad \mathbb{A}_2 = \begin{pmatrix} -1 & -1 \\ 0 & 1 \end{pmatrix}$$

$$\mathbb{A}_3 = \begin{pmatrix} 1 & 1 \\ 0 & -1 \end{pmatrix} \quad \mathbb{A}_4 = \begin{pmatrix} 1 & 1 \\ 0 & 1 \end{pmatrix}. \tag{14.11}$$

---

### 14.1.3 The sigmoid function

We will, from now on, make some basic assumptions about the nature of our input and output data, and motivate a choice for the function $g(p)$. There are a variety of such choices, and we are picking the simplest for reasons that we will outline. Our

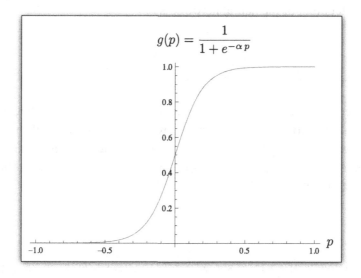

Figure 14.1 An example of $g(p)$ defined according to (14.12) – the value of $\alpha$ determines the steepness of the transition from 0 to 1. For this plot, $\alpha = 10$, giving a nonlinear transition for $p \in [-1, 1]$.

primary motivation comes from our desire to simplify and automate the process of setting the entries of the matrix $\mathbb{A}$.

We'll start by assuming that all entries in our output data are scaled so that they lie in $(0, 1)$, i.e. $z_j \in (0, 1)$ for $j = 1 \to m$. Our function $g(p)$ must then return values on $(0, 1)$, and, for reasons that will be clear in a moment, should be continuous and differentiable. It is easiest to use the sigmoid function:

$$g(p) = \frac{1}{1 + e^{-\alpha p}}. \tag{14.12}$$

An example of this function is shown in Figure 14.1 – it is $s$-shaped, defined for all values of $p$, and returns values that asymptotically approach 0 and 1. The parameter $\alpha$ sets the steepness of the transition from 0 to 1. We will set $\alpha$ based on the size of the input vector and the data it contains, so that we can probe the nonlinear transition of $g(p)$ from 0 to 1 (an alternative is to fix $\alpha = 1$, and then modify the input data encoding, and you can try that in Problem 14.1).

We can write the action of our neural network, defined with this $g(p)$ in terms of the summation implicit in the matrix-vector multiplication. That is, we have:

$$(\mathbb{A}\mathbf{x})_i = \sum_{j=1}^{n} A_{ij} x_j \quad \text{for } i = 1 \longrightarrow m, \tag{14.13}$$

so that the $i$th output of the neural network is

$$z_i = g\left[(\mathbb{A}\mathbf{x})_i\right] = g\left[\sum_{j=1}^{n} A_{ij}x_j\right]. \tag{14.14}$$

Because $z_i$ comes from $g$ applied to the elements of $\mathbb{A}\mathbf{x}$, we know that $z_i \in (0, 1)$. To ensure that the components of $\mathbb{A}\mathbf{x}$ run over the whole range of the nonlinear transition, we assume that $\mathbb{A}$ has entries of order one, and then the maximum element of $\mathbb{A}\mathbf{x} \sim n\max|x_i|$, i.e. the dimension of the vector $\mathbf{x}$ times its largest element. Then taking

$$\alpha = \frac{10}{n \max |x_i|} \tag{14.15}$$

allows the elements of $\mathbb{A}\mathbf{x}$ to be spread across the transition range for $g(p)$.

To use the neural network model, we must first find the coefficients $A_{ij}$ given a set of input/output data – that process is called "training." Then we'll use $\mathbf{z} = g(\mathbb{A}\mathbf{x})$ with those coefficients to generate output given an input $\mathbf{x}$; this is the operational phase of the neural network.

## 14.2 Training

Given our function $g(p)$, we have a fully defined neural network model. We now need to figure out how to set the coefficients in the matrix $\mathbb{A}$. Assume we are provided with $\{\mathbf{x}^j, \mathbf{y}^j\}_{j=1}^{T}$ as training data. To distinguish between the training pairs $\{\mathbf{x}^j, \mathbf{y}^j\}_{j=1}^{T}$, the $T$ associated pairs we will use to generate the matrix $\mathbb{A}$, and the action of the neural network on a generic vector $\mathbf{x}$, we'll write $\mathbf{z} = g(\mathbb{A}\mathbf{x})$ in general, then during training, we will demand that $g(\mathbb{A}\mathbf{x}^j) = \mathbf{y}^j$ for all $j = 1 \longrightarrow T$ (or some acceptable subset).

Start with a matrix $\mathbb{A}$ that has random entries in $[-1, 1]$, and then our goal is to update these entries so as to minimize (to within some user-specified tolerance $\epsilon \approx 0$) the difference between the output $g(\mathbb{A}\mathbf{x}^j)$ and the target $\mathbf{y}^j$ – define the function to be minimized as:

$$f(A_{ij}) = \|g(\mathbb{A}\mathbf{x}^k) - \mathbf{y}^k\|^2 \quad \text{for } k = 1 \longrightarrow T, \tag{14.16}$$

so that $f(A_{ij})$ is the length squared (in the Pythagorean sense) of the residual vector $g(\mathbb{A}\mathbf{x}^k) - \mathbf{y}^k$. We write $f(A_{ij})$ to emphasize that this function depends on the matrix coefficients. By minimizing this function, evaluated over the training data, we will find a set of coefficients $A_{ij}$.

### *14.2.1 Setting coefficients*

We'll use the simplest procedure available, in the interest of transparency and speed. We have a function $f(A_{ij})$ to minimize – the steepest descent method from Section 12.3.1 will suffice – it requires only that we evaluate the derivative of $f(A_{ij})$ with respect to (all of) the $A_{ij}$, a total of $m \times n$ parameters. The approach is to iteratively update the entries according to:

$$A_{pq} = A_{pq} - \eta \frac{\partial f}{\partial A_{pq}}, \qquad (14.17)$$

for all $p$ and $q$ (all entries of $\mathbb{A}$), with $\eta$, some adjustable step size. From the explicit form (14.14), we can construct $\frac{\partial f}{\partial A_{pq}}$ without too much difficulty. Writing out $f(A_{ij})$, we have

$$f(A_{ij}) = \sum_{i=1}^{m} \left[ g \left( \sum_{j=1}^{n} A_{ij} x_j \right) - y_i \right]^2, \qquad (14.18)$$

with $\mathbf{x} \equiv \mathbf{x}^k, \mathbf{y} \equiv \mathbf{y}^k$ the $k$th input/output pair. Define

$$b_i \equiv \sum_{j=1}^{n} A_{ij} x_j \qquad (14.19)$$

$$z_i \equiv g(b_i)$$

then

$$f(A_{ij}) = \sum_{i=1}^{m} (z_i - y_i)^2. \qquad (14.20)$$

Taking the derivative of $f$ with respect to the element $A_{pq}$, we have

$$\frac{\partial f}{\partial A_{pq}} = \sum_{i=1}^{m} 2(z_i - y_i) \frac{\partial z_i}{\partial A_{pq}}, \qquad (14.21)$$

and working our way down:

$$\frac{\partial z_i}{\partial A_{pq}} = g'(b_i) \frac{\partial b_i}{\partial A_{pq}}$$

$$\frac{\partial b_i}{\partial A_{pq}} = \sum_{j=1}^{n} \frac{\partial A_{ij}}{\partial A_{pq}} x_j = \sum_{j=1}^{n} \delta_{ip} \delta_{jq} x_j = \delta_{ip} x_q, \qquad (14.22)$$

where we note that the amount of $A_{pq}$ in $A_{ij}$ is zero unless $i = p$ and $j = q$, hence the deltas that collapse the sum. Now we work our way back up – the choice of a nice $g(p)$ allows for simplifications as we go – the derivative, in particular, works out well:

$$g'(p) = \frac{\alpha e^{-\alpha p}}{(1 + e^{-\alpha p})^2} = \alpha g(p)(1 - g(p)). \tag{14.23}$$

Inserting this, together with our expression for $\frac{\partial b_i}{\partial A_{pq}}$ back in to the right-hand side of $\frac{\partial z_i}{\partial A_{pq}}$ gives

$$\frac{\partial z_i}{\partial A_{pq}} = \alpha g(b_i)(1 - g(b_i)) \delta_{ip} x_q = \alpha z_i (1 - z_i) \delta_{ip} x_q, \tag{14.24}$$

and then

$$\frac{\partial f}{\partial A_{pq}} = \sum_{i=1}^{m} 2(z_i - y_i) \alpha z_i (1 - z_i) \delta_{ip} x_q = 2\alpha (z_p - y_p) z_p (1 - z_p) x_q. \tag{14.25}$$

Here, we have a simple iterative scheme that involves relatively few computations – given a set of $\{\mathbf{x}^k\}_{k=1}^T$ and $\{\mathbf{y}^k\}_{k=1}^T$, we take each pair, $\{\mathbf{x}^k, \mathbf{y}^k\}$, and perform steepest descent iteration until $f(A_{ij}) < \epsilon$, a tolerance provided by the user. The procedure is shown in Algorithm 14.1, where we assume that the training pairs, a function $g(p)$, and the step size $\eta$ have been provided.

---

**Algorithm 14.1** Steepest descent for neural networks

$\mathbb{A} \leftarrow$ random initial values
**for** $k = 1 \rightarrow T$ **do**
  $\mathbf{z} = g(\mathbb{A}\mathbf{x}^k)$
  **while** $(\mathbf{z} - \mathbf{y}^k) \cdot (\mathbf{z} - \mathbf{y}^k) > \epsilon$ **do**
    **for** $i = 1 \rightarrow m$ **do**
      **for** $j = 1 \rightarrow n$ **do**
        $\Delta A_{ij} = 2\alpha(z_i - y_i^k)z_i(1 - z_i)x_j^k$
        $A_{ij} = A_{ij} - \eta\Delta A_{ij}$
      **end for**
    **end for**
    $\mathbf{z} = g(\mathbb{A}\mathbf{x}^k)$
  **end while**
**end for**

---

A few implementation details necessary to run Algorithm 14.1: We must fix the value of $\alpha$ prior to training using the maximum of all the input data. The desired tolerance, $\epsilon$, is generally based on some acceptable maximum error for any individual element of the residual (squared). The step size $\eta$, used in the steepest descent portion of the algorithm, can be fixed "experimentally" (try to get a small subset of the training data to converge quickly), or you can do additional line minimization as discussed in Section 12.5.

At this point, you might ask, given the nonlinear nature of both the neural network mapping, and the iterative procedure that is used to set the internal coefficients: How do we know, once the matrix has been adapted to a particular pair $\{\mathbf{x}^k, \mathbf{y}^k\}$ that it still returns the correct values for any previous pair? You don't (and in general, it won't) – the only way to reasonably assure that the neural network will function well for all training data is to provide these known examples multiple times and in random order, then test the accuracy of the network on the training data set. So, given a set of training data of size $t$, we might pick $T = 10t$ samples at random, and use those as the input to Algorithm 14.1. After running those through, we would check the action of the resulting $\mathbb{A}$ on all $t$ examples from the data set, and if we still do not have good agreement, run another $10t$. It is possible that we will never achieve good agreement even on training data, and we'll address extensions that allow for broader matches later on.

After the neural network is sufficiently trained, i.e. the matrix $\mathbb{A}$ correctly identifies all input and output pairs (or some acceptable percentage), we can use the matrix on new data by generating $g(\mathbb{A}\mathbf{x})$ for some input $\mathbf{x}$. Thus, training aside, the neural network is cheap to run, requiring only $m \times n$ operations (the matrix-vector multiplication is the expensive part) to generate output.

## 14.3 Example and interpretation

As a simple example that demonstrates some of the structure of the neural network, our test problem will be: Given a string of ten numbers, randomly drawn from $\{-0.4, -0.3, -0.2, -0.1, 0, 0.1, 0.2, 0.3, 0.4\}$, return the tenth number. This is a strictly linear operation, so the use of a nonlinear procedure is massive overkill.

Suppose we take $t = 10$, so that we will use ten known sets of input and output data, then we'll mix those ten together randomly, $T = 100t$ times, to generate the entire training set, $\{\mathbf{x}^k, \mathbf{y}^k\}_{k=1}^{T}$. We send those through the training routine described by Algorithm 14.1. If we take the resulting $\mathbb{A} \in \mathbb{R}^{1 \times 10}$ matrix and apply $g(\mathbb{A}\mathbf{x})$ for each of the ten original data pairs, we get residual $\mathbf{r}^k \equiv g(\mathbb{A}\mathbf{x}^k) - \mathbf{y}^k$ for $k = 1 \rightarrow t$ shown in Figure 14.2.

Notice, in this example, that if we restrict the output of the neural network to the known set of numbers that make up each of the entries of $\mathbf{y}^k$, we get perfect agreement, since no residual value is greater than 0.1. That's a lesson on its own: The more information you have about the encoding of your problem, the better off the result will be.

Now that the neural network is trained, it's time to use it on real data – we can see the result of applying $g(\mathbb{A}\mathbf{x})$ to a hundred different input vectors $\mathbf{x}$. This time,

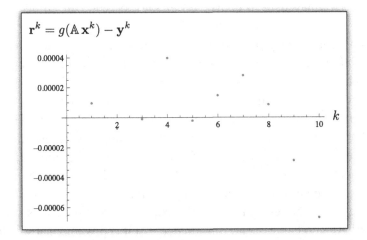

Figure 14.2  Using the neural network to predict the tenth entry in a random list of numbers – this is a plot of the residual of the neural network output and the actual output. For the ten sets of training data, the residual is much smaller than the data resolution.

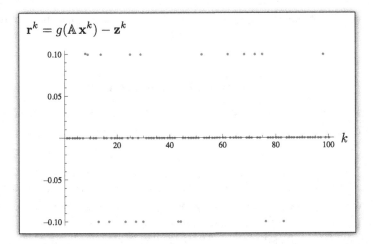

Figure 14.3  The result of our trained neural network applied to 100 new inputs. Here we show the residual $\mathbf{r}^k = g(\mathbb{A}\mathbf{x}^k) - \mathbf{z}^k$ for the correct output set $\{\mathbf{z}^k\}_{k=1}^{100}$. For 80 of these new inputs, the output is correct, and for the remaining 20 the neural network is only one off.

we have performed all the rounding, so we assume that we know the neural network output can be rounded to the values $[-0.4, 0.4]$ in steps of 0.1. In Figure 14.3, we see the results – they are not great as far as picking out the tenth digit of a string of numbers is concerned, although they are within $\pm 0.1$ for all the data, and perfect 80% of the time.

If we think about the information content of $\mathbb{A}$, there are relatively few adjustable parameters – the input vectors are of length $n = 10$, and the output vectors are of length $m = 1$, so there are only ten matrix entries to set. Forgetting about the function $g(p)$ for a moment, what we would expect for the matrix $\mathbb{A}$ is something like:

$$\mathbb{A} \doteq (0 \quad 0 \quad 0 \quad 0 \quad 0 \quad 0 \quad 0 \quad 0 \quad 0 \quad 1), \qquad (14.26)$$

that would provide the correct output for any $\mathbf{x}$ if we just took $\mathbb{A}\mathbf{x}$ as the output. It is interesting, then, that the neural network training matrix has a similar form:

$$\mathbb{A} \approx (-0.03 \quad -0.08 \quad 0.17 \quad 0.12 \quad 0.12 \quad -0.09 \quad 0.06 \quad -0.01 \quad -0.1 \quad 2.05)$$

$$(14.27)$$

with the tenth entry much larger than the rest.

There are a variety of ways to improve the performance of our neural network model for this example, but given the artificial nature of the example, there is not much point. In general, if you find you can predict the results by hand, or have a complete definition for the map you are trying to get the neural network to model, you shouldn't be using a neural network (we have to violate this rule to demonstrate the setup, of course).

## 14.4 Hidden layers

One thing should be clear from the above – we are constrained in the number of adjustable parameters we can set by the size of the input and output vectors. We would like to introduce freedom in the process, so that even if we have ten inputs and one output, we have more than ten parameters in the neural network.

The simplest way to introduce additional parameters is to add a matrix, and to keep things nonlinear, an additional application of the nonlinear function $g(p)$ – call it $\bar{g}(p)$ since it has its own parameter $\bar{\alpha}$ not necessarily equal to $\alpha$. We'll use a "hidden layer" of dimension $k$, and then proceed as follows – let $\mathbb{A} \in \mathbb{R}^{m \times k}$, and let $\mathbb{B} \in \mathbb{R}^{k \times n}$, these two matrices will provide the parametric content of the neural network model. Given a vector $\mathbf{x} \in \mathbb{R}^n$, construct the output $\mathbf{z} \in \mathbb{R}^m$ as follows:

$$\tilde{\mathbf{z}} = g(\mathbb{B}\mathbf{x}) - \frac{1}{2}$$
$$\mathbf{z} = \bar{g}(\mathbb{A}\tilde{\mathbf{z}}). \qquad (14.28)$$

The constant offset in the definition of $\tilde{\mathbf{z}}$ is to shift the output of $g(\mathbb{B}\mathbf{x})$, necessarily in $(0, 1)$ down to the range $(-\frac{1}{2}, \frac{1}{2})$ – that's to probe the nonlinear transition of the second sigmoid, $\bar{g}(p)$. Since we know the data range for the input to $\bar{g}$, we can estimate the size needed for $\bar{\alpha}$ – by analogy with (14.15), set $\bar{\alpha} = 10/(\frac{1}{2}k)$.

The advantage of the "hidden" layer (associated with the matrix $\mathbb{B}$, and represented internally by $\tilde{\mathbf{z}}$) is that the choice of $k$ is independent of the size of the input and output. We now have a parameter for the neural network itself, allowing us to train using $k(m + n)$ total matrix elements.

This hidden layer introduces a new set of matrix entries in our minimization – we need to (re-)compute the gradient of $f(A_{ij}, B_{ij})$ with respect to all $A_{ij}$ and $B_{ij}$. If we write out the $\ell$th element of $\mathbf{z}$ explicitly, then we can take the relevant derivatives:

$$
\tilde{z}_i = g\left(\sum_{j=1}^{n} B_{ij} x_j\right) - \frac{1}{2}
$$
$$
z_\ell = \bar{g}\left(\sum_{s=1}^{k} A_{\ell s}\tilde{z}_s\right).
$$
(14.29)

From these, repeated application of the chain rule on the definition of $f$:

$$
f(A_{\ell s}, B_{ij}) = \sum_{\ell=1}^{m}(z_\ell - y_\ell)^2
$$
(14.30)

will give

$$
\frac{\partial f}{\partial A_{pq}} = 2\bar{\alpha}\left(z_p - y_p\right) z_p\left(1 - z_p\right)\tilde{z}_q.
$$
(14.31)

We can define the portion of the result that does not depend on $q$ to be $\sigma_p$ – that is

$$
\sigma_p \equiv 2\bar{\alpha}(z_p - y_p)z_p(1 - z_p).
$$
(14.32)

Working out the derivatives with respect to the entries of the matrix $\mathbb{B}$:

$$
\frac{\partial f}{\partial B_{pq}} = \sum_{\ell=1}^{m}\sigma_\ell A_{\ell p}\alpha\left(\frac{1}{2} + \tilde{z}_p\right)\left(\frac{1}{2} - \tilde{z}_p\right) x_q.
$$
(14.33)

As with the single layer case, we update the matrices, for a given pair $\mathbf{x}^k$ and $\mathbf{y}^k$ by iteration until $f \leq \epsilon$ using the update

$$
A_{pq} = A_{pq} - \eta\frac{\partial f}{\partial A_{pq}}
$$
$$
B_{pq} = B_{pq} - \eta\frac{\partial f}{\partial B_{pq}}.
$$
(14.34)

The modification of Algorithm 14.1 is straightforward, we just replace the update for $\mathbb{A}$ with updates for $\mathbb{A}$ and $\mathbb{B}$. This doesn't significantly change the amount of time required to calculate the updates, although the number of iterations to convergence will change, since we now have more parameters.

## 14.5 Usage and caveats

Neural networks are fun to try out on various problems, they are easy to implement, and straightforward to apply to almost anything. Given "appropriate" sizes for the matrices $\mathbb{A}$ and $\mathbb{B}$, the combination

$$\bar{g}(\mathbb{A}(g(\mathbb{B}\mathbf{x}) - 1/2)) \tag{14.35}$$

with our choice of $g(p)$ can approximate any nonlinear function – that is the content of the "Cybenko theorem," carefully stated, and proven in [13]. So in a theoretical sense, the neural network can be used to model any underlying behavior, describable as a nonlinear function of the input.[1]

As a rule-of-thumb, you can use neural networks successfully when you believe there is an underlying pattern to a set of data, but you cannot "see" that pattern clearly, either because it is described by equations too difficult to solve (or ones whose timescales make even numerical solution untrustworthy), or because the model that makes the pattern clear is incomplete – then use the network as an unbiased set of eyes. People use neural networks in weather prediction – we know there is an underlying set of physics governing the motion of weather systems, but the fluid dynamics equations we use to model this physics are highly nonlinear, and very sensitive to initial conditions, making numerical solution for prediction difficult. On the other hand, we have tons of weather data, and that can be used to train a neural network.

Yet neural networks are not a cure-all – they are highly dependent on the way in which a particular problem is encoded. Take an example from data analysis – suppose we have a piece of noisy data, and we know that underlying the data is a single frequency which we would like to find using a neural network. One way we could use the network is to generate a bunch of artificial data at different frequencies, and then have, as output, a pure sinusoidal signal with the noise stripped away – from that pure signal, we could find the frequency (by, for example, taking the Fourier transform of the neural network output and picking the largest peak). But we could also take the same artificial input data, and output a number, $f$, the frequency of the underlying pure signal. Which is better?

---

[1] The Cybenko theorem does not tell us how to construct the matrices $\mathbb{A}$ and $\mathbb{B}$ that define the correct approximation to a particular nonlinear function, unfortunately.

Aside from careful encoding, and testing different encodings, there are difficulties in the choice of size, $k$, for the hidden layer (see [28], for example). You don't want to make the hidden layer too big (whatever that means). Remember that the function $f(A_{ij}, B_{ij})$ is defined in a high-dimensional parameter space – we can think of the process of training as finding a minimum of the function $f$ in the large space – but we don't know that we are at a "global" minimum – we could be at a local minimum. Once we're there, we will be trapped, like a mass on a spring. One way to keep the coefficients $A_{ij}$ and $B_{ij}$ from too narrow a region of the overall space is to periodically, during training, add some random noise to the matrix values – this can allow a particular set of coefficients to jump over barriers and flow downhill towards a "better" minimum.

The process of introducing a little noise now and then is the basis for a type of "genetic" approach to neural networks. One can start training a bunch of networks (each one described completely by the matrices $\mathbb{A}$ and $\mathbb{B}$), see which ones perform best (either on the training data itself, or on actual (or artificial) data), and "kill" neural networks that do not meet some fitness criteria. The remaining ones can be culled and trained further, perhaps with the introduction of some noise.

Finally, the single hidden layer "perceptron" model we have studied in this chapter is just the beginning – there are a variety of switching functions $g(p)$ that can be used, various different minimization routines (replacing steepest descent, for example), data pre-processing/filtering, and better matrix initialization (for starters) that can improve the behavior of these models. The goal in this chapter is to highlight neural network models as an interesting approach to approximating complicated functions – but, just as the functions they approximate are complicated, so too are the different options available within the neural network model umbrella.

## 14.6 Signal source

Let's do a relevant example – suppose we know, on physical grounds, that radiation of one of four frequencies is being detected by an antenna on the Earth. The signal is noisy – to model this situation, we take the signal

$$s(t) = \cos(2\pi f t) + 5\xi \tag{14.36}$$

where $\xi$ represents a random number chosen from a flat distribution on $[-1, 1]$ – the magnitude of the noise, then, could be as much as five times greater than the magnitude of the signal in this example. We'll take $t_n = (1/100)n$, so that we are sampling in time from $0 \to 1$ with a hundred (and one) data points. The frequency $f$ takes on four values: $f = 1, 2, 3,$ or $4$ Hz. A typical instantiation of the random noise, together with the pure signal that underlies the data, is shown in Figure 14.4. We'll take the signal as input, and for output, we'll return $f/5$ (that lies on $(0, 1)$).

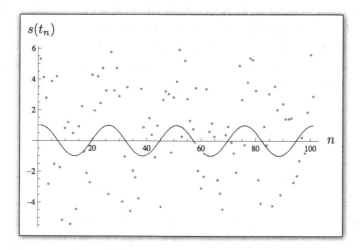

Figure 14.4 An example of a test signal, generated with $f = 4$ Hz. The points are the data from (14.36), while the curve is the underlying cosine with no noise added.

We'll start by determining, experimentally, a reasonable size for the hidden layer. In all that follows, we take $\bar{\alpha} = 20/k$ for hidden layer size $k$, $\epsilon = 10^{-12}$, so that we require close to perfect agreement in each training pair, and $\eta = 0.5$ for the descent step.

If we take 100 training pairs, and set the hidden layer size to $k = 2$, then the result of the neural network – appropriately rounded so that it returns one of the known frequency values, acting on the training data is:

$$
\begin{array}{ll}
\text{perfect} & 55\% \\
\pm 1 & 37\% \\
\pm 2 & 6\% \\
\pm 3 & 2\%
\end{array}
\tag{14.37}
$$

Setting $k = 4$, the numbers improve (using the same training set):

$$
\begin{array}{ll}
\text{perfect} & 65\% \\
\pm 1 & 29\% \\
\pm 2 & 5\% \\
\pm 3 & 1\%
\end{array}
\tag{14.38}
$$

For $k = 8$:

$$
\begin{array}{ll}
\text{perfect} & 65\% \\
\pm 1 & 32\% \\
\pm 2 & 3\%
\end{array}
\tag{14.39}
$$

We'll stop at $k = 8$, since there is not much change with respect to the $k = 4$ case. Now the training data is just a set of instantiations of the noise, and since we don't have a real data set, we'll take *different* instantiations of the random noise as our artificial data – data that the network has never seen. The numbers are significantly worse:

$$
\begin{array}{ll}
\text{perfect} & 41\% \\
\pm 1 & 45\% \\
\pm 2 & 13\% \\
\pm 3 & 1\%
\end{array}
\tag{14.40}
$$

How should we evaluate the network? The perfect score is still above 25%, so we are doing better than random guessing. The percentage of the time we are just one off is 86% which seems acceptable. Of course, we should ask how far off the result would be if we just looked for the maximum peak in the power spectrum of the data (or even better, the auto-correlation of the power spectrum).

### Further reading

1. Anderson, James A. *An Introduction to Neural Networks*. The MIT Press, 1995.
2. Gurney, Kevin. *An Introduction to Neural Networks*. UCL Press Limited, 1997.
3. Haykin, Simon. *Neural Networks: A Comprehensive Foundation*. Prentice Hall, 1999.

### Problems

**Problem 14.1**
The data that is sent in to a neural network can be pre-processed to put it in a good range for the sigmoid function with a fixed parameter $\alpha$. For $\alpha = 1$, we'd like the data to have zero mean and standard deviation $10/n$ (where $n$ is the length of the data) – that way, we probe the nonlinearity of $g(p)$ (try plotting $g(p)$ for $\alpha = 1$ – remember that $\mathbf{x}$ gets multiplied by $\mathbb{A}$ before it is sent in to $g(p)$)). Write a function that takes an input vector $\mathbf{x}$ and generates $\bar{\mathbf{x}} = A\mathbf{x} + B$ so that:

$$
\frac{1}{n} \sum_{j=1}^{n} \bar{x}_j = 0 \qquad \frac{1}{n} \sum_{j=1}^{n} \bar{x}_j^2 = \frac{100}{n^2},
\tag{14.41}
$$

and have the function return $A$ and $B$. This would allow you to rewrite the neural network functions from the chapter notebook with $\alpha = 1$.

## Problem 14.2
For

$$f(\mathbb{A}, \mathbb{B}) = \sum_{\ell=1}^{m} (z_\ell - y_\ell)^2, \tag{14.42}$$

with

$$z_\ell = \bar{g}(b_\ell) \qquad b_\ell \equiv \sum_{i=1}^{n} A_{\ell i} \tilde{z}_i$$

$$\tilde{z}_i = g(c_i) - \kappa \qquad c_i \equiv \sum_{j=1}^{n} B_{ij} x_j, \tag{14.43}$$

show that

$$\frac{\partial f}{\partial A_{pq}} = \underbrace{2\bar{\alpha} z_p (1 - z_p)(z_p - y_p)}_{\equiv \sigma_p} \tilde{z}_q$$

$$\frac{\partial f}{\partial B_{pq}} = \sum_{\ell=1}^{n} \sigma_\ell A_{\ell p} \alpha (\tilde{z}_p + \kappa)(1 - \tilde{z}_p - \kappa) x_q. \tag{14.44}$$

## Problem 14.3
Write a function that generates a table of data, given a frequency, $f$, and variance $v$
according to:

$$d(t) = \cos(2\pi f t) + \sqrt{v} G(0, 1) \tag{14.45}$$

where $G(0, 1)$ represents a random number drawn from a normal distribution of mean
zero, variance one (the built-in command `NormalDistribution` can be used to
provide these numbers). Your table should be of size $N = 128$, and start at $t = 0$ with the
final time $T = 10$, in steps, then, of $\Delta t = \frac{T}{N-1}$.

## Problem 14.4
Write a function that takes a table of data, like the one described in the previous problem,
computes the power spectrum of the data, and returns the frequency of the maximum
peak. Check your function by sending in data with $f = 0.4$ Hz, $v = 0$ that you generate –
you should recover 0.4 Hz as the location of the peak. Test your peak-finder for increasing
value of $v$ as follows: For $v = 1, 2, 3$, etc. generate a hundred sets of data each at $f = 0.4$
Hz, and run them through your peak finder. At what value of $v$ do you lose the ability to
correctly locate the signal within the noise (to be quantitative, find the value of $v$ for
which your peak finder works less than 25% of the time, so that if we had four known
frequency sources, random guessing between these four frequencies would be better)?

**Problem 14.5**

Suppose we have a function like $u(x) = \frac{1}{2}(\sin(2\pi x) + 1)$ (so that $u(x) \in [0, 1]$) – we know the infinite series expansion for sine, that's

$$\frac{1}{2}(\sin(2\pi x) + 1) = \frac{1}{2} + \sum_{j=0}^{\infty}(-1)^j \frac{(2\pi x)^{2j+1}}{2(2j + 1)!}. \qquad (14.46)$$

Now consider a neural network that is meant to produce $u(x)$ given $x$ (remember, the Cybenko theorem says such approximations can be accomplished):

(a) If we use a single layer, then there is a single coefficient to set,

$$g(\mathbb{A}x) = \frac{1}{1 + e^{-\alpha A_{11}x}}. \qquad (14.47)$$

Compute the Taylor expansion of this function (use the built-in command `Series`, if you like) – how many terms of the sine expansion can you match with this single coefficient (remember that $\alpha$ is fixed)?

(b) Try using a minimal hidden layer of size $k = 2$, so that there are two entries in $\mathbb{A}$ and $\mathbb{B}$ – now the output looks like:

$$\bar{g}(\mathbb{A}(g(\mathbb{B}x) - 1/2)) = \left[1 + e^{-\bar{\alpha}\left[\left(A_{11}\left(\frac{1}{1+e^{-\alpha B_{11}x}} - \frac{1}{2}\right) + A_{12}\left(\frac{1}{1+e^{-\alpha B_{21}x}} - \frac{1}{2}\right)\right)\right]}\right]^{-1}. \qquad (14.48)$$

Find the Taylor expansion of this function (definitely use the built-in command `Series`) – now how many terms of the sine expansion can you match?

## Lab problems

**Problem 14.6**

We'll test the neural network model against the power spectrum peak finder using data as in Problem 14.3.

(a) Start by making a function that randomly chooses between $f = 0.2, 0.4, 0.6$, and $0.8$ hertz and constructs the appropriate data, together with the correct frequency $f$: use your function from Problem 14.3 with $v$ from Problem 14.4 (i.e. construct data that will "fool" your peak finder). Using your function, generate five thousand input output pairs – make a table X containing the five thousand input vectors, and Y with the correct output. Set $\alpha = \frac{10}{128|x|}$ as in (14.15).

(b) Using your data sets, train (that is, find the matrices $\mathbb{A}$ and $\mathbb{B}$ using the function `TrainOnData` from the chapter notebook) a neural network with a hidden layer of size $k = 4$, use $\bar{\alpha} = 10/(4\frac{1}{2})$, $\eta = 2.5$, and $\epsilon = 10^{-4}$. Test the resulting matrices on your training data (you may take the output of $\bar{g}(\mathbb{A}g(\mathbb{B}x) - 1/2)$ and round it to the nearest 0.2 to capture the idea that we know the frequencies must be one of four specific ones). What percentage of the training data do you capture correctly?

(c) Now generate 100 sets of new data (you can use your function from part a) together with the correct source frequencies. Run the neural network on this "real" data, again rounding to the nearest 0.2 Hz, and compare that with the power spectrum peak finder you wrote in Problem 14.4. What percentage of completely correct data comes from each method? Check that random guessing gives around 25% correct results. The point is, you have data that random guessing gets right a quarter of the time, the "rational" power spectrum method is worse than this (slightly), and the neural network model is better.

## Problem 14.7

In this problem, we'll try to get a neural network model to capture sine. Start by writing a function u[x] that takes $x$ and returns the value of $\sin(2\pi x)$ appropriately scaled so that the minimum is at 0.1 and the maximum at 0.9 (this is to keep the output on (0, 1)). To generate the training data, take the interval $[-1, 1]$ for $x$ and make a grid with $\Delta x = 2/100$, that provides your $x$ data – for the correct output, give u[x] (so you have a hundred and one correctly associated data pairs $\{x_i, u[x_i]\}$ for the grand point $x_i$). Mix these pairs together randomly to get a total of 20 000 training examples (we're providing each example multiple times). Train using a neural network with a hidden layer of size $k = 32$. Use $\alpha = 10$ (from (14.15) with one input and a maximum value, for $x$, of one) and $\bar{\alpha} = 10/(\frac{1}{2}32)$ – take $\eta = 2.5$ and $\epsilon = 10^{-5}$.

(a) Once you've trained the network, generate one hundred additional pairs $\{x, u[x]\}$ using random $x \in [-1, 1]$ – calculate the residual of the correct output and your neural network's prediction. What is the mean of that residual?
(b) Plot the function: $\bar{g}(\mathbb{A}(g(\mathbb{B}x) - 1/2))$ (for $x \in [-1, 1] - x \in \mathbb{R}$ here).

## Problem 14.8

In this problem, we'll look at how one might use a neural network model to predict stock price.

(a) Get the financial data for ten years of Apple stock; you can use the built-in command FinancialData with the stock "APPL," this returns a list, each element includes a date, and the share price at market's close for that date. Strip out the date section of the data, and scale the remaining prices so that the scaled data is in $[-1, 1]$ – this will be the input data. Make another version of the data that is scaled to be in $[0.05, 0.95]$ – this is the output data.
(b) As our inputs, we'll take one hundred consecutive days of stock prices (from the input data), and the output will be the stock price on the hundred and first day (taken from the output data). Write a function that takes the ten-year data, picks a random point in that data, and returns one hundred consecutive prices from that point – it should also return the 101st price, that represents the correct outcome for the neural network.
(c) Using your function from part (b), generate 5000 input/output pairs – the inputs are of length one hundred (one hundred days of price data), the outputs of length one (stock price on the 101st day). Train a neural network with this data, take a hidden layer with $k = 32$, and use $\eta = 1.5$ and $\epsilon = 0.001$ (use $\alpha = 10/100$ from (14.15), and $\bar{\alpha} = 10/(\frac{1}{2}32)$).

(d) Test the trained neural network model on one hundred sets of data – use your training-data-generator from part (b) to make the "real" data. For each data pair, calculate the difference between the actual stock price on the 101st day, and your neural network's output. Make a plot of this residual data, and find its mean. The accuracy is not worth quitting one's day-job – but there's a lot of room for improvement – this problem has been set up to run relatively quickly – more data, a larger hidden layer, these would slow down the training process but improve results.

# 15

# Galerkin methods

In Chapter 4, we saw how to turn linear partial differential equations into finite differences, reducing, for example, the Poisson problem to a matrix inverse problem. There were two properties we exploited there – the first was linearity (that was what led to the linear matrix-vector equations like (4.46)), and the second was the boundary-value formulation of the Poisson problem (where the solution is given on the boundary of some domain). Then, in Chapter 5, we extended our numerical solutions to nonlinear problems in initial value formulation using separate discretizations in time and space. The missing piece, then, is boundary-value formulations of *non*-linear problems.

We can solve these, and introduce a new way to solve nonlinear initial value problems as well, using a class of techniques called "Galerkin methods".[1] Whereas finite difference, whether linear or not, requires a set of grids (spatial, for the Poisson problems, and spatial and temporal for nonlinear time-dependent problems), Galerkin methods are grid-free. The idea is to start with an expansion, like:

$$u(x, t) = \sum_{j=0}^{\infty} a_j(t) \sin\left(\frac{j\pi x}{L}\right) \tag{15.1}$$

for a solution to some PDE, $u(x, t)$, that vanishes at $x = 0$ and $L$ (the boundary conditions). Any solution can be written in this form; it's a Fourier sine series with time-varying coefficients. The approximation comes in truncating the sum at some value $N$ – then we use the PDE itself to develop ODEs governing the coefficients $a_j(t)$ and solve those. If we use, for example, a Runge–Kutta method with adaptive step size to find the $a_j(t)$, then even the temporal grid goes away, and we have access to the value of $u(x, t)$ at any $x$ and $t$. For PDEs that involve two spatial variables, the resulting ODEs (for $a_j(y)$, say) will be in boundary-value formulation, and we can use the shooting techniques from Chapter 3 to solve them.

---

[1] These are related to the "finite element" methods for solving PDEs [47].

The approach sketched above can be carried out analytically in some cases, and we will look at this series solution method applied to the wave equation, Schrödinger's equation and the heat equation in that context. But the real utility of the Galerkin method comes from its application to nonlinear problems, and these are not as prevalent in the physical systems we have studied so far, so we'll spend some time setting up a few different nonlinear PDEs related to interesting systems.

## 15.1 Physical motivation

For linear field theories, like E&M, we have field equations that relate sources ($\rho$ in that case) to fields ($V$) usually via some differential operator – in $\nabla^2 V = -\rho/\epsilon_0$, the source is $\rho$ and the operator is $\nabla^2$. When the field equation is nonlinear, the distinction between source and differential operator is blurred. Suppose, for example, we had a physical theory where the source $\rho$ was related to the field $V$ by

$$\nabla^2 V = -\frac{\rho}{\epsilon_0} - \alpha V^2 = -\frac{1}{\epsilon_0}\left(\rho + \alpha\epsilon_0 V^2\right), \tag{15.2}$$

for some $\alpha$. The $V^2$ appears on the right, and combines with the physical sources (charge density, $\rho$). So we could say that $V$ is "sourcing itself."

But, we could just as easily put the $V^2$ term on the left, making it part of the differential operator:

$$(\nabla^2 + \alpha V)V = -\frac{\rho}{\epsilon_0}, \tag{15.3}$$

and as we study nonlinear equations, we will use whichever point of view is appropriate.

It is true that, aside from direct modification, nonlinearity appears in, for example, general relativity as a mathematical manifestation of the idea that the fields of general relativity act as their own sources. This self-sourcing is easy to see – in a relativistic theory of gravity, mass and energy act as sources (since they are equivalent). As in most field theories, the field in general relativity carries energy, and therefore acts as its own source. This is different from E&M, for example, where the source of **E** and **B** is charge and moving charge – the fields themselves are neutral, so do *not* act as their own sources.

We'll consider three specific cases of nonlinear theories – first, Born–Infeld E&M, a modification of Maxwell's equations that leads to a finite cutoff for the electric field of a point charge. Then we'll take the massive Klein–Gordon equation and add a self-source term, and finally, we'll look at the motion of an extensible string (the wave equation applies to an inextensible string).

### *15.1.1 Born–Infeld E&M*

Maxwell's equations are linear, and their predictive power has been verified for well over a hundred years. There is not much compelling experimental evidence that they require correction, and their applicability spans both classical and quantum descriptions. Yet, there is a theoretical problem with their solution for point sources: The energy stored in the electric field of a point charge is infinite, meaning that to make a point charge requires an infinite amount of energy. This turns out to be manageable, but leads to some restrictions on the availability of true point charges in classical E&M. We can either content ourselves with pre-made point charges (ones that were generated by some external agent, and float around interacting electromagnetically), or we can agree that all classical point charges have finite, very small, radius.

There is another way to save point charges: we can change the field equations that provide their solution. This is the approach taken by Born and Infeld [10] (see [48] for discussion). The source of the energy problem is the point charge electric field. For a point charge $q$ sitting at the origin, the field is

$$\mathbf{E} = \frac{q}{4\pi\epsilon_0 r^2}\hat{\mathbf{r}}. \tag{15.4}$$

The energy density (energy per unit volume) associated with this field is $e = \frac{1}{2}\epsilon_0 E^2$, or

$$e = \frac{q^2}{32\pi^2\epsilon_0 r^4} \tag{15.5}$$

and the total energy, obtained by integrating $e$ over all space is:

$$E = \int_0^\infty \int_0^{2\pi} \int_0^\pi e r^2 \sin\theta \, d\theta \, d\phi \, dr = \frac{q^2}{8\pi\epsilon_0} \int_0^\infty \frac{1}{r^2} dr = \infty. \tag{15.6}$$

There's the problem, now for the Born–Infeld fix – the divergence of the integral comes from the $1/r^2$ divergence of the electric field as $r \longrightarrow 0$. If the electric field of a point charge instead had finite value as $r \longrightarrow 0$, then we could save the total energy. How should we modify:

$$\nabla \cdot \mathbf{E} = \frac{\rho}{\epsilon_0} \tag{15.7}$$

so that the point source density $\rho = q\delta^3(\mathbf{r})$ leads to an electric field with finite value everywhere?

Think of the other unacceptable infinity we encounter early on in physics – particle speeds. Newton's second law: $\frac{d(m\dot{x})}{dt} = F$ can lead to speeds $|\dot{x}(t)| > c$, violating one of the postulates of special relativity. The resolution of that problem is to introduce relativistic momentum, then Newton's second law takes the form

$$\frac{d}{dt}\left[\frac{m\dot{x}}{\sqrt{1-\frac{\dot{x}^2}{c^2}}}\right] = F, \tag{15.8}$$

and this equation enforces $|\dot{x}(t)| < c$ for any $F$ (barring pathological cases).

We can copy the form of (15.8), and apply it to (15.7), giving the modified equation[2]

$$\nabla \cdot \left[\frac{\mathbf{E}}{\sqrt{1-\frac{E^2}{b^2}}}\right] = \frac{\rho}{\epsilon_0}. \tag{15.9}$$

This should have the same cutoff structure as (15.8) by construction, with $E < b$ for all space. That will lead to a finite energy of assembly for point charges in this modified theory.

Let's see how it works out by finding the electric field for a point charge. We can apply the divergence theorem to (15.9): take a sphere of radius $r$, and assume that the electric field (governed by (15.9) now) is spherically symmetric, so that $\mathbf{E} = E(r)\hat{\mathbf{r}}$ as always. Then using this updated Gauss's law, we get a relation for the magnitude of the electric field for a point charge sitting at the origin:

$$\oint \frac{\mathbf{E} \cdot d\mathbf{a}}{\sqrt{1-\frac{E^2}{b^2}}} = \frac{q}{\epsilon_0} \longrightarrow \frac{E}{\sqrt{1-\frac{E^2}{b^2}}} = \frac{q}{4\pi\epsilon_0 r^2} \tag{15.10}$$

which we can invert to find $E(r)$:

$$E(r) = \frac{\frac{q}{4\pi\epsilon_0 r^2}}{\sqrt{1+\left(\frac{q}{4\pi\epsilon_0 r^2}\right)^2 \frac{1}{b^2}}} = \frac{1}{\sqrt{\left(\frac{4\pi\epsilon_0 r^2}{q}\right)^2 + \frac{1}{b^2}}}. \tag{15.11}$$

The second equality provides a nice expression for checking the limiting behavior – for $r = 0$, $E(0) = b$, the cutoff we introduced in (15.9). As $r$ gets large, (15.11) matches the Maxwell solution (15.4). The only remaining issue is the maximum field value, $b$. There is no a priori way to set the value of the cutoff, although we can experimentally constrain it. The Born proposal was to set the (now finite) total energy associated with this solution equal to the rest energy of an electron, $m_e c^2$.

Regardless of the details, what we have, in (15.9), is a spatial nonlinear PDE – the solutions are unknown for all but the simplest configurations (the dipole field, for example, is elusive). We want a numerical method that will allow us to make some progress in solving (15.9) given boundary conditions.

---

[2] In the case of (15.9), we take the nonlinearity to be part of the left-hand side, a modification of the operator applied to the field $\mathbf{E}$.

### 15.1.2 Nonlinear scalar fields

Next, we'll introduce nonlinearity in the Klein–Gordon equation – here, we are using additional dependence on the scalar field to mimic precisely the self-coupling present in the tensor field of general relativity. The Klein–Gordon equation for a massive scalar field is

$$-\frac{1}{c^2}\frac{\partial^2 \phi(x,t)}{\partial t^2} + \frac{\partial^2 \phi(x,t)}{\partial x^2} - \frac{m^2 c^2}{\hbar^2}\phi(x,t) = 0. \tag{15.12}$$

This is just the wave equation with an additional term that gives the field mass $m$. That interpretation comes from the wave-like solutions to (15.12) (see [15, 24], for example), we take: $\phi(x,t) = Ae^{i(kx-\omega t)}$, then $k$ and $\omega$ are constrained by the field equation:

$$-\frac{1}{c^2}(-i\omega)^2 \phi(x,t) + (ik)^2 \phi(x,t) - \frac{m^2 c^2}{\hbar^2}\phi(x,t) = 0 \tag{15.13}$$

which gives

$$\hbar^2\omega^2 - \hbar^2 k^2 c^2 = m^2 c^4. \tag{15.14}$$

If we identify $E$ with $\hbar\omega$ (think of light), then $\hbar k$ is like a momentum, and we have an energy–momentum relation of the form

$$E^2 - p^2 c^2 = m^2 c^4, \tag{15.15}$$

precisely what we expect for a particle of mass $m$ in special relativity. For this reason, the field $\phi(x,t)$ is associated with a massive particle, and the Klein–Gordon equation plays a role analogous to Schrödinger's equation in relativistic quantum mechanics.

Starting from the "free" field equation (15.12), we can introduce interactions – we'll take $\hbar = c = 1$, and consider the "self-coupled" form:

$$-\frac{\partial^2 \phi(x,t)}{\partial t^2} + \frac{\partial^2 \phi(x,t)}{\partial x^2} = \alpha^4 \phi(x,t)^3, \tag{15.16}$$

where $m = 0$, and $\phi(x,t)^3$ represents a typical polynomial modification (coming from the so-called "$\phi$ to the fourth" Lagrangian) – we have put the term on the right, so that $\phi^3$ is acting as a source. This particular case could be solved numerically using the Lax–Friedrichs method from Section 5.3.1, for example. But think of the generalization to higher spatial dimension, and assume a static form, then we have

$$\nabla^2 \phi - \alpha^4 \phi^3 = 0, \tag{15.17}$$

with an implicit boundary condition. Now we have a nonlinear PDE in boundary-value formulation, a problem that requires a new method for its numerical solution.

### 15.1.3 Nonlinear wave equation

For a string with tension that varies spatially, the one-dimensional wave equation for transverse motion is

$$\mu \frac{\partial^2 u(x,t)}{\partial t^2} - \frac{\partial}{\partial x}\left(T(x)\frac{\partial u(x,t)}{\partial x}\right) = 0, \tag{15.18}$$

where $\mu$ is the mass density (constant, here), $u(x,t)$ represents the height of the string at location $x$ (and time $t$), and $T(x)$ is the tension at $x$. Suppose we allow the tension to change based on the stretching of the string. A simple, linear model for the tension can be developed by assuming constant $T_0$ for $u = 0$ (so the string is under tension prior to any transverse motion), and when a portion of the string is extended upwards by an amount $du$, we pick up a correction linear in the difference $ds - dx$ for $ds \equiv \sqrt{du^2 + dx^2}$:

$$T = T_0 + \frac{K(ds - dx)}{dx}, \tag{15.19}$$

where $K$ is a constant with the dimension of force (we made a ratio out of $ds - dx$ and $dx$ for convenience, to leave us with this constant), and is set by the material parameters of the string. Geometrically, the configuration is shown in Figure 15.1. This model comes from a Hooke's law response of the string's tension under (transverse-only) stretching (see, for example [38]). 

We can simplify, and highlight the role of $u(x,t)$ by noting that

$$ds - dx = dx\left[\sqrt{1 + \left(\frac{\partial u}{\partial x}\right)^2} - 1\right], \tag{15.20}$$

so the tension is

$$T = T_0 + K\left[\sqrt{1 + \left(\frac{\partial u}{\partial x}\right)^2} - 1\right]. \tag{15.21}$$

If we take $\frac{\partial u}{\partial x} \equiv u'$ to be small (this is the small angle approximation), then the tension is approximately:

$$T \approx T_0 + \frac{1}{2}Ku'^2, \tag{15.22}$$

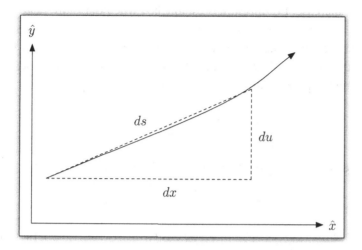

Figure 15.1 A string is stretched due to its transverse motion. Instead of having length $dx$, it has length $ds = \sqrt{dx^2 + du^2}$ where $du$ is the infinitesimal height of the string.

and putting this into the wave equation (15.18) gives (using dots to denote partial derivatives w.r.t. time):

$$\mu \ddot{u} - T_0 u'' = \frac{3}{2} K u'^2 u''. \tag{15.23}$$

Notice that if $u' \ll 1$, then the right-hand side is effectively zero, and we recover the wave equation with no source. As it stands, what we have is a nonlinearly sourced form of the wave equation for the transverse waves on the string.

### 15.1.4 The heat equation

The heat equation is another linear PDE, like the wave equation and Schrödinger's equation, but one we have not discussed previously. For $u(x, y, z, t)$, the temperature at the point $(x, y, z)$ at time $t$, we have the following second-order PDE:

$$\frac{\partial u}{\partial t} - \alpha \nabla^2 u = 0, \tag{15.24}$$

where $\alpha$ is a constant (to set the dimension).

There are "dissipative" solutions to this PDE, meaning solutions that decrease in magnitude over time. This makes sense when we think of the dissipation of heat in a solid – given a metal plate, if we heat the center to a temperature $T_0$, then over time, the temperature at the center will decrease while the temperature over the rest of the plate will increase. Equilibrium is established when the entire plate is the same temperature, and the heat equation governs the dynamics of this process.

Let's treat the separable solutions to see the idea – if we take, in one spatial dimension, the ansatz: $u(x, t) = X(x)T(t)$, then (15.24) becomes (setting $\alpha = 1$ by appropriate temporal scaling)

$$\frac{1}{T}\frac{dT}{dt} = \frac{1}{X}\frac{d^2X}{dx^2} \qquad (15.25)$$

and both sides must be constant, call it $-\omega^2$. The solution is

$$u(x, t) = e^{-\omega^2 t}\left[A\sin(\omega x) + B\cos(\omega x)\right], \qquad (15.26)$$

and we can see the decrease in amplitude clearly. If we start with a pure sine wave (setting $B = 0$ above), then the amplitude of the sine wave is $Ae^{-\omega^2 t}$, decreasing over time, starting at $A$. The heat equation bears significant resemblance to the Schrödinger equation – in the quantum mechanical case, "dissipation" is not allowed – remember that the integral of $\Psi(x, t)^*\Psi(x, t)$ over all space must be a constant, and it is the factor of $i$ in the Schrödinger equation that restores this "amplitude conservation."

Suppose we want to describe a bar of metal, and we hold one end at fixed temperature $T_\ell$, the other at temperature $T_r$ – if we provide the initial temperature field $T(x, 0)$ taking us across the bar, how does temperature change in time? This is, again, a linear problem, and can be solved using the finite difference methods of Chapter 5, but this is an interesting problem for our Galerkin method, as well (particularly if we allow $\alpha$ to vary across the bar).

## 15.2 Galerkin method

Galerkin methods refer to a general strategy for solving nonlinear PDEs. Suppose we have a PDE governing the behavior of a function $u(x, t)$ (so that here, we think of a function of two variables, one temporal, one spatial), let's write the generic PDE as

$$\mathcal{L}u(x, t) = F(x, t) \qquad (15.27)$$

where $\mathcal{L}$ is some differential operator that acts on $u(x, t)$, and $F(x, t)$ is a specified function of position and time. We want to solve this equation given appropriate spatial/temporal boundary/initial conditions (the number and type depend on the details of $\mathcal{L}$). The operator $\mathcal{L}$ is just shorthand for the portion of the equation that acts on $u$, it could depend on $u(x, t)$ (if the PDE is nonlinear) – here are a few examples:

$$\mathcal{L}u(x, t) = \left(-\frac{1}{v^2}\frac{\partial^2}{\partial t^2} + \frac{\partial^2}{\partial x^2}\right)u(x, t)$$

$$\mathcal{L}u(x, t) = \left(-\frac{\hbar^2}{2m}\frac{\partial^2}{\partial x^2} + V(x, t) - i\hbar\frac{\partial}{\partial t}\right)u(x, t) \qquad (15.28)$$

$$\mathcal{L}u(x, y) = \left(\frac{\partial^2}{\partial x^2} + \frac{\partial^2}{\partial y^2} - \alpha u(x, y)^3\right)u(x, y),$$

where the operators are, in order, the wave operator, the Schrödinger differential operator (both linear), and a fictitious nonlinear operator (a purely spatial example).

The idea is to expand $u(x, t)$ in a complete orthonormal basis for the spatial components with time-dependent coefficients, and then use the PDE itself to generate ODEs for those time-dependent coefficients. The choice of basis leads to a proliferation of names (Chebyshev, pseudo-spectral, etc.; see [11]), but they all share the same motivation. Unlike finite difference methods applied to these problems, where we generate a grid in time and space, using the Galerkin approach, we are dispensing with a spatial grid in favor of functions that form a continuous basis over some domain of interest. Then we pick up a bunch of ODEs, and these must be solved (in general) using Runge–Kutta (for example), either in initial value or boundary-value form (in which case, shooting is required).

Take a discrete basis of functions $\{\phi_j(x)\}_{j=0}^{\infty}$ that satisfy the boundary conditions (for $x \in [0, L]$) required by the operator $\mathcal{L}$.[3] As a basis for functions, we want the set to be orthogonal,

$$\int_0^L \phi_j(x)\phi_k(x)dx = \delta_{jk}, \qquad (15.29)$$

and complete, meaning that any function (of $x \in [0, L]$) can be written as a linear combination of the $\{\phi_j(x)\}_{j=0}^{\infty}$.

Using that completeness, the solution to (15.27), $u(x, t)$, can be decomposed with time-varying coefficients as

$$u(x, t) = \sum_{j=0}^{\infty} a_j(t)\phi_j(x). \qquad (15.30)$$

---

[3] Any basis set will do, although a good choice is to use the eigenfunctions of a relevant (to the nonlinear problem) linear operator, so that we might pick the eigenfunctions of $\nabla^2$ for use in the third example in (15.28).

The differential operator $\mathcal{L}$, applied to $u(x, t)$ acts on the entire sum, equation (15.27) becomes:

$$\mathcal{L}\left[\sum_{j=0}^{\infty} a_j(t)\phi_j(x)\right] = F(x, t). \tag{15.31}$$

Multiply both sides of (15.31) by $\phi_k(x)$ and integrate as in (15.29) to obtain the infinite family of equations

$$\int_0^L \phi_k(x)\mathcal{L}\left[\sum_{j=0}^{\infty} a_j(t)\phi_j(x)\right] dx = \int_0^L \phi_k(x)F(x, t)dx \quad k = 0 \longrightarrow \infty. \tag{15.32}$$

---

**Example – the wave equation**

Let's take the wave operator, $\mathcal{L} = -\frac{1}{v^2}\frac{\partial^2}{\partial t^2} + \frac{\partial^2}{\partial x^2}$, and $F(x, t) = 0$ to see that (15.32) reduces as we expect. Assume we are given the boundary conditions $u(0, t) = u(L, t) = 0$. We'll pick basis functions that support these. Let $\phi_k(x) = \sqrt{2/L}$ $\sin(k\pi x/L)$, the normalized functions that make up the sine series decomposition of a spatial function (in this case, it is natural to start the sum (15.30) with $j = 1$ rather than zero). Then, since the operator $\mathcal{L}$ is linear, it acts on each term in the sum separately,

$$\mathcal{L}[a_j(t)\phi_j(x)] = -\frac{1}{v^2}\ddot{a}_j(t)\phi_j(x) - a_j(t)\left(\frac{j\pi}{L}\right)^2\phi_j(x), \tag{15.33}$$

and inserting this in (15.32) gives

$$\sum_{j=0}^{\infty} \int_0^L \phi_k(x)\phi_j(x)\left[-\frac{1}{v^2}\ddot{a}_j(t) - \left(\frac{j\pi}{L}\right)^2 a_j(t)\right] dx = 0. \tag{15.34}$$

The orthogonality of the basis functions introduces the Kronecker delta, collapsing the sum on the left, and returning the target (infinite) set of ODEs. These we can immediately solve:

$$-\frac{1}{v^2}\ddot{a}_k(t) - \left(\frac{k\pi}{L}\right)^2 a_k(t) = 0 \longrightarrow a_k(t) = A_k \cos\left(\frac{vk\pi}{L}t\right) + B_k \sin\left(\frac{vk\pi}{L}t\right). \tag{15.35}$$

The coefficients $A_k$ and $B_k$ will be set by the initial conditions, $u(x, 0)$ and $\dot{u}(x, 0)$.

---

So far, all we have done is separation of variables in an infinite sum, with a specified basis for the spatial dependence. The approximation lies in the truncation of the sum (15.30); we'll take:

$$u_N(x, t) \equiv \sum_{j=0}^{N} a_j(t)\phi_j(x), \tag{15.36}$$

for some $N$. Then instead of an infinite sum in (15.32), we have $j = 0$ up to $N$. What we'll do now is restrict $k$ to run from 0 to $N$ as well – that will give us the $N + 1$ ODEs we need to solve for the coefficients $\{a_j(t)\}_{j=0}^{N}$. We can solve these to obtain the complete, approximate solution.

One way to think about the process is to define the residual:

$$r_N(x, t) \equiv \mathcal{L}u_N(x, t) - F(x, t) \tag{15.37}$$

obtained by applying the operator $\mathcal{L}$ to $u_N$ and subtracting the "source" function $F(x, t)$. As $N \longrightarrow \infty$, $r_N$ will go to zero, so that (15.27) is satisfied. For finite $N$, we will require that all of the error we are making by truncating the sum (15.30) lies in the portion of the sum that has index greater than $N$. This notion is enforced quantitatively by the requirement:

$$\boxed{\int_0^L \phi_k(t)r_N(x, t)dx = 0 \quad k = 0 \longrightarrow N} \tag{15.38}$$

or in words: "the residual has no component in the $\phi_k(x)$ 'direction'." The error occurs beyond our truncation level, the same sentiment we used in the Krylov subspace approach to matrix inversion that led to conjugate gradient in Section 11.3.4 (the analogous equation was (11.44)).

---

**Example – time-dependent potentials**
We'll work out the Galerkin approach for a quantum mechanical problem with time-varying potential. Schrödinger's equation is

$$-\frac{\hbar^2}{2m}\frac{\partial^2\Psi(x, t)}{\partial x^2} + V(x, t)\Psi(x, t) = i\hbar\frac{\partial\Psi(x, t)}{\partial t}, \tag{15.39}$$

so that our $\mathcal{L}$ operator is:

$$\mathcal{L} \doteq \left(-\frac{\hbar^2}{2m}\frac{\partial^2}{\partial x^2} + V(x, t) - i\hbar\frac{\partial}{\partial t}\right), \tag{15.40}$$

and let's once again specialize to the case $\Psi(0, t) = \Psi(L, t) = 0$, the wave function is localized. We'll use the $N$-decomposition in the orthogonal sine basis:

$$\Psi_N(x, t) = \sqrt{\frac{2}{L}} \sum_{j=1}^{N} a_j(t) \sin\left(\frac{j\pi x}{L}\right). \tag{15.41}$$

Then the residual requirement (15.38) (or equivalently (15.32) for $k = 1 \longrightarrow N$ instead of infinity) will yield:

$$
\begin{aligned}
0 = {} & \frac{\hbar^2}{2m}\left(\frac{k\pi}{L}\right)^2 a_k(t) - i\hbar \dot{a}_k(t) \\
& + \sum_{j=1}^{N}\left[\frac{2a_j(t)}{L}\int_0^L V(x, t)\sin\left(\frac{j\pi x}{L}\right)\sin\left(\frac{k\pi x}{L}\right)dx\right].
\end{aligned}
\tag{15.42}
$$

Now we can see that the coefficients $a_k(t)$ are going to inherit time-dependence from the time-dependence of the potential, and will be coupled from the spatial dependence of the potential (imagine writing the potential itself in terms of a time-dependent sum of spatial basis functions). We again have $N$ ODEs in time, and we must be given the initial configuration of the wave function in order to solve these ODEs.

The content of (15.38) is a set of ODEs that we have to solve with, in this case, initial conditions to pin down the constants that appear. Even though we have been using the language of time and space in formulating the decomposition for $u(x, t)$, the approach is general. For $u(x, y)$, a function of two spatial variables, we might take:

$$u_N(x, y) = \sum_{j=0}^{N} a_j(y)\phi_j(x), \tag{15.43}$$

where we expand in the basis for the $x$-variable and leave $\{a_j(y)\}_{j=0}^{N}$ as our unknown functions (we could let the $y$-dependence be encapsulated in a basis function, and find equations for $\{a_j(x)\}_{j=0}^{N}$ instead). Then (15.38) will give ODEs in $y$ rather than $t$. But these ODEs are still coupled and finite (in number), so can be solved using standard methods once boundary values have been specified (necessitating, in some cases, shooting – if, for example, $u(x, 0) = u(x, b) = 0$ for $y \in [0, b]$, we would need to use an RK method with, say, bisection, as in Chapter 3).

### 15.2.1 Nonlinear Klein–Gordon

Even though it is a somewhat painful process once you move from linear examples to nonlinear applications, we will form $u_N$, and find the ODEs explicitly for (15.17).

The full problem we will solve, via Galerkin, is (PDE plus initial/boundary conditions, now):

$$\mathcal{L}u(x,t) \equiv -\ddot{u}(x,t) + u''(x,t) + \alpha^4 u(x,t)^3 = 0$$

$$u(0,t) = u(L,t) = 0$$

$$u(x,0) = u_0\sqrt{\frac{2}{L}}\sin\left(\frac{\pi x}{L}\right)$$

$$\dot{u}(x,0) = 0.$$

(15.44)

Expand in the basis functions $\phi_j(x) = \sqrt{2/L}\sin(j\pi x/L)$ as usual. To keep the algebra to a minimum, we'll let $N = 3$, so the starting sum is

$$u_3(x,t) = a_1(t)\phi_1(x) + a_2(t)\phi_2(x) + a_3(t)\phi_3(x),$$ (15.45)

and then the residual (15.37) is

$$r_3 = -\phi_1(x)\left(\ddot{a}_1(t) + \frac{\pi^2}{L^2}a_1(t)\right) - \phi_2(x)\left(\ddot{a}_2(t) + \frac{4\pi^2}{L^2}a_2(t)\right)$$

$$- \phi_3(x)\left(\ddot{a}_3(t) + \frac{9\pi^2}{L^2}a_3(t)\right)$$ (15.46)

$$+ \alpha^4(a_1(t)\phi_1(x) + a_2(t)\phi_2(x) + a_3(t)\phi_3(x))^3.$$

The first two lines are just the content of the wave equation, and the third line is the nonlinearizing cubic term.

We will get three equations by requiring that $\int_0^L \phi_k(x)r_3 dx = 0$ for $k = 1, 2,$ and 3. The wave-like portions of $r_3$ are independent, but the cubic term will couple modes together. At the end of the integration, we have:

$$0 = -\left(\ddot{a}_1(t) + \frac{\pi^2}{L^2}a_1(t)\right)$$

$$+ \alpha^4\frac{3}{2L}\left(a_1(t)^3 - a_1(t)^2 a_3(t) + a_2(t)^2 a_3(t) + 2a_1(t)\left(a_2(t)^2 + a_3(t)^2\right)\right)$$

$$0 = -\left(\ddot{a}_2(t) + \frac{4\pi^2}{L^2}a_2(t)\right)$$

$$+ \alpha^4\frac{3}{2L}a_2(t)\left(2a_1(t)^2 + a_2(t)^2 + 2a_1(t)a_3(t) + 2a_3(t)^2\right)$$

$$0 = -\left(\ddot{a}_3(t) + \frac{9\pi^2}{L^2}a_3(t)\right)$$

$$+ \alpha^4\frac{1}{2L}\left(-a_1(t)^3 + 3a_1(t)a_2(t)^2 + 6\left(a_1(t)^2 + a_2(t)^2\right)a_3(t) + 3a_3(t)^3\right)$$

(15.47)

(where you can clearly see the nonlinear coupling introduced by $\alpha$). This triplet of second-order ODEs can be turned into six first-order ones in the usual way (from Section 2.4.2), and we can solve that vector set using RK with the initial conditions:

$$a_1(0) = u_0 \quad a_2(0) = a_3(0) = 0 \quad \dot{a}_1(0) = \dot{a}_2(0) = \dot{a}_3(0) = 0. \quad (15.48)$$

The Galerkin solution of this problem is complete once you have solved the ODEs for $a_1(t)$, $a_2(t)$ and $a_3(t)$.

### 15.2.2 Nonlinear string

As an example of the Galerkin approach for another nonlinear problem, take the nonlinear string equation (15.23). Suppose our string problem has been posed with fixed boundary conditions on a (stretched) string of length $L$, so that we want to solve:

$$\mu \ddot{u}(x, t) - T_0 u''(x, t) = \frac{3}{2} K u'(x, t)^2 u''(x, t)$$

$$u(0, t) = u(L, t) = 0 \quad\quad\quad (15.49)$$

$$u(x, 0) = u_0 \sin\left(\frac{\pi x}{L}\right)$$

$$\dot{u}(x, 0) = 0.$$

We again use the sine series, appropriate for the boundary conditions, and truncate at $N$:

$$u_N(x, t) = \sum_{j=1}^{N} a_j(t) \sin\left(\frac{j \pi x}{L}\right). \quad (15.50)$$

The residual is:

$$r_N(x, t) = \mathcal{L} u_N(x, t) - F(x, t) = \mu \ddot{u}_N - T_0 u_N'' - \frac{3}{2} K u_N'^2 u_N''. \quad (15.51)$$

We are left with $N$ temporal ODEs from (15.38) – this time, even writing them out is an unilluminating chore. Once we have the ODEs, we need to solve them with the correct initial conditions, here only $a_1(0)$ is non-zero. That type of observation can provide an estimate for $N$ – if the initial waveform can be decomposed in a sine series with smaller and smaller coefficients, then we can use a truncation of the initial waveform (take as many terms of the sum as necessary to capture its rough shape, say) to get an idea of how large $N$ must be. For example, if $u(x, 0) = \sin(50 \pi x / L)$, we would use $N > 50$.

For the current problem, the solution starts "from rest" in the first mode (i.e. $u(x, 0) = \phi_1(x)$), so we'll take $N = 5$ to capture the initial configuration and

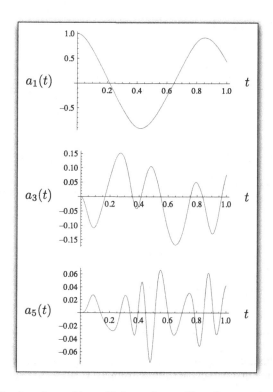

Figure 15.2 The first few odd coefficients for an $N = 5$ approximate solution to the nonlinear wave equation. Here, we set $L = 1$ m, $\mu = 1$ kg/m, $K = 1$ N and $T_0 = 0.5$ N.

include some higher modes. For the linear wave equation, the full solution would be a standing wave with a single, appropriate period; no other modes would be involved. In the nonlinear case, all odd modes are coupled, and contribute in different amounts as time goes on. The odd solution coefficients, as functions of time, are shown in Figure 15.2 – as a check, we computed the even coefficients as well, they are all zero (as they should be, given these initial conditions) to within numerical accuracy. The solution itself $u_N(x, t)$ can be constructed, and some snapshots of the trajectory are shown in Figure 15.3.

## Further reading

1. Boyd, John P. *Chebyshev and Fourier Spectral Methods*. Dover, 1999. Available for free at `http://www-personal.umich.edu/~jpboyd/`.
2. Franklin, Joel. *Advanced Mechanics and General Relativity*. Cambridge University Press, 2010.
3. Griffiths, David J. *Introduction to Elementary Particles*. Wiley-VCH, 2008.

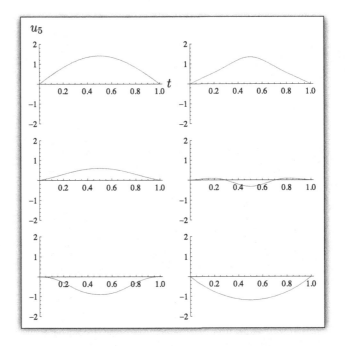

Figure 15.3 The approximation $u_N(x, t)$ for $N = 5$ – these are equally spaced snapshots over the first half cycle starting from $u_5(x, 0) = \phi_1(x)$ and $\dot{u}_5(x, 0) = 0$.

4. Pang, Tao. *An Introduction to Computational Physics.* Cambridge University Press, 2006.
5. Zienkiewicz, O. C. & K. Morgan. *Finite Elements & Approximation.* John Wiley & Sons, 1983.
6. Zwiebach, Barton. *A First Course in String Theory.* Cambridge University Press, 2004.

## Problems

**Problem 15.1**

Use (15.9) to find the Born–Infeld electric field a distance $s$ from an infinite line of charge with uniform charge-per-unit-length, $\lambda$.

**Problem 15.2**

For a string with both longitudinal and transverse wave components, we let $u(x, t)$ refer to the height of the wave at $x$ (time $t$), and $v(x, t)$ is the longitudinal displacement (from equilibrium) of the mass at $x$ – then the wave equation reads, for each component (see [38] and references therein):

$$\mu \frac{\partial^2 u}{\partial t^2} - \frac{\partial}{\partial x}(T(x) \sin \theta) = 0 \qquad \mu \frac{\partial^2 v}{\partial t^2} - \frac{\partial}{\partial x}(T(x) \cos \theta) = 0, \qquad (15.52)$$

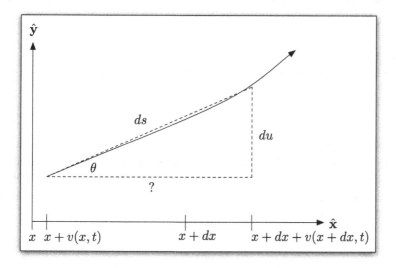

Figure 15.4 A portion of string, with displacement in both the horizontal and vertical directions.

where $\theta$ is the angle of the tangent to the string w.r.t. horizontal – the configuration is shown in Figure 15.4. Find the lengths $ds$ and ? shown in Figure 15.4, and express both $T(x)$ (from (15.19), using $\frac{ds-?}{dx}$ instead of $\frac{ds-dx}{dx}$) and $\theta$ in terms of $\frac{\partial u}{\partial x}$ and $\frac{\partial v}{\partial x}$ (assume that $dx$ is small and Taylor expand the expression $v(x + dx, t)$).

**Problem 15.3**
Show that an infinite square well (extending from zero to $L$) with bottom that moves up and down sinusoidally according to $V(t) = V_0 \sin(2\pi f t)$ cannot cause "transitions" from one energy eigenstate to another – if we start at $t = 0$ in the state $\psi(x) = \psi_j(x) = \sqrt{2/L} \sin(j\pi x/L)$, there is no time at which $\psi(x, t)$ has spatial component $\psi_k(x)$ for $k \neq j$ (i.e. show that $\int_0^L \psi(x, t)^* \psi_k(x) dx = 0$ for all $t$).

**Problem 15.4**
Solve the wave equation, in one spatial dimension,

$$-\frac{1}{v^2}\frac{\partial^2 u}{\partial t^2} = \frac{\partial^2 u}{\partial x^2} \tag{15.53}$$

subject to the boundary conditions, $u(0, t) = 0$, $\frac{\partial u}{\partial x}|_{x=1} = 0$ and the initial conditions: $u(x, 0) = u_0 x(2 - x)$, $\frac{\partial u}{\partial t}|_{t=0} = 0$. Start by finding the basis functions $\{\phi_j\}_{j=1}^\infty$ that satisfy the appropriate boundary conditions (a good choice is the set of eigenfunctions of the spatial differential operator), then expand $u(x, t)$ in this basis and find the time-dependence of the decomposition coefficients. With the full solution assembled, truncate the sum at $N = 100$, and make a movie of $u_N(x, t)$ with $u_0 = 1$, $v = 1$, for $t = 0 \longrightarrow 10$.


**Problem 15.5**
Solve the heat equation, in one spatial dimension,

$$\frac{\partial u}{\partial t} = \alpha \frac{\partial^2 u}{\partial x^2} \tag{15.54}$$

with $u(0, t) = u(1, t) = 0$ and $u(x, 0) = u_0 x(1 - x)(x + 1/8)$ using the Galerkin approach – identify the basis functions $\{\phi_j\}_{j=1}^\infty$ that satisfy the appropriate boundary conditions, then expand $u(x, t)$ in this basis and solve the resulting temporal ODEs governing the decomposition coefficients. Truncate the sum at $N = 100$, and make a movie of the solution with $\alpha = 1$, $u_0 = 1$, for $t = 0 \longrightarrow \frac{1}{2}$.


**Problem 15.6**
Suppose a wave travels in a medium with spatially varying speed $v(x)$ (in one dimension) according to:

$$-\frac{\partial^2 u}{\partial t^2} + v(x)^2 \frac{\partial^2 u}{\partial x^2} = 0. \tag{15.55}$$

Show that this equation can be solved with separation of variables by writing out the spatial equation (which will depend on the arbitrary function $v(x)$) and solving the temporal one.


**Problem 15.7**
Referring to (15.55), suppose we take the speed to be $v(x) = v_0 x / L$ – develop the temporal ODEs from (15.38) that will be solved numerically in the Galerkin method. Assume the boundary conditions support the use of $\phi_j(x) = \sqrt{2/L} \sin(j\pi x / L)$ as the spatial basis. Write out the ODEs explicitly for the case $N = 3$.


**Problem 15.8**
Show that, in source-free regions (where $\rho = 0$), the Born–Infeld field equations (15.9) can be written (in terms of $\tilde{\mathbf{E}} \equiv \frac{\mathbf{E}}{b}$) as

$$\nabla \cdot \tilde{\mathbf{E}}(1 - \tilde{E}^2) + \sum_{i,j=1}^D \tilde{E}_i \tilde{E}_j \frac{\partial}{\partial x^i} \tilde{E}_j = 0 \tag{15.56}$$

in $D$ dimensions, with $\{x^i\}_{i=1}^D$ the $D$ (Cartesian) coordinates. Introduce the potential in the usual way, $\tilde{\mathbf{E}} = -\nabla V$, and consider the Galerkin ansatz, truncated at $N = 2$,

$$V_2(x, y) = a_1(y) \sin\left(\frac{\pi x}{L}\right) + a_2(y) \sin\left(\frac{2\pi x}{L}\right) \tag{15.57}$$

appropriate to a two-dimensional, square region with side length $L$ (with $V_2 = 0$ at $x = 0$, $L$). Find the equations for $a_1''(y)$ and $a_2''(y)$ by requiring that (15.38) hold (feel free to use Mathematica to perform this relatively unilluminating final task).

## Problem 15.9

The Galerkin approach works even when the basis set is itself discrete (i.e. the functions $\{\phi_j\}_{j=1}^\infty$ are vectors $\{\mathbf{v}_j\}_{j=1}^n$). Consider a sourced wave equation (think of the potential associated with a time-varying charge distribution) in two dimensions:

$$-\ddot{u} + \nabla^2 u = -f(x, y)\cos(\omega t), \tag{15.58}$$

where $f(x, y)$ is some spatial function, and the source's time dependence is given explicitly. Take $\{\mathbf{v}_j\}_{j=1}^n$ to be the eigenvectors of the discrete Laplacian matrix, $\mathbb{D}$ (from, for example, Section 4.3), with eigenvalues $\{\lambda_j\}_{j=1}^n$ so that: $\mathbb{D}\mathbf{v}_j = \lambda_j \mathbf{v}_j$. Since these eigenvectors are complete, we know that the function $f(x, y)$, evaluated on our two-dimensional grid, can be written as:

$$\mathbf{f} = \sum_{j=1}^n \beta_j \mathbf{v}_j, \tag{15.59}$$

for some set of coefficients $\{\beta_j\}_{j=1}^n$. The discretized PDE is: $-\ddot{\mathbf{u}} + \mathbb{D}\mathbf{u} = -\mathbf{f}\cos(\omega t)$. Solve for the temporal evolution of the coefficients in the (finite) Galerkin expansion

$$\mathbf{u} = \sum_{j=1}^n \alpha_j(t)\mathbf{v}_j, \tag{15.60}$$

given the coefficients of the decomposition at $t = 0$, $\alpha_j(0) \equiv \alpha_{j0}$, and assuming we begin "from rest," so that $\dot{\alpha}_j(0) = 0$. The boundary conditions, implicit in the definition of $\mathbb{D}$, are $u = 0$ for all four sides of the two-dimensional domain.

## Problem 15.10

Using the Cartesian finite difference matrix-making-functions from the Chapter 4 notebook, make the discrete Laplacian matrix for $N_x = 40$, $N_y = 40$ with $\Delta x = \Delta y = \frac{1}{N_x+1}$.

(a) Find the eigenvalues and eigenvectors of this matrix, and make contour plots of the first four eigenvectors (corresponding to the largest four eigenvalues, these will be the last four if you use Eigensystem) using the technique for plotting the solutions to the Poisson problem in the Chapter 4 notebook. These eigenvectors are discrete approximations to the eigenfunctions associated with the two-dimensional, quantum mechanical, infinite square box.

(b) Note that the second and third eigenvalues are the same, call the associated eigenvectors $\mathbf{v}_2$ and $\mathbf{v}_3$. Remember from Problem 10.4 that the sum of two eigenvectors with the same eigenvalue is also an eigenvector, and make a contour plot of the eigenvectors $\mathbf{v}_2 \pm \mathbf{v}_3$.

**Problem 15.11**

We can find the spectrum of the finite difference matrix approximating the Laplacian
without resorting to numerical evaluation. Set:

$$u_{jk} = u_0 e^{i\pi jq \Delta x} e^{i\pi kr \Delta y} \tag{15.61}$$

for integer $q$, $r$, and insert this approximation to $u(x_j, y_k)$ directly in the finite difference
equation:

$$\frac{u_{(j+1)k} - 2u_{jk} + u_{(j-1)k}}{\Delta x^2} + \frac{u_{j(k+1)} - 2u_{jk} + u_{j(k-1)}}{\Delta y^2} = \lambda u_{jk} \tag{15.62}$$

representing the eigenvalue problem for the discrete form. Find the resulting values of $\lambda$,
which will be indexed by $q$ and $r$. Now make the finite difference matrix from the
previous problem, with $N_x = N_y = 40$, $\Delta x = \Delta y = \frac{1}{N_x+1}$, and find the eigenvalues.
Construct the $\lambda_{pq}$ for $p = 1 \longrightarrow 40$, $q = 1 \longrightarrow 40$, and generate a sorted list of these
values. They should be the "same" as the eigenvalues of the matrix computed using
`Eigenvalues`, verify that this is the case by subtracting your analytically determined
values from the numerical ones and plot the result. From the exact form, $\lambda_{qr}$, you can see
the source of the degeneracy noted in the previous problem.

## Lab problems

**Problem 15.12**

Solve the three ODEs for the cubic Klein–Gordon field theory from (15.47) – use your
RK4 ODE solver from Chapter 2 on the coupled ODEs, with $L = 1$.

(a) The odd coefficients are coupled to one another, so that if you start with
$u(x, 0) = \phi_1(x)$, for example, only $a_1(t)$ and $a_3(t)$ will be non-zero – try setting
$\alpha = 1.5$, with $u(x, 0) = \phi_1(x)$ and $\frac{\partial u}{\partial t}|_{t=0} = 0$ – use $N_s = 1000$ steps with $\Delta t = \frac{2\pi}{N_s}$.
Plot $a_1(t)$, $a_2(t)$ and $a_3(t)$, and make a movie of the solution $u(x, t)$ for $t = 0 \longrightarrow 2\pi$
in steps of $10\Delta t$.

(b) Adding in a little initial $\phi_2(x)$ allows us to probe the full nonlinear solution. Set
$\alpha = 1$, and take $u(x, 0) = \phi_1(x) + \phi_2(x)$, again starting from rest. Use the same RK4
parameters as in part (a), and plot all three coefficients and make a movie of the
output. Try the same conditions with $\alpha = 0$ to see the Galerkin solution to the *linear*
problem.

**Problem 15.13**

We can also solve the cubic field equations using a finite difference approach as
in Chapter 5 – take (15.16) in one spatial dimension, and suppose we introduce a fixed
grid in time and space, $t_n = n\Delta t$, $x_j = j\Delta x$, with $\phi_j^n \approx \phi(x_j, t_n)$. Then the PDE can be
approximated by:

$$\left[ \frac{\phi_j^{n+1} - 2\phi_j^n + \phi_j^{n-1}}{\Delta t^2} \right] = \left[ \frac{\phi_{j+1}^n - 2\phi_j^n + 2\phi_{j-1}^n}{\Delta x^2} \right] - \alpha^4 (\phi_j^n)^3 . \tag{15.63}$$

Use this update scheme, and take $\phi(x, 0) = \phi_1(x)$, with $\frac{\partial \phi}{\partial t}|_{t=0} = 0$ (so that you can set $\phi_j^{-1} = \phi_j^0$ for all $j$). For boundary conditions, set $\phi(0, t) = \phi(1, t) = 0$ (for a spatial grid extending from zero to $L = 1$). Make the $x$-grid with $N_x = 100$, $\Delta x = \frac{1}{N_x+1}$, then set $\Delta t = \frac{1}{2}\Delta x$ (so that $\frac{\Delta t}{\Delta x} < 1$). Run your solution with $\alpha = 1.5$ for a thousand (time) steps and make a movie by plotting the output every ten steps. Try using the initial condition from part (b) of the previous problem, and compare with your solution from that problem.

## Problem 15.14

For the wave equation (15.55) with $v(x) = v_0 x/L$ as in Problem 15.7 (with $v_0 = 1$ m/s) – take $N = 3$ and the initial conditions: $u_3(x, 0) = \phi_1(x)$, $\dot{u}_3(x, 0) = 0$ in a Galerkin approach. You should have the three ODEs as part of your solution to Problem 15.7 – use your Runge–Kutta routine to find approximations to $a_1(t)$, $a_2(t)$, and $a_3(t)$ with $\Delta t = 0.01$, and 1000 steps. Make plots of your approximations, and a movie of $u_3(x, t)$.

## Problem 15.15

We'll solve for the truncated Born–Infeld potential, $V_2(x, y)$ from Problem 15.8 using an ODE solver with shooting. Set up your RK4 routine for the problem, using your solution to Problem 15.8 to define the vector function **G**. For the boundary conditions, use $V(x, 0) = 0.01 \sin(\pi x)$ and $V(x, 1) = 0.01 \sin(2\pi x)$, or, in terms of the coefficients, $a_1(0) = 0.01$, $a_2(0) = 0$, and $a_1(1) = 0$, $a_2(1) = 0.01$. Write a function u that takes $a_1'(0)$ and $a_2'(0)$ as its arguments, runs your ODE solver for $N_s = 100$ steps at $\Delta y = 0.01$, and returns the approximation $\sqrt{a_1(1)^2 + (a_2(1) - 0.01)^2}$ – that will be the function we want to minimize. Use steepest descent as in Algorithm 3.1. You must provide a function du that computes the gradient of u – use a step size $\Delta = 0.0001$ for those (finite difference) derivative approximations, and return the gradient as a unit vector. Starting from the origin, run your steepest descent routine with $\eta = 0.0001$ for the first hundred steps, then divide $\eta$ by ten for the next hundred, and so on, until u $\leq 10^{-4}$. What values of $a_1'(0)$ and $a_2'(0)$ must you use to achieve this accuracy? Using those values, generate $V_2(x, y)$ and make a contour plot of it on the unit square.

## Problem 15.16

Using the eigenvectors of the finite difference approximation to the Laplace operator, construct the solution to the forced wave equation as in Problem 15.9 – i.e. write a coefficient function $a_j(t)$ that contains the $j$th coefficient's solution, and use that in the sum (15.60) to generate the **u**(t). For your grid, take $N_x = N_y = 29$ and $\Delta x = \Delta y = \frac{1}{N_x+1}$. As the time-varying source, set $f(x_j, y_k) = 1$ if $x_j = 15$ and $y_k = 15$, else zero. Start with $u(x, y, 0) = 0$, and use frequency $\omega = 1$. Plot $a_1(t)$ for $t = 0 \longrightarrow 1$ and $a_{841}(t)$ for $t = 0 \longrightarrow 10$ to get an idea of the time scales involved, then make a contour plot movie of your solution (refer to you contour plots from Chapter 4 for a reminder of the indexing involved) for $t = 0 \longrightarrow 2$ in steps of $\Delta t = 0.01$.

**Problem 15.17**

We can treat the heat equation in the same, fully discrete manner as we did for the wave equation in Problem 15.16. Using

$$\mathbf{u} = \sum_{j=1}^{n} \alpha_j(t)\mathbf{v}_j \qquad (15.64)$$

for the eigenvectors $\{\mathbf{v}_j\}_{j=1}^{n}$ of the discrete Laplacian operator, find the "exact" solution to $\mathbb{D}\mathbf{u} = \dot{\mathbf{u}}$ (with the heat coefficient set to unity) – you will again need to solve for the $\{\alpha_j(t)\}$ analytically, as in Problem 15.9. As your initial condition, take $u(x, y, 0) = \sin(2\pi x/L)\sin(3\pi y/L)$ for a square grid with $L = 1$. Use the same discretization parameters as in Problem 15.16, namely $N_x = N_y = 29$, $\Delta x = \Delta y = \frac{1}{N_x+1}$. Plot the coefficients $\alpha_1(t)$ and $\alpha_{841}(t)$ as functions of time, and make a movie of the three-dimensional plots of the solution $\mathbf{u}(t)$ for $t = 0 \longrightarrow 0.03$ in steps of $\Delta t = 0.001$.

# References

[1] Allen, M. P. & D. J. Tildesley. *Computer Simulations of Liquids*. Oxford University Press, 2001.

[2] Anderson, James A. *An Introduction to Neural Networks*. The MIT Press, 1995.

[3] Arfken, George B. & Hans J. Weber. *Mathematical Methods for Physicists*. Academic Press, 2001.

[4] Baker, Gregory L. "Control of the chaotic driven pendulum," *American Journal of Physics* (1995) **64** 9, 832–838.

[5] Baker, Gregory L. & Jerry P. Gollub. *Chaotic Dynamics: An Introduction*. Cambridge University Press, 1990.

[6] Batchelor, G. K. *An Introduction to Fluid Dynamics*. Cambridge University Press, 1967.

[7] Baumgarte, Thomas W. & Stuart L. Shapiro. *Numerical Relativity: Solving Einstein's Equations on the Computer*. Cambridge University Press, 2010.

[8] Bender, Carl M. & Steven A. Orszag. *Advanced Mathematical Methods for Scientists and Engineers*. McGraw-Hill Book Company, 1978.

[9] Boas, Mary. *Mathematical Methods in the Physical Sciences*. Wiley, 2005.

[10] Born, M. & L. Infeld. "Foundations of the new field theory," *Proceedings of the Royal Society of London A* (1934) **144**, 425–451.

[11] Boyd, John P. *Chebyshev and Fourier Spectral Methods*. Dover, 1999. Available for free at http://www-personal.umich.edu/~jpboyd/.

[12] Cormen, Thomas H., Charles E. Leiserson, & Ronald L. Rivest. *Introduction to Algorithms*. The MIT Press, 1990.

[13] G. Cybenko, "Approximation by superpositions of a sigmoidal function," *Mathematics of Control, Signals, and Systems* (1989) **2**, 303–314.

[14] Demmel, James W. *Applied Numerical Linear Algebra*. Siam, 1997.

[15] Franklin, Joel. *Advanced Mechanics and General Relativity*. Cambridge University Press, 2010.

[16] Giordano, Nicholas J. *Computational Physics*. Prentice Hall, 1997.

[17] Goldberg, Abraham, Harry M. Schey, & Judah L. Schwartz, "Computer-generated motion pictures of one-dimensional quantum-mechanical transmission and reflection phenomena," *American Journal of Physics* (1967) **35**, 177–186.

[18] Goldstein, Herbert, Charles Poole, & John Safko. *Classical Mechanics*. Addison-Wesley, 2002.

[19] Golub, Gene H. & Charles F. Van Loan. *Matrix Computations*. The Johns Hopkins University Press, 1996.

[20] Golub, Gene H. & James M. Ortega. *Scientific Computing and Differential Equations: An Introduction to Numerical Methods*. Academic Press, 1992.

[21] Gould, Harvey, Jan Tobochnik, & Wolfgang Christian. *Computer Simulation Methods: Applications to Physical Systems*. Pearson, 2007.

[22] Griffiths, David J. *Introduction to Electrodynamics*. Prentice Hall, 1999.

[23] Griffiths, David J. *Introduction to Quantum Mechanics*. Pearson Prentice Hall, 2005.

[24] Griffiths, David J. *Introduction to Elementary Particles*. Wiley-VCH, 2008.

[25] Gutzwiller, Martin C. *Chaos in Classical and Quantum Mechanics*. Springer-Verlag, 1990.

[26] Gurney, Kevin. *An Introduction to Neural Networks*. UCL Press Limited, 1997.

[27] Gustaffson, Bertil, Heinz-Otto Kreizz, & Joseph Oliger. *Time Dependent Problems and Difference Methods*. John Wiley & Sons, 1995.

[28] Haykin, Simon. *Neural Networks: A Comprehensive Foundation*. Prentice Hall, 1999.

[29] Hinch, E. J. *Perturbation Methods*. Cambridge University Press, 1991.

[30] Infeld, Eryk & George Rowlands. *Nonlinear Waves, Solitons and Chaos*. Cambridge University Press, 2000.

[31] Isaacson, Eugene & Herbert Bishop Keller. *Analysis of Numerical Methods*. Dover, 1994.

[32] Koonin, Steven E. & Dawn C. Meredith. *Computational Physics: Fortran Version*. Westview Press, 1990.

[33] Knuth, Donald. *Art of Computer Programming Vol 1. Fundamental Algorithms*. Addison-Wesley, 1997.

[34] Knuth, Donald. *Art of Computer Programming Vol 2. Semi-Numerical Algorithms*. Addison-Wesley, 1997.

[35] Knuth, Donald. *Art of Computer Programming Vol 3. Sorting and Searching*. Addison-Wesley, 1998.

[36] LeVeque, Randal J. *Numerical Methods for Conservation Laws*. Birkhäuser, 1992.

[37] Milne-Thomson, L. M. *Theoretical Hydrodynamics*. Dover, 2011.

[38] Murthy, G. S. "Nonlinear character of resonance in stretched strings," *Journal of the Acoustical Society of America* (1965) **38**, 461–471.

[39] Pang, Tao. *An Introduction to Computational Physics*. Cambridge University Press, 2006.

[40] Press, William H., Saul A. Teukolsky, William T. Vetterling, & Brian P. Flannery. *Numerical Recipes in C*. Cambridge University Press, 1996.

[41] Riley, K. F., M. P. Hobson, & S. J. Bence. *Mathematical Methods for Physics and Engineering*. Cambridge University Press, 2002.

[42] Stoer, J. & R. Bulirsch. *Introduction to Numerical Analysis*. Springer-Verlag, 1993.

[43] Strogatz, Steven H. *Nonlinear Dynamics and Chaos*. Addison-Wesley, 1994.

[44] Stuart, A. M. & A. R. Humphries. *Dynamical Systems and Numerical Analysis*. Cambridge University Press, 1998.

[45] Trefethen, Loyd N. & David Bau III. *Numerical Linear Algebra*. Siam, 1997.

[46] Wilson, James R. & Grant J. Matthews. *Relativistic Numerical Hydrodynamics*. Cambridge University Press, 2003.

[47] Zienkiewicz, O. C. & K. Morgan. *Finite Elements & Approximation*. John Wiley & Sons, 1983.

[48] Zwiebach, Barton. *A First Course in String Theory*. Cambridge University Press, 2004.

# Index